The Environment and the European Public Sphere: Perceptions, Actors, Policies

THE ENVIRONMENT AND THE EUROPEAN PUBLIC SPHERE: PERCEPTIONS, ACTORS, POLICIES

edited by

Christian Wenkel, Eric Bussière, Anahita Grisoni and Hélène Miard-Delacroix

The White Horse Press, The Old Vicarage, Main Street, Winwick, Cambridgeshire, UK

Set in 11 point Adobe Caslon Pro and Lucida Sans
Printed by Lightning Source

British Library Cataloguing in Publication Data
A catalogue record for this book is available from the British Library

ISBN 978-1-912186-14-3

TABLE OF CONTENTS

Contents

CONTRIBUTOR BIOGRAPHIES

Sophie Baziadoly is a lawyer, specialised in the fields of European Union law, public finances and taxation, as well as being an assistant professor at the law department of the University of Cergy-Pontoise, interested in trans-sectoral issues, like European environmental law. She has published several books on this topic.

Eric Bussière is Professor Emeritus at Sorbonne University and Jean Monnet Chair in the history of European integration; and was founding director of the Laboratoire d'excellence 'Ecrire une histoire nouvelle de l'Europe' (2012–2018). A member of several scientific councils, he is also corresponding member of the Academy of History of Portugal. Author and editor of numerous publications on economic history and the history of European integration, he is editor of *Histoire, Economie et société* (HES) and co-editor of *The European Commission – History and Memories of an Institution 1958–2000* (3 vols, 2007–2019).

Michel Dupuy is an associated researcher at the Institute of Modern and Contemporary History in Paris. Following his Ph.D. on the history of forest ecology in France and in Germany, he held a fellowship at the Centre Marc Bloch in Berlin in order to investigate the environmental history of the GDR under various aspects, such as air pollution, hunting or forest recreation. His research also focuses on the media treatment of environmental issues.

Marjolein van Eerd is advisor on water quality and international collaboration at Rijkswaterstaat, the public works department of the Dutch Ministry of Infrastructure and Water. From 2013 to 2018, she was a research fellow at the Radboud University Nijmegen, where she contributed to the Dutch 'Knowledge for Climate' research project. In her own research, she has dealt with transboundary climate adaptation governance and the reloading of implementation experiences in EU water governance.

Christopher Fabre holds a Ph.D. in history of energy, is a graduate of CentraleSupélec and attended IHEDATE and the XXVII Class of the French Ecole de Guerre as an auditor. After starting his career in public communication, then completing his thesis at ADEME, he joined the Public Affairs Department of the electricity distribution company Enedis in 2014. Since 2018 he has been Head of the Private Office of the Secretary General.

Liesbeth van de Grift is associate professor in history of International Relations at Utrecht University. She specialises in the history of political representation through the lens of rural and environmental governance in the twentieth century. She is currently leading the project 'Consumers on the March', which studies the rise of consumer governance and the role of consumer representatives within the European Community.

Giorgio Grimaldi holds a Ph.D. in history of federalism and European integration from the University of Pavia and is currently fellow researcher and adjunct professor at the University of Genoa, where he teaches history and institutions of the Middle East countries and holds supplementary courses on green political thought, history of global and EU environmental governance and green parties in Europe. His research interests and publications include green parties, history of European integration, EU environmental policy and global environmental governance.

Anahita Grisoni is a French sociologist and urban planner, affiliated to the research unit 'Environnement, ville, société' at the University of Lyon. Author of several books about green transition, she currently leads the French think tank 'La fabrique écologique'.

Emilie van Haute is chair of the Department of Political Science at the Université libre de Bruxelles (ULB) and researcher at the Centre d'étude de la vie politique (Cevipol). Her research interests focus on party membership, intra-party dynamics, elections and voting behaviour. Her research has been published in *West European Politics*, *Party Politics*, *Electoral Studies*, *Political Studies*, *European Political Science*, or *Acta Politica*. She is also co-editor of *Acta Politica*.

Karena Kalmbach is assistant professor in history at Eindhoven University of Technology. Her areas of expertise include social and cultural history of technology and the environment (with a particular focus on nuclear history), politics of memory and social studies of science and technology. She has done extensive research on the question of how national and international nuclear politics have influenced the debate on the health effects of Chernobyl.

Astrid Mignon Kirchhof is a senior researcher and project manager at the Saxon Academy of Science and Humanities in Leipzig, following positions at Deutsches Museum in Munich, Georgetown University in Washington, DC, and at Humboldt University of Berlin. Her research covers transnational nineteenth to twenty-first century German history in its social, ecological, technological and urban dimensions, especially the histories of nature protection, technology and energy, conflict and social movements as well as philanthropic and gender history. Recently she co-edited (with J.R. McNeill) *Nature and the Iron Curtain. Environmental Policy and Social Movements in Communist and Capitalist Countries 1945–1990* (2019).

Duncan Liefferink is assistant professor in the Environmental Governance and Politics Group at the Institute for Management Research, Radboud University Nijmegen, the Netherlands. His main research fields are European and comparative environmental politics, with a particular interest in the dynamic interrelationship between national and EU environmental policy making and implementation.

Charles-François Mathis is senior lecturer in history at Bordeaux Montaigne University. He specialises in the environmental history of Britain and France in the nineteenth and twentieth centuries. He is currently preparing a cultural and material history of coal in Britain from the 1840s to 1940. Former chairman of the French network of environmental historians, he is also a member of the board of the ESEH and is responsible for an environmental series for Champ Vallon publishing house. His *Greening the City: Nature in French Towns from the 17th Century*, co-authored with Emilie-Anne Pépy and translated from the French original, was published by The White Horse Press in 2020.

Silke Mende is a contemporary historian and deputy director at Centre Marc Bloch (CMB) in Berlin. In 2009, she received her doctorate with a thesis on the emergence of the West-German green movement in the 1970s and 1980s. In 2018 she qualified with a habilitation thesis on the history of Francophonie across the nineteenth and twentieth centuries. Her current research deals with the European history of democracy and parliamentarism in the last third of the twentieth century.

Jan-Henrik Meyer is a senior researcher at the Max Planck Institute for European Legal History in Frankfurt/Main, conducting research on the emergence of the environmental law and policy in the European Union; and an associate researcher at the Leibniz Centre for Contemporary History in Potsdam. His research interests include transnational environmental and energy history, notably nuclear history, and the history of the public sphere. He recently published the chapter on environmental policy in *The European Commission 1986–2000 – History and Memories of an Institution*.

Contributor Biographies

Hélène Miard-Delacroix is an historian and currently professor at the Department of German Studies at Sorbonne University in Paris. She is a member of the research group SIRICE (Sorbonne - Identités, relations internationales et civilisation de l'Europe). Her research focuses on the history of Germany since 1945, in particular the history of the Franco-German relationship and of the German contribution to the European integration process. Among her various publications is a transnational Franco-German history since 1963 published in French and in German (2011). She is also member of the Academic Committee of the House of European History in Brussels.

Daniela Neubacher studied Journalism and Communications in Graz and Saint Petersburg. After working in both fields, she completed a Masters in Central European Studies at the Andrássy University in Budapest. Currently, she is a Ph.D. candidate in history, focusing on cross-border protest movements and environmentalism in Central Eastern Europe during the 1980s. Since 2018 she has been working as a research associate at the Institute of the Danube Region and Central Europe (IDM).

Eva Oberloskamp is research associate at the Leibniz Institute for Contemporary History (Munich). She has published on the history of intellectuals, the history of European integration and the history of counterterrorism in Europe. At present, her main research interest is the influence of environmental thinking on energy policies in West Germany and the United Kingdom since the 1970s.

Hans Rodenburg is a staff member at Wetenschappelijk Bureau GroenLinks, the think tank of the Dutch Green Party. Previously, he worked as an historian at the Centre for Parliamentary History at the Radboud University in Nijmegen. His research focuses on the political history and future of the welfare state.

Andrew S. Tompkins is lecturer in modern European history at the University of Sheffield. He has previously worked at the University of Oxford and the Humboldt University of Berlin. He is the author of *Better Active than Radioactive! Anti-Nuclear Protest in 1970s France and West Germany* (2016). His current research examines transnational relations in everyday life along the French-West German and Polish-East German borders after 1945.

François Walter has been professor of early modern and modern history at the University of Geneva since 1986. His research focuses on Swiss history, urban history, landscape and territory history, the construction of cultures of risk, and the production of social identities. His numerous publications include *Les figures paysagères de la nation* (2004), *Catastrophes: une histoire culturelle* (2008), which was translated into Italian and German, and *Hiver: histoire d'une saison* (2014).

Christian Wenkel is associate professor of contemporary history at Artois University. After receiving a Ph.D. from the University of Munich and Sciences Po Paris, he was a senior research fellow at the German Historical Institute in Paris and at the Laboratoire d'excellence 'Ecrire une histoire nouvelle de l'Europe' at Sorbonne University. His research interests cover the Franco-German relationship, the Cold War, European integration and phenomena of Europeanisation in the long twentieth century. He recently co-edited a special issue of *Comparativ* on transnational spheres of communication in Europe from 1914 to 1945 (2018).

Guus Wieman is a Ph.D. candidate in the Department of History and Civilisation at the European University Institute in Florence. He is interested in processes of quantification and economic knowledge production within international organisations. His current research focuses on the history of Eurostat, the Statistical Office of the European Communities.

Anthony R. Zito is professor of European Public Policy at Newcastle University. He is currently co-director of the Jean Monnet Centre for Excellence at Newcastle University and co-editor of *Environmental Politics*. His research focuses on environmental governance and European Union decision-making. He has authored and co-authored numerous journal articles as well as *Creating Environmental Policy in the European Union* (2000) and *Environmental Governance in Europe* (2013).

LIST OF ABBREVIATIONS

ADLT	*Les Amis de la Terre*
AFP	Agence France-Presse
AP	Associated Press
ARD	*Arbeitsgemeinschaft der öffentlich-rechtlichen Rundfunkanstalten der Bundesrepublik Deutschland* (Germany)
BBC	British Broadcasting Corporation
BBU	*Bundesverband Bürgerinitiativen Umweltschutz* (Germany)
BEUC	Bureau européen des unions de consommateurs
BME	*Budapesti Műszaki és Gazdaságtudományi Egyetem* (Hungary)
BRD	*Bundesrepublik Deutschland* (FRG)
BUND	*Bund für Umwelt und Naturschutz Deutschland* (Germany)
CAAG	Clean Air Action Group (Hungary)
CDA	*Christen-Democratisch Appèl* (The Netherlands)
CDU	*Christlich Demokratische Union* (Germany)
CEGRP	Coordination of Green and Radical Parties in Europe
CFSP	Common Foreign and Security Policy
CIS	Common Implementation Strategy
CJEU	Court of Justice of the European Union
CO_2	Carbon dioxide
CSU	*Christlich-Soziale Union* (Germany)
DDR	*Deutsche Demokratische Republik* (GDR)
DG XI/ENV	Directorate-General for Environment (European Commission)
EAP	Environment Action Programme
EC	European Community or European Communities
ECGP	European Coordination of Green Parties
ECOROPA	Ecological European Action Group
ECPA	European Crop Protection Association
EDF	*Electricité de France*
EEA	European Environmental Agency
EEB	European Environmental Bureau
EEC	European Economic Community
EELV	*Europe Écologie Les Verts* (France)
EFGP	European Federation of Green Parties
EGP	European Green Party
ELTE	*Eötvös Loránd Tudományegyetem* (Hungary)
EMAS	Eco-Management and Audit Scheme (EU)
EMU	European Monetary Union
ENEL	*Ente nazionale per l'energia elettrica* (Italy)

EP	European Parliament
ER-EER	*Erakond Eestimaa Rohelised* (Estonia)
ESF	European Social Forum
ETS	Emissions Trading Scheme (EU)
EU	European Union
Euratom	European Atomic Energy Community
EUREAU	European Federation of National Associations of Water Services
FBR	Fast Breeder Reactor
FD	Floods Directive (EU)
FDP	*Freie Demokratische Partei* (Germany)
FoE	Friends of the Earth International
FR3	*France Régions 3*
FRG	Federal Republic of Germany
FV	*Federazione dei Verdi* (Italy)
G	The Greens (EP)
G/EFA	The Greens/European Free Alliance (EP)
GATT	General Agreement on Tariffs and Trade
GDP	Gross domestic product
GDR	German Democratic Republic
GGC	Global Greens Coordination
GGEP	Green Group in the EP
GJM	Global Justice Movement
GL	Green League (Finland)
GLP	*Grünliberale Partei* (Switzerland)
GMO	Genetically modified organism
GNV	*Gabčíkovo-Nagymarosi Vízlépcső* (Hungary)
GPA	*Groen Progressief Akkoord* (The Netherlands)
GPEW	Green Party of England and Wales (UK)
GPS	*Grüne Partei der Schweiz* (Switzerland)
GRAEL	Green-Alternative European Link
GW	Gigawatt
HNF	*Hazafias Népfront* (Hungary)
HSZ	*Hrvatska Stranka Zelenih* (Croatia)
IAEA	International Atomic Energy Agency
IBB	*Internationales Bildungs- und Begegnungswerk* (Germany)
ICMESA	*Industrie Chimiche Meda Società Azionaria* (Italy)
ICPR	International Commission for the Protection of the Rhine River
IEA	International Energy Agency
IFRM	Integrated flood risk management
IGC	Intergovernmental conference
IGO	Intergovernmental organization

List of Abbreviations

IISH	International Institute of Social History (The Netherlands)
IKT	*Ifjusági Környezetvédelmi Tanács* (Hungary)
IMF	International Monetary Fund
ISO	International Organization for Standardization
IUCN	International Union for Conservation of Nature and Natural Resources
JEF	*Junge Europäische Föderalisten*
KB	*Kommunistischer Bund* (Germany)
KISZ	*Kommunista Ifjúsági Szövetség* (Hungary)
KPD	*Kommunistische Partei Deutschlands* (Germany)
kT	Product of the Boltzmann constant (k or kB) and the temperature (T)
kWh	Kilowatt-hour
LIFE	*L'instrument financier pour l'environnement* (EU)
LMP	*Magyarország Zöld Pártja* (Hungary)
LVZS	*Lietuvos valstiečių liaudininkų sąjunga* (Lithuania)
LZP	*Latvijas Zaļā partija* (Latvia)
LZP	*Lietuvos žaliųjų partija* (Lithuania)
MEP	Member of the European Parliament
MER-FER	*Mişcarea Ecologista din România- Federaţia Ecologistă din România*
MP	*Miljöpartiet de gröna* (Sweden)
MP	Member of parliament
Mtoe	Megatons of oil equivalent
MZP	*Magyarországi Zöld Párt* (Hungary)
NABU	*Naturschutzbund Deutschland* (Germany)
NATO	North Atlantic Treaty Organization
NERSA	*Centrale nucléaire européenne à neutrons rapides SA*
NGO	Non-Governmental Organization
OCT	*Organisation communiste des travailleurs* (France)
OECD	Organisation for Economic Co-operation and Development
ÖH	*Österreichische Hochschüler_innenschaft* (Austria)
OKTH	*Országos Környezet- és Természetvédelmi Hivatal* (Hungary)
OOA	*Organisationen til Oplysning om Atomkraft* (Danmark)
OP	*Oikologi Prasinoi*
PER	*Partidul Ecologist Român* (Romania)
PEV	*Partido Ecologista 'Os Verdes'* (Portugal)
PPP	Polluter pays principle
PSUC	*Partido Socialista Unificado de Cataluña* (Spain)
PV-MVDA	*Partidul Verde-Mişcarea Verzilor-Democraţi Agrarieni* (Romania)
R&D	Research and development
RAF	*Rote Armee Fraktion* (Germany)
RAP	Rhine Action Programme
RAPF	Rhine Action Plan on Floods

RBC	River Basin Commission
RBMK	*Reaktor bolshoy moshchnosti kanalnyy* (reactor type)
RBW	Rainbow Group (EP)
REACH	Registration, Evaluation, Authorisation and Restriction of Chemicals (EU)
RIAS	Radio in the American Sector (Germany)
RSPB	British Royal Society for the Protection of Birds
RWE	*Rheinisch-Westfälisches Elektrizitätswerk* AG (Germany)
SCAPA	Society for Checking the Abuses of Public Advertising
SCPRI	*Service central de protection contre les rayonnements ionisants* (France)
SEA	Single European Act
SED	*Sozialistische Einheitspartei Deutschlands* (GDR)
SGC	Stichting Greenpeace Council
SMS-Zeleni	*Stranka mladih Slovenije-Zeleni* (Slovenia)
SO_2	Sulphur dioxide
SPD	*Sozialdemokratische Partei Deutschlands* (Germany)
SPÖ	*Sozialdemokratische Partei Österreichs* (Austria)
SPPF	*Société pour la protection des paysages de France*
SPV	*Sonstige Politische Vereinigung Die Grünen* (Germany)
SZ	*Strana Zelených* (Czech Republic)
TEU	Treaty on the European Union (Maastricht Treaty, 1992; Treaty of Lisbon, 2007)
TFEU	Treaty on the Functioning of the European Union (Treaty of Rome, 1957)
TF1	*Télévision française 1* (France)
TMI	Three Mile Island (US)
TRS	*Trajnostni Razvoj Slovenije* (Slovenia)
UK	United Kingdom
UN	United Nations
UPI	United Press International
US/USA	United States of America
USSR	Union of Soviet Socialist Republics
VA	Voluntary agreement
VGÖ	*Vereinte Grünen Österreichs* (Austria)
VVD	*Volkspartij voor Vrijheid en Democratie* (The Netherlands)
WFD	Water Framework Directive (EU)
WGGP	West German Green Party
WHO	World Health Organization
WIR	*Wirtschaftsring* (Germany)
WSF	World Social Form
WTO	World Trade Organization
WWF	World Wide Fund for Nature

ZA	*Zelena Alternativa* (Slovenia)
ZA-ZDS-ZB	*Zöld Alternativa-Zöld Demokraták Szövetsége-Zöld Baloldal* (Hungary)
ZDF	*Zweites Deutsches Fernsehen* (Germany)
ZH	*Zeleni Hrvatske* (Croatia)
ZL-ORaH	*Zelena lista- Održivi razvoj Hrvatske* (Croatia)
ZPB	*Zelena Partija Bulgaria*
ZS	*Zelena Stranka* (Croatia)
ZS	*Zeleni Slovenije* (Slovenia)

INTRODUCTION

Is it child's play to formulate environmental policies today? In any case, it seems that the young initiators of and participants in the 'Fridays for Future' movement have understood one of the main mechanisms of environmental policies in Europe for more than fifty years. The success of the green parties in the May 2019 European elections and resulting debates in some of the other parties currently in power in several EU member states correspond to a setting of the strike students' demands on the political agenda. This movement, which mobilises young people from all European countries, and even beyond,[1] around the Swedish girl Greta Thunberg on the issue of climate change, is part of a long-term evolution, marked by the emergence of a new environmental consciousness within the European public sphere, which is emerging at the same time. This evolution includes the gradual institutionalisation of environmental movements, the placing of their themes on the political agenda and, above all, the formulation of environmental policies, following a growing convergence of debates within this European public and political sphere.

The history of environmental policies since the 1970s enables us to better understand the transformations of the political field in Europe in general and illustrates most notably the entrance of new actors, who are investing the political and public sphere, as well as the growing importance of the European level. Both phenomena call for a renewal of historiography. The existence of a link between the formation of the environment as a political object and that of Europe as a political actor is reflected by a certain parallelism between the two trends, an overlap that became noticeable especially during the first direct elections to the European Parliament in 1979 – a key moment, both for the construction of Europe and for the institutionalisation of environmental movements.[2]

The transformation of political culture and of the political field took place in the aftermath of the 1968 upheavals. Previously clearly circumscribed to and shared between a limited number of political actors, the political field has since then become much more complex. This transformation, which concerns first and foremost the decision-making process at all levels, has

1 It is a movement that is developing on a global level, but it originated in Europe and is more present there than anywhere else.

2 See the contributions to this volume of Emilie van Haute, Silke Mende and Giorgio Grimaldi.

been amplified by a new era of globalisation that began in the 1970s, as well as by the end of the Cold War in 1989–91.

But while the public sphere is becoming increasingly important in the decision-making process, research on the history of European integration, much influenced by the methods of the history of international relations and those of political history in general, continues to be based mainly on the study of the executive, considering the public sphere only as a secondary factor. However, it seems particularly worthwhile to reverse the perspective on the decision-making process by starting with a study of the public sphere and its long-term dynamics, particularly at the European level.

While the existence of national, regional or local public spheres and their importance for decision-making in European democracies is generally acknowledged, the question of the existence of a European public sphere is a matter for debate – a debate as old as European integration itself.[3] In fact, any public sphere exists only as a corollary to a political entity. Thus, the creation of a new political entity necessarily calls for the emergence of a new public sphere. As far as the European Community is concerned, the unfinished state of its public sphere corresponds clearly to the unfinished state of the Union itself, at least from a political point of view. However, linguistic diversity and different political cultures within the Community are significant impediments too. Indeed, the European public sphere does not have a clearly identifiable existence, but rather presents itself as a possibility whose future contours are perceptible through a multitude of public spheres, at different – especially cross-border – levels, or even of communication spheres, which together foster an increasing convergence of debates. On environmental issues, one of the first communication spheres was initiated on a European level towards the end of the nineteenth century thanks to a few individual actors and their simultaneous presence in several national

3 On European public sphere, see R. Frank, H. Kaelble, M.-F. Lévy et al. (eds), *Building a European Public Sphere: from the 1950s to the Present* (Brussels: Peter Lang, 2010); J. Habermas, *Structural Transformation of the Public Sphere: An Inquiry into a Category of Bourgeois Society* (Cambridge: Polity, 1989); A. Mercier (ed.), *Vers un espace public européen? Recherches sur l'Europe en construction* (Paris: L'Harmattan, 2003); C. Doria and G. Raulet (eds), *Questioning the European Public Sphere. L'espace public européen en question* (Brussels: Peter Lang, 2016); C. Doria, 'Espace public et projet européen', in *Encyclopédie pour une histoire nouvelle de l'Europe* (2015) https://ehne.fr/node/49 (accessed 11 Aug. 2019); J. Requate and M. Schulze-Wessel (eds), *Europäische Öffentlichkeit. Transnationale Kommunikation seit dem 18. Jahrhundert* (Frankfurt: Campus, 2002); J.-H. Meyer, *The European Public Sphere. Media and Transnational Communication in European Integration 1969–1991* (Stuttgart: Franz Steiner, 2010); more recently on the European citizen, see H. Kaelble, *Der verkannte Bürger. Eine andere Geschichte der europäischen Integration seit 1950* (Frankfurt a. M.: Campus, 2019).

communication spheres.[4] The existence of shared communication spheres or axes such as the continent's major rivers, facilitates this evolution, as do shared memories of a common history.[5]

The emergence of such transnational communication spheres in Europe, no longer limited to elites as in previous centuries and preceding the European integration process, was not called into question by the two world wars. By contrast, the wars and the economic or demographic crises of the first half of the twentieth century produced European experiences, especially through the phenomenon of a more or less forced migration of millions of Europeans, thus facilitating the development of networks and social ties at a transnational level.[6] However, the emergence of a genuine European communication sphere, as described by Hartmut Kaelble, started only after the reconstruction of Europe following World War II, the first steps in European integration and the establishment of a climate of détente in East-West relations.[7] This decisive transformation during the 1970s can also be seen as a first stage on the road to a genuine European public sphere.

The 1970s are characterised by numerous changes in terms of perceptions, political practices and institutions and were indeed a decisive decade, not only for the emergence of this European public sphere and the construction of the European Community, but also for the constitution of the environment as a political object.[8] This decade can be considered as a bridge between,

4 See the contribution to this volume of Charles-François Mathis.

5 On European memories, see E. François and Th. Serrier (eds), *Europa notre histoire* (Paris: Les Arènes, 2017); P. den Boer, H. Duchhardt, G. Kreis and W. Schmale (eds), *Europäische Erinnerungsorte. Mythen und Grundbegriffe des europäischen Selbstverständnisses* (Munich: Oldenbourg Verlag, 2011).

6 On transnational communication spheres in Europe during the first half of the 20th century, see B. Lambauer and Ch. Wenkel (eds), 'Entstehung und Entwicklung transnationaler Kommunikationsräume in Europa zu Kriegszeiten, 1914–1945'. Special issue of *Comparativ* **28** (1) (2018).

7 H. Kaelble, 'Das europäische Selbstverständnis und die europäische Öffentlichkeit im 19. und 20. Jahrhundert', in H. Kaelble, M. Kirsch and A. Schmidt-Gernig (eds), *Transnationale Öffentlichkeiten und Identitäten im 20. Jahrhundert* (Frankfurt: Campus Verlag, 2002), pp. 85–109; see also H. Kaelble, *Auf dem Weg zu einer europäischen Gesellschaft? Eine Sozialgeschichte Westeuropas 1889–1980* (München: Beck, 1987); H. Kaelble, 'Die gelebte und gedachte europäische Gesellschaft', in H. Kaelble and J. Schriewer (eds), *Gesellschaften im Vergleich. Forschungen aus Sozial-und Geschichtswissenschhaften* (Frankfurt: Peter Lang, 1998), pp. 343–351; H. Kaelble (ed.) 'European public sphere and European identity in 20th century history'. Special issue of *Journal of European Integration History* **8** (2) (2002).

8 The literature on the 1970s is quite abundant. See, for instance, G. Migani and A. Varsori (eds), *Europe in the International Arena during the 1970s. L'Europe sur la scène internationale dans les années 1970* (Brussels: Peter Lang, 2011); F. Bösch, *Zeitenwende 1979. Als die Welt von heute began* (Munich: C.H. Beck, 2019); D. Hellema, *The Global 1970s. Radicalism, Reform, and Crisis* (New York: Routledge, 2018).

on the one hand, the beginnings of a movement for a return to nature, for the reform of life ('*Lebensreformbewegung*') and for the protection of nature – a movement that was Europeanised following a major industrialisation and technological wave at the end of the nineteenth century – and today's European environmental policies on the other hand. The density of changes in environmental issues during the 1970s, described by Joachim Radkau in terms of a 'great chain reaction',[9] is of crucial importance to further evolution in this field during the following decades. The present book therefore takes this period into particular consideration.

The new conception of nature in the long nineteenth century serves as a starting point for a patrimonialisation of nature as an integral part of the construction of national identities, in analogy with the patrimonialisation of culture.[10] The link with the nation promotes the transformation of nature protection into a political object throughout Europe. International meetings and publications, such as Hugo Conwentz's book *The Care of Natural Monuments with Special Reference to Great Britain and Germany* (1909), both contributed to the emergence of a first transnational communication sphere around these issues, and led to a first wave of legislation in this field and the creation of national parks.[11] These developments are fostered in particular by the spread of a life reform movement ('*Lebensreformbewegung*') with roots in Germany and Switzerland. If this movement evolved on the ground of a new conception of nature, widespread in Europe during the nineteenth century, it served itself as an ideological basis for the emergence of a new environmental consciousness from the 1960s and 1970s onwards.

The environmental movement was interrupted by the two world wars and their respective post-war periods. It regained momentum when growth began to decline in the Western world during the 1960s. With the increasing difficulties of the United States in winning the war in Vietnam and the battle for the Great Society within its own borders, both accelerating its economic crisis, the decline was looming not only of American power but above all of an American socio-economic model, with significant repercussions for

9 J. Radkau, 'The great chain reaction. The "ecological revolution" in and around 1970', in J. Radkau. *The Age of Ecology* (Cambridge: Polity Press, 2014), pp. 79–113.

10 On the conceptualising of heritage, see A. Swenson, *The Rise of Heritage. Preserving the Past in France, Germany and England, 1789–1914* (Cambridge: Cambridge University Press, 2013).

11 H. Conwentz, *The Care of Natural Monuments with Special Reference to Great Britain and Germany* (Cambridge: The University Press, 1909); see also the contribution of Charles-François Mathis in this volume.

Western Europe.[12] Crises on various levels, whether political, economic, financial, oil, environmental or even cultural, overlapped during the 1970s and caused a major change in mentalities within European societies. In particular, the two oil crises of the 1970s contributed to raising awareness among Europeans of a new environmental reality. Simultaneously, energy security became a major challenge for Western European states in the same way as military or monetary security. This challenge was even more important because the United States' protection in these three security domains was no longer as unconditional as before. While the European NATO member states were called upon to participate in the United States' military protection of Western Europe (i.e. burden sharing), they were mainly confronted with the needs to organise their own monetary protection (resulting in the creation of the European Monetary System in the 1970s) and to find alternative energy suppliers (as shown by the construction of a gas pipeline to transport Soviet gas from the early 1980s onwards).

The new environmental consciousness results first of all from an awareness of the finiteness of natural resources, put at the centre of the debate by the reports of the Club of Rome, and of a new vulnerability to environmental disasters, which have become more visible through the latest mass-media developments. But this consciousness also emerges against the backdrop of the questioning of a model of almost eternal growth and constant technological progress, a questioning that goes hand in hand with the questioning of the political system by new social movements. In the specific Cold War context that fed the fear of a nuclear apocalypse, a generalised feeling of crisis thus gradually developed, generating fears of all kinds about the future.[13] The environmental issue, however, seems to have been at the core of this widespread concern.

In the same way, the development of this new environmental consciousness was fostered by the emergence of the whole set of new social movements in the 1960s and 1970s, as they provided the environmental movements, initially rather limited, with a much broader base. A new attitude towards the environment was a constitutive element of almost all these new social

12 P. Melandri, *Le siècle américain. Une histoire* (Paris: Perrin, 2016); Edward D. Berkowitz, *Something Happened. A Political and Cultural Overview of the Seventies* (New York: Columbia University Press, 2006).

13 On how to deal with the future, see L. Hölscher, *Die Entdeckung der Zukunft* (Göttingen: Wallstein Verlag, 2106); M. Giraudeau and F. Graber (eds), *Les projets. Une histoire politique (16e–21e siècles)* (Paris: Presses des Mines, 2018).

movements, thus contributing to a generalisation of various convictions and patterns of perception. And beyond the regional or national framework, the transnational links available to these different movements also facilitated the establishment of transnational networks between environmental movements across Western Europe.[14]

The growth of environmental consciousness across territorial and linguistic borders has also resulted from several environmental disasters and ensuing media coverage, which have left particularly strong images in the collective imaginary over the past few decades. Thus, the images of the shipwrecks of the *Torrey Canyon* in 1969 and the *Amoco Cadiz* in 1979, causing oil spills respectively in the United Kingdom and France, played a particularly strong role. So did technological accidents such as that of Seveso in 1976, polluting Northern Italy with dioxin; or that of Chernobyl in 1986, irradiating large regions in the Soviet Union – and sending a radioactive cloud to the Western part of the continent as a threat all the more treacherous because it was invisible. While there have always been technological accidents and failures, their impact on the environment as well as the perception of this impact has changed considerably since the 1960s for technological, demographic and also media reasons, further reinforcing the feeling of a worsening state of the environment and that of growing vulnerability to such accidents.[15] But not only do environmental disasters demonstrate the vulnerability of the environment in concrete terms, as well as the limits of a widespread belief in technological progress at the time; they also reduce space because environmental damage no longer stops at territorial or linguistic borders, and neither do fears of a possible proliferation of the danger. These disasters are indeed increasingly perceived as phenomena involving the European sphere as a whole, which creates a certain congruence between geographical space and communication sphere.

The democratisation of colour television and the use of colour photographs in the tabloid press[16] during these years encouraged a spatial concentration of communication spheres revolving around these environmental disasters.

14 On the transnational connections of the new social movements, see for instance M. Klimke, *The Other Alliance Student Protest in West Germany and the United States in the Global Sixties* (Princeton: Princeton University Press, 2010); M. Klimke and J. Scharloth (eds), *1968 in Europe. A History of Protest and Activism, 1956–1977* (Basingstoke: Palgrave Macmillan, 2008).

15 See, in particular, F. Walter, *Catastrophes: une histoire culturelle XVIe–XXIe siècle* (Paris: Ed. du Seuil, 2008); U. Beck, *Risk Society. Towards a New Modernity* (New York: SAGE Publications, 1992); see also the contributions of François Walter and Karena Kalmbach in this volume.

16 Colour television was gradually introduced in Western Europe from 1967. The first colour photo appeared in the West German *BILD* newspaper on 21 July 1969.

The public had a front row seat to watch them, as well as the horrors of the war in Vietnam. Distant accidents turned into disasters for an environment which is potentially the same as that of each spectator, even those far from the affected places. The end of the Cold War and the development of new communication technologies accelerated this spatial concentration of communication spheres from the 1990s onwards.

If one of the obstacles to the emergence of a European public sphere is the linguistic, and consequently cultural, diversity in Europe, images of environmental disasters helped to build a communication sphere on a European level, by linking those at lower levels, separated in principle by the use of different languages. These images thus served as a focal point for environmental debates across Europe. The circulation of images facilitates the circulation of certain concepts, such as 'peak of oil', '*marée noire*' or '*Waldsterben*', which structured the debates through their omnipresence. A common vocabulary, necessary for the emergence of a Europe-wide communication sphere, found its origins in some key publications, such as *Silent Spring* (1962) by the American biologist Rachel Carson, translated into many languages and selling more than two million copies all over the world. Another type of transnational, and more apocalyptic, vocabulary was provided by the Club of Rome, a think-tank created in 1968 by scientists, economists and national and international officials, but also entrepreneurs, whose reports resonate in a semantic context that reflects Cold War patterns of perception.[17]

The circulation of ideas and concepts between different communication spheres and therefore the increasing convergence of debates may be particularly facilitated by the heritage of a centuries-long common history. One of the best examples is the Franco-German one, characterised by particularly dense and deep links, which played a twofold role in the emergence of the European public sphere as well as in the emergence of environmental movements in Europe.[18] These links were, for instance, at the origin of the formation of a green list for the European elections in 1979 and facilitated the formation of a cross-border

17 See, for instance, B. Greiner, Ch. Th. Müller and D. Walter (eds), *Angst im Kalten Krieg* (Hamburg: Hamburger Edition, 2009).

18 On the Franco-German example, see M. Espagne and M. Werner (eds), *Transferts. Les relations interculturelles dans l'espace franco-allemand (XVIIIe–XIXe siècles)* (Paris: Editions Recherche sur les civilisations, 1988); or the Franco-German History, especially the last volume: H. Miard-Delacroix, *Le défi européen. Histoire franco-allemande de 1963 à nos jours* (Villeneuve d'Ascq: Presses du Septentrion, 2011); the relevance of Franco-German history for the development of transnational networks is also confirmed by numerous articles in this volume.

anti-nuclear movement in the Rhine Valley.[19] It is therefore not surprising that the perceptions, debates, movements and environmental policies within this Franco-German area have been particularly well researched. For this volume, it served as a starting point while a firmly European perspective is adopted.

In any case, the environment is a challenge and a political object that in most cases cannot be dealt with at the national level and is addressed either at the local or regional level or most likely at a supranational or global level. And, for some aspects of the environment, only the global level really matters. Climate is thus an archetype of a global public good. The research conducted so far reflects this situation by examining environmental issues mainly on a global or a regional level. Yet very few studies are interested at the European level or adopt a European perspective to study the formation of the environment as a political object. However, such a shift of perspective seems to be justified precisely by the emergence of a new communication sphere at the European level since the 1970s.

As far as the European Community is concerned, the environmental policy framework has been developing in a discreet but effective way since the 1970s. While some European countries set up the first Environment Ministries in the early 1970s, the European Commission only created a unit for environmental issues within the Directorate-General for Industry in 1973. Transformed into an independent Directorate-General in 1981, it was constantly increasing in importance and number of staff.[20] The influence of the European Economic Community was twofold during the last five decades: firstly, in instigating environmental policies of the Member States; and secondly and mainly by developing ecological and environmental standards whose application even goes far beyond the Community's framework itself. One of the best examples of this second kind of action is the famous Seveso Directives in response to the technological accident in the ICMESA chemical plant on 10 July 1976, causing a toxic cloud that impacted several municipalities in the Lombardy plain, including Seveso. Since 1982, the successive so-called Seveso Directives have required Member States to classify all industrial sites that are potentially dangerous for the environment and to put in place preventive policies to anticipate any possible risks.[21]

19 See the contributions of Andrew Tompkins and Giorgio Grimaldi in this volume.

20 M. Dumoulin et al. (eds), *The European Commission 1973–86. History and Memories of an Institution* (Luxembourg: Publication Office of the European Union, 2014).

21 See the contribution of Sophie Baziadoly in this volume.

The European level in this field seems in fact much more important than it appears at first sight. It even seems possible to invoke the emergence of Europe through the perception of environmental problems and the suggested solutions. The methodological challenge is therefore to grasp this level against a global issue on the one hand and the predominance of the national perspective in the political debate and its influences on political history on the other. The aim of this research is to reveal European characteristics of the way the environment is perceived, in order to identify the parameters of a specific European environmental consciousness and those of distinct European policies in this field.

By focusing on the simultaneous emergence of the European public sphere and the environment as a political object across Europe, this volume aims to contribute to a renewal of European history, which too often remains compartmentalised by a national prism. The theme provides an opportunity to contribute to a history of the Europeanisation of the continent beyond political turning points and limits. The aim is a European history of Europe that is not confined to any division, as for example that of the Cold War, but is rather based on long-term dynamics, transcending any project of integration or disintegration of the European continent, and shaped by the global challenges of our times.

While the concept of Europeanisation offers a broader vision of the history of Europe in the twentieth century,[22] the theme of the volume makes it possible to study Europeanisation as a phenomenon at three levels: first of all, institutional Europeanisation, namely at the political, economic and legal levels, through the process of European integration; then, structural Europeanisation, resulting from intra-European transport, communication or migration flows and producing convergences of perceptions, representations, discourses and also values; and finally the interdependencies between these two forms of Europeanisation. Supra-national issues, such as the environment or the vision of the future, seems particularly suitable for this. In contrast to other topics, such as social issues, this suitability for the study of Europeanisation dynamics seems to depend not least on the relative novelty of the environment as a political object.

22 On the use of the concept of Europeanisation for portraying European history in the 20th century, see H. Kaelble, 'Europäisierung', in M. Middell (ed.) *Dimensionen der Kultur- und Gesellschaftsgeschichte* (Leipzig: Leipziger Universitätsverlag, 2007); M. Conway and K.K. Patel (eds), *Europeanization in the Twentieth Century. Historical Approaches* (Basingstoke: Macmillan, 2010); M. Osmont, E. Robin-Hivert, K. Seidel, M. Spoerer and Ch. Wenkel (eds), *Européanisation au 20e siècle: un regard historique. Europeanisation in the 20th century: the historical lens* (Brussels: Peter Lang, 2012).

This volume also aims to contribute to a renewal of the historiography, especially in the field of international relations, but also beyond, by providing an analytical framework for the decision-making process that corresponds to long-term transformations in the political field since the 1970s and the arrival of new actors within that field. Based on a study of the convergences of debates within a European communication sphere, the volume examines the influence of such convergences on the formation of political objects and their setting on the agenda. The influence of converging public opinion on the formulation of internal or external policies is certainly not a new phenomenon in Europe, but the dimensions have changed with each new media development and especially since the emergence of a European public sphere. Decision-making and the formulation of policies in the environmental field seems particularly suitable for such an analytical framework. Further research is needed to determine whether this approach is also suitable to historical analysis of decision-making within other policy fields.

Taking the environment as its object and example, the volume offers to retrace the different stages of this very process, starting with the convergence of perceptions and debates that are gradually taking root at a European level between various regional or national communication spheres as a result of this global challenge, and ending with the setting of regional, national or European policies on the agenda in reaction to these converging debates at all levels. While the contributions in this book examine each step separately, taken together they provide a general understanding of the conditions and timelines of this process. Thus, the temporal dimension of the process is at the very centre of the overall analysis, although it is a subject at the crossroads of history, sociology, law and political science. The importance of long-term phenomena, such as the *Lebensreformbewegung*, and of connections with other long-term processes, such as industrialisation, globalisation of trade or technologisation, in order to understand the basis for possible convergences, make the historical dimension of the approach predominant.

The volume, with its four parts, follows the process of the emergence of a European public sphere and its impact on decision-making through environmental issues, focusing first on perceptions, then on actors before dealing with the policies themselves.

At the centre of the first part is the question of the convergence of perceptions and debates about the environment. These convergences occur in principle on a global level, but they are more substantial at a Western level, given that environmental consciousness nowadays is mainly formed in response to a high degree of industrialisation and technological development. But within this Western area, these convergences are even more clearly established on a European level, which can be related to the formation of a European public sphere and to the common heritage of this sphere. The differences between the European and North American regions in this field appeared as early as in the nineteenth century, visible for example through the different modalities at stake in the protection of nature on either side of the Atlantic. While American influences, particularly those of American discourses, should not be neglected, as shown by the history of the diffusion of Rachel Carson's book *Silent Spring* in Europe and its influence in the emergence of a new environmental consciousness since the late 1960s, it is much more the cultural heritage that constitutes a favourable terrain for convergence in Europe.

In his chapter, Charles-François Mathis reveals the very first convergences on environmental issues at a European level at the beginning of the twentieth century. At a series of international conferences, leading experts from all over Europe met and formulated initial standards for the protection of nature, which were soon adopted by the legislation of various European countries. Mathis illustrates some form of Europeanisation of environmental issues prior to the beginning of World War I and we can thus state that a European communication sphere for landscape preservation was in the making at this very moment. François Walter's chapter goes back much further in time, exploring the historical origins of a specifically European way of dealing with environmental disasters. To investigate this question, he is interested in the perception and memory of such disasters from the seventeenth century, especially in the long-term relevance of a religious conditioning of these perceptions and memories, but also in the development of disaster research in Europe since the late twentieth century. More specifically, he argues for an examination of European 'risk culture(s)', based on specific historical experience, and explains the transition from a 'prevention society' to a 'risk society' as a result of changes during the 1970s, in particular the first oil crisis, and of the end of the Cold War. The formation of a 'risk society', as described by Ulrich Beck,[23] is also the subject of Karena Kalmbach's contribution, which describes the characteristics of a specific European risk culture, taking the example of the Chernobyl accident of 1986, its perceptions, experiences and memories. Although this environmental disaster did not immediately

23 Beck, *Risk Society*.

trigger the emergence of a specific European public sphere, as Kalmbach shows, it was, in the long run, a decisive moment for the emergence of a communication sphere not only on radioactivity or on nuclear risks, but on modern risks in general throughout Europe, even beyond East-West divisions. The emergence of a pan-European communication sphere that was not limited by the Iron Curtain and in which environmental issues were tackled is also the theme of Michel Dupuy's contribution. He examines the Western European perception of environmental degradation in Eastern Europe, in particular when it affected the environment on the other side. The emergence of this pan-European communication sphere, which can be considered as a prelude to a European public sphere, is particularly evident if German history is taken as an example: Dupuy shows how East German dissidents could use West German media in the 1980s to draw the attention of the West as well as the East German public to environmental pollution in the GDR, which was largely concealed by East German officials.

In the context of an increasing convergence of perceptions and debates at a European level, environmental movements, which initially arose mostly on national, regional or local levels, are becoming more and more transnationalised, transcending national, linguistic and sometimes even ideological boundaries throughout Europe. The essential role of environmental movements in the formulation of environmental consciousness on a European level is the subject of the second part of the book. The plurality of social forms and action repertoires related to the environment, as developed on the continent, creates a complex picture of engagement and reveals different aspects of Europe, broadening the contours of an institutional Europe.

This Europeanisation of environmental consciousness and ecological action by social movements is taking shape in quite different ways. Astrid Mignon Kirchof's work reveals a cultural and social history of environmental protection in the East that has hitherto been poorly documented. Based on an in-depth study of the biographies of two East German environmentalists, her contribution tells us about the importance of individuals acting as mediators of currents and repertoires of action between different times and places; i.e. between different periods of the twentieth century, thus linking the *Lebensreformbewegung* to the new social movements of the 1970s; between East and West, especially between the two Germanies; but also within the countries of Eastern Europe. The Europe that emerges from this portrayal is far more geographical and cultural than institutional and political. It is a Europe made up of individual convictions and exemplary action, from one border to another. This history of environmentalism in socialist Central Europe is also described by Daniela Neubacher in her chapter 'Wetlands of Protest'. She finely demonstrates how militant mobilisation around the Danube is creating new links between Hungary and Austria, beyond the Iron Curtain and official contacts, thus contributing to the emergence of a Europe at grassroots level. In the West, the political system was in favour of the emergence of social movements with solid foundations, allowing even cross-border organisation to become sustainable. Andrew Tompkins describes this type of organisation in the chapter 'Towards a Europe of Struggles?' about the taking shape of anti-nuclear mobilisation in France, Germany and the United Kingdom. This type of transnational movement undergoing an evolutionary institutionalisation is also at the very centre of the study conducted by Liesbeth van de Grift, Hans Rodenburg and Guus Wieman on Greenpeace. Although the story of an NGO working at an international level and having to adapt to the European public sphere is a radical and militant one, the degree of organisation of this structure, created in 1971 in Canada, and the power of its action at the international level, have conferred on it a quasi-institutional standing within Europe, albeit not without difficulties of adjustment.

The beginnings of institutionalisation of environmental movements and the establishment of green parties in many Western European states took place in parallel with a major push for the institutionalisation of the European Community following the 1969 Hague Summit. Subsequently, this process of institutionalising environmental movements took great advantage of the European framework and in particular of the institutional one provided by the European Parliament since it was directly elected in 1979. Thus, green MEPs entered the European Parliament even before green parties were represented in most of the national parliaments in Western Europe. The third part of this book is therefore focused on the emergence of political parties dealing mainly with environmental issues in Europe and the introduction of such issues in parliamentary debates, a crucial link between public sphere and political decision-making. The perspective is a multi-level one, including a focus on the national level, a comparative case study and an approach that encompasses all EU countries.

Emilie van Haute emphasises the obstacles and opportunities specific to green parties in various European countries from the 1970s to the present, enabling us to understand the potential for affirming a new organisational model but also the reasons for its limitations. Giorgio Grimaldi traces the development of the various European green parties throughout the federalist turning point of the 1990s. Silke Mende then sheds light on how the West German Greens dealt with European issues during their formative phase in the 1970s and 1980s, and thus shows how the European level has been adopted by one of the parties at the forefront of environmental presence at the Community level. Through a comparative approach, involving France, the United Kingdom and Federal Germany, Eva Oberloskamp examines how environmental issues were established in parliamentary debates between the 1970s and 1990s. This case study reflects the comparable importance of energy issues in the emergence of the environment as a topic for parliamentary debate within all three countries, a phenomenon that can be regarded as a characteristic of the emergence of a European public sphere. However, the highly contrasting picture reveals the limits of the phenomenon, as national specificities continue to dominate both in terms of energy security and of national representation in parliaments. All in all, this section underlines the extent to which nation states and their specific ways of organisation have hampered citizens' representation and political decision-making in the emergence of this common European sphere of communication and action.

Starting from reflections on the growing convergence of perceptions and debates as well as on the environmental movements emerging in this context since the 1970s and their institutionalisation at the parliamentary level, the fourth and final part of the book deals with the question of how these convergences determine the political agenda, both at the national level throughout Europe and at the supranational level in Brussels.

According to Jan-Henrik Meyer, the contribution of the European institutions in shaping European environmental policy has long been underestimated. Yet, this role has emerged from internal movements within Europe that are both interconnected and influenced by the internationalisation of debates in frameworks such as that provided by the Stockholm conference in 1972. The development of internal EU legislation is leading to a set of regulatory tools that will for example pave the way for major reforms in Central Europe in the context of the enlargement during the 2000s. Meyer analyses the origins of the construction of a public sphere through a series of very different case studies – Rhine pollution, the nuclear issue and the protection of birds – and explores the different ways in which public spheres function and are influenced by environmental movements. By presenting the various stages and the constitution of a European environmental law from the very first communication and the first action plan on the subject up to the Lisbon Treaty and the last milestones of this policy, Sophie Baziadoly shows how much the environmental question has become one of the central issues in European policies, especially from the decisive stage of the Single Act onwards. She highlights two main driving forces: internally, the role of civil society through the central place given to the citizens in environmental policies in Europe; externally, the global nature of the issue, which is also an element of policy impetus. She thus demonstrates the effectiveness of citizen action at several levels as well as the central position of the European regional level in the way a global issue is dealt with politically. Marjolein van Eerd and Duncan Liefferink point out the role of the management of large river basins in the rise of a European environmental policy, referring to the Rhine and the International Commission for the Protection of the Rhine River (ICPR). They show that we are thus moving from a functionalist approach based on de facto interdependencies and their necessary common management, to the convergence of experiences and thence to an active contribution in defining the EU's common policies. The presence of the European Commission within the ICPR and the adoption of directives of general interest that are of interest to the Union as a whole is a fundamental step in this politicisation.

The key element of the environmental challenge for European policies is the energy constraint, in particular through greenhouse gas emissions. The EU energy and climate change package implemented from 2008 onwards reflects the convergence of European energy and environmental policies. Drawing on the German and French cases, Christopher Fabre analyses how this convergence has developed by highlighting first the importance of the economic dimension (price) in reducing energy consumption and second the gradual empowerment of the environmental dimension of energy policies even beyond the oil counter-shock of the 1980s. He shows that Franco-German structural convergence is in fact part of the growing importance of a European policy that affects the entire EU. The analyses proposed by Antony R. Zito make it possible to examine the ways in which environmental policies are implemented and to identify some of the specific features of the European Union. Beyond the common guidelines that pass through the global level (UN) or the Western level (OECD), the European Union has long been distinguished by the regulatory dimension of its policies and by a political culture that is conducive to building consensus, as in the Netherlands or in Germany. The trend towards the use of economic instruments such as taxes is mainly due to the implementation of the Single Act. The example of the United Kingdom clearly shows to what extent the evolution of British environmental policies is determined by its accession to the EU, thereby demonstrating the effects of integration on the country and indirectly how difficult it might be to undo them.

This volume is the result of a three-year research seminar, a couple of workshops and an international conference organised by the Research Center of Excellence 'Ecrire une histoire nouvelle de l'Europe' (Sorbonne University, with funding from the French *Agence nationale de la recherche*) and the German Historical Institute in Paris between 2014 and 2017. The editors want to thank all the participants in these various scientific events as well as those involved in their organisation and the subsequent editorial process, in particular Arby Gharibian for the translation of several contributions in French.[24]

24. Chapters 1, 2, 4, 14 and 16.

PART I

THE EMERGENCE OF A EUROPEAN PUBLIC SPHERE ON ENVIRONMENTAL ISSUES

CHAPTER 1.

THE FIRST INTERNATIONAL CONGRESS FOR THE PROTECTION OF LANDSCAPES: A EUROPEAN CONVERGENCE?

Charles-François Mathis

'Nineteenth-century man entered nature like an executioner': this quotation from the Russian writer and journalist Menshikov opens the article by Henri Cazalis (otherwise known as Jean Lahor), the French doctor and writer who in 1901 called for the creation of a society to protect French landscapes, which came into existence that same year.[1] Such criticism of the relations between human societies and the natural environment during the century of industrialisation and urbanisation emerged at various paces throughout Europe. It was based on a patrimonial conception of nature that expressed aesthetic, spiritual and patriotic values, and was itself inscribed within a broader movement of concern for the preservation of historical, artistic and increasingly natural heritage, which actively developed and became internationalised during the last third of the nineteenth century.[2] A desire to protect landscapes emerged everywhere, as they became the 'beloved face of the homeland' in the Briton John Ruskin's memorable phrase, which has been repeated over and over again since.[3] In his reference work on the topic, the historian François Walter evokes 'the landscape figures of the nation', which especially took form during the nineteenth century thanks to artists, scholars and intellectuals seeking to contain the ravages of modernity.[4]

This concern was not specific to Europe, for it was present in all territories where industrialisation and urbanisation had grown in scope and increasingly seemed to threaten landscape spaces, the United States in particular.[5] The

1 Jean Lahor, 'Une société à créer pour la protection des paysages français', excerpt taken from the *Revue des Revues*, 1 Mar. 1901.

2 Astrid Swenson, *The Rise of Heritage. Preserving the Past in France, Germany and England, 1789–1914* (Cambridge: CUP, 2013).

3 See Jean Astié, *La Protection des paysages* (Lyon: Legendre, 1912), p. 14.

4 François Walter, *Les figures paysagères de la nation* (Paris: EHESS, 2004).

5 Also in Canada, Australia, New Zealand and Argentina at the turn of the twentieth century. See for instance Maria D. Rivarola, Daniel Simberloff and Christy Lepannen, 'History of protected areas

creation of Yosemite National Park in 1864 and later Yellowstone National Park in 1872 was a pioneering act, and invented a model that was broadly commented on in European countries. While this precedent offered inspiration, there was wariness toward strict imitation, and considerable variance from one country to another. This protection was, of course, also part of a movement of national construction in the United States and was accompanied by an aestheticising of nature. It nevertheless differed from European intentions through its desire to replay within national parks the original encounter between pioneers and the wilderness, and to emphasise the supposedly wild natural heritage whose wonders distinguished the American continent from an Old World marked by history.[6] For all that, sensitivity to these differences varied considerably from one space to another: the Alps or large forests of Northern Europe could more easily resemble those seemingly untouched American spaces than other territories apparently more marked by human activity – for instance Fontainebleau forest in France, over 1,000 of whose hectares were protected in 1861. The American example thus served more as a reference point than a model, one that inspired landscape protection movements in unequal ways depending on the location.

These movements emerged with force throughout Europe, taking the form of associations such as the National Trust in England, founded in 1894–95; the Associazione Nazionale per i Paesaggi, established in Italy in 1906; and the Danish society for the preservation of natural beauty, which appeared in 1911. They gave rise to protective laws – the most famous being the law of 21 April 1906 by the Frenchman Charles Beauquier on 'the protection of natural sites and monuments of an artistic nature' – although they were also present, for instance, in Norway in 1910. They also led to preservation activities, as the first European national parks appeared in Sweden in 1909, also the year in which Ravenna's pinewoods were protected in Italy, with Switzerland also creating a national park in 1914.[7]

Luigi Piccioni has underscored this surprising European convergence, and believes that this movement of landscape patrimonialisation emerged in

in Argentina: A seesaw of shifting priorities and policies in a developing country', *Environment and History* (online first 2019).

6 Charles-François Mathis, '1864. Création du parc de Yosemite', in Pierre Singaravélou and Sylvain Venayre (eds), *Histoire du Monde au XIXe siècle* (Paris: Fayard, 2017), pp. 303–306.

7 Swenson, *The Rise of Heritage*, pp. 274–279.

the very early twentieth century, at the same time as its internationalisation.[8] One of the most visible manifestations of this internationalisation was the First International Congress for the Protection of Landscapes, which took place in Paris in 1909. It has not been the subject of many specific studies, and is only mentioned in passing.[9] This relative lack of interest can firstly be explained by the very minor historical traces it left behind,[10] and by the fact that it was quickly forgotten, as the organisers of the major international events that followed in 1923 and 1931 hardly mentioned it.[11] It is precisely this discrepancy between the immediate satisfaction of an at-first-glance successful meeting and the weakness of its historical impact that offers an interesting angle for exploring how the internationalisation - or more precisely Europeanisation, as the United States went it alone on this issue - of environmental issues took place before the First World War around a few conceptions of nature, and what its successes and limits were.[12] Did this internationalisation of concerns surrounding landscapes and nature provide an opportunity for the emergence and development of a European space of communication[13] and action on these questions, revolving around a civil society that was increasingly aware of these issues?

8 Luigi Piccioni, *Il volto amato amato della Patria* (Trento: Temi, 2016), pp. 124–129. To be published in English by The White Horse Press as *The Beloved Face of the Homeland* (2020)

9 See for example John McCormick, *The Global Environmental Movement: Reclaiming Paradise* (Hoboken: John Wiley, 1992), who only mentions this congress, then concentrating on the post-war period. The exception is the recent work by Caroline Ford, in which she offers a detailed presentation of this congress, but without lingering on it too long: Caroline Ford, *Natural Interests* (Cambridge: Harvard University Press, 2016), pp. 116–117.

10 Nothing remains aside from a report by the organisers, and commentaries in the bulletin of the society that organised it.

11 See, for example, Raoul de Clermont, Fernand Cros-Mayrevieille and Louis de Nussac, *Premier Congrès international pour la Protection de la nature, faune et flore, sites et monuments naturels* (Paris: Guillemot et de Lamother, 1926), p. vi.

12 My linguistic abilities limited me to English and French sources: this article is based essentially on the archives of the Society for Checking the Abuses of Public Advertising (SCAPA), created in 1893 to combat abusive advertising, which became the primary contact in the United Kingdom for other associations in Europe; on the archives of the Société pour la protection des paysages de France (SPPF); and the publications surrounding the 1909 congress. This research must of course be extended to archives of German societies in particular, as well as those of the other European countries involved.

13 Here, once again, the role of the United States in this public sphere should be explored (although this would be the subject of a separate article); as stated earlier, its model was often signalled but rarely truly followed, with personal bonds apparently being weaker.

An international movement under formation

This European public sphere can be seen in the international coordination that was implemented during the early years of the twentieth century, which enabled the successful organisation of the 1909 congress. This European movement, made possible by the shared ideology of the patrimonialisation of nature, was based on numerous personal contacts between actors in landscape protection, along with constant sharing of activities and international meetings.

The transmission and sharing of experiences

It is striking to observe that each association for the protection of nature emphasised the achievements of its European neighbours, whether during general meetings, conferences or the publication of journals. Associations had the dual goal of invigorating their national movement, which was supported by the existence of foreign counterparts, and of suggesting new means for effective action, notably institutional and legislative ones. They consequently engaged in monitoring, which enabled them regularly to inform members about what was taking place beyond their borders. The *Bulletin de la Société pour la protection des paysages de France* (SPPF), for instance, had a 'foreign news' section that explored the laws and actions conducted in Germany, Switzerland, the UK, Belgium, etc.[14] The same thing was true of the magazine *A Beautiful World*, published by the Society for Checking the Abuses of Public Advertising (SCAPA), or the one published by the *Selborne Society*.[15] Closer links were sometimes established between foreign associations, for instance when SCAPA requested to become an associate correspondent of the SPPF, and subscribed to its magazine;[16] the German society *Heimatschutz* mentions relations with Austria, Switzerland, Belgium, Sweden and Holland.[17] This search for information was even encouraged by the British government, as the eminent environmental activist and member of Parliament James Bryce transmitted a request in 1903 from SCAPA, asking embassies in France, Germany, Austria, Belgium, Switzerland, Italy,

14 This interest was even highlighted in the report from the 1909 congress, which indicates all the issues relating to foreign countries: de Clermont et al., *Le Premier Congrès*, p. 34.

15 For example, *A Beautiful World*, no. X, Sept. 1909: 136–160, London Metropolitan Archives, A/SCA/V/3/2.

16 Letter from Anselme Changeur to Richardson Evans, 12 Oct. 1912, in London Metropolitan Archives, A/SCA/III/2/10 (Europe).

17 De Clermont et al., *Le Premier Congrès*, p. 43.

the Netherlands and the United States to report on the state of legislation regarding advertising.[18] In similar fashion, in Germany Hugo Conwentz, a central figure in the patrimonialisation movement, took advantage of his position as the head of a Prussian administration dedicated to the protection of nature to collect all possible information on the subject in a library that drew people from long distances.[19] Similar research was conducted for the drafting and application of the French law of 1906, and the reflections it prompted.[20] The French case was incidentally central to the reflections conducted in Italy during the same period.[21]

Personal bonds

These exchanges were facilitated, and sometimes simply made possible, by personal links established between actors in the environmental struggle. The outings conducted by the growing number of Alpine Clubs and Touring Clubs facilitated these encounters on both the national and international levels.[22] They were sometimes undertaken simply to create a network, as when Conwentz went on tours of Europe that took him to Sweden, the UK, Austria and Denmark, for instance.[23] A correspondence was thus initiated between these figures, including Hugo Conwentz in Germany, Richardson Evans in the UK,[24] and Charles Beauquier, Anselme Changeur and Raoul de Clermont[25] in France: the environmental movement cannot of course be reduced to this handful of individuals, but they nevertheless played a driving role in leading it. Moreover, by becoming the essential intermediaries for

18 *A Beautiful World*, no. X, 1909: 136, London Metropolitan Archives, A/SCA/V/3/2.

19 Anna-Katharina Wöbse, 'Les liaisons sinueuses: les relations franco-allemandes en matière de protection de la nature dans la première moitié du XXᵉ siècle', in Mathis and Jean-François Mouhot (eds), *Une Protection de l'environnement à la française?* (Seyssel: Champ Vallon, 2013), p. 110.

20 Numerous legal works explore the question of landscape protection by drawing a parallel between France and its European neighbours: Jean Astié, *La Protection des paysages* (Lyon: Legendre, 1912), ch. 6; Louis Gassot de Champigny, *La Protection des sites et paysages* (Paris: Michalon, 1909), preface and ch. 1; Lucien Sorel, *La protection des paysages naturels et des perspectives monumentales*, Ph.D. thesis in law from l'Université de Caen, 1932, pp. 195–201.

21 Luigi Piccioni, 'L'influence de la France dans la protection de la nature en Italie au début du XXᵉ siècle', in Mathis and Mouhot (eds), *Une Protection*, pp. 97–107.

22 See, for example, Catherine Bertho Lavenir, *La roue et le stylo* (Paris: Odile Jacob, 1999); Olivier Hoibian, *Les Alpinistes en France, 1870–1950* (Paris: L'Harmattan, 2000).

23 Wöbse, 'Les liaisons sinueuses', 109–110.

24 The founder and secretary-general of SCAPA.

25 Respectively the president, secretary-general and steering committee member of the SPPF.

other nations, they also sought to impose themselves in their own country. For instance, SCAPA enjoyed an aura abroad that its activity in the UK itself did not fully justify.[26] Strong links also existed between Conwentz and Jean Massart, one of the actors in landscape protection in Belgium, as well as Paul Sarasin, a central figure in the protection of nature in Switzerland.[27] Very concretely, this gradually expanding network also helped determine whom to seek out in order to obtain information, extend an invitation, etc.[28]

The first international meetings on the European level

These personal contacts were of course supported by, and sometimes established during, international meetings attended by various actors in the movement at the turn of the twentieth century.[29] In 1900, a conference was held in London on wildlife preservation, but did not include the primary actors from the landscape protection movement. However, to consider only the case of the French, in 1905 Raoul de Clermont presented a paper at the Congrès International d'Art Public in Liège, which focused on the protection of nature.[30] In 1908, the SPPF also participated in the French-British exhibition held in London to present its bulletins, the text of the Beauquier law, paintings of protected areas, etc.[31] This participation was explicitly designed to establish contacts in view of strengthening cooperation between France and Great Britain in matters of landscape protection, something that the SPPF considered to be part of the *Entente cordiale*![32] Finally, once again in 1908, the SPPF granted its support to the English branch of the Ligue pour la conservation de la Suisse pittoresque, which opposed the construction of a railway on the Matterhorn.[33]

It was the existence of this network and international movements with similar objectives and comparable ideology that made it possible to envi-

26 It was no more than a cog in a broader movement led by more talented organisers. See Charles-François Mathis, *In Nature We Trust* (Paris: Presses Universitaires Paris-Sorbonne, 2010), pp. 369–400.

27 Walter, *Les figures paysagères*, p. 276.

28 Letter from Fritz Koch to Richardson Evans, 26 Apr. 1912, London Metropolitan Archives, A/SCA/III/2/10 (Europe).

29 The congresses held earlier focused chiefly on wildlife, birds in particular (the first congress for the protection of birds took place in Paris in 1895).

30 De Clermont et al., *Premier Congrès*.

31 *Bulletin de la SPPF*, no. 26, 15 Apr. 1908: 36.

32 *Bulletin de la SPPF*, no. 27, 15 July 1908: 62.

33 *Bulletin de la SPPF*, no. 25, 15 Jan. 1908: 304.

sion a major international congress for the protection of landscapes on the European level.[34]

The first International Congress for the protection of landscapes

The goals of the Congress

It was France, by way of the SPPF, which took the initiative in 1907, at a time when the association was enjoying the success of the Beauquier law from the preceding year.[35] The invitations that were extended help grasp the specific objectives.

They were firstly sent to all French and foreign societies, whether environmental, scholarly, agricultural or artistic, and 'tending, like [SPPF] to protect in each country the great artistic heritage represented by its forests, rock formations, beautiful and sublime landscapes, along with its picturesque sites and some of the monuments accompanying them'.[36] The SPPF remained faithful to its patrimonial conception, and therefore did not exclude any field of protection, combining the natural and the historical, the scientific and the artistic.

It established two types of goals for itself: to awaken public opinion on the broader environmental question, and to lay the foundation for a public sphere on the international level, one that was in fact first and foremost European:

> In the presence of the constantly growing movement in favour of conserving [artistic heritage] in the general interest, this Congress seeks to inform public opinion regarding the serious danger that such outrageous excess and destruction represent for these invaluable elements of the nation's wealth.[37]

The congress thus had a role as a pathfinder, as a guide for public opinion seeking information and advice: the aim was to address everyone across Europe who was concerned by these threats and destruction. Without of course specifying how it would be implemented, there was a desire to spark

34 There was no emphasis placed at the time on a European singularity that distinguished it from American vision or practice, as the United States remained a model whose achievements were praised. In other respects, there were indeed singularities between the approaches pursued by different countries of the Old World.

35 'Pour les paysages de France', *Le Figaro*, 9 Oct. 1909.

36 *Bulletin de la SPPF*, no. 26, 15 Apr. 1908: 17–18.

37 Ibid.

a European debate regarding the protection of nature, by providing people with arguments and examples supporting landscape preservation. While there was a strong European dimension, the space of action remained firmly national, as it was within each country where associations had to act, albeit on the basis of arguments developed during the congress.

As it happens, these arguments were founded on a moderate approach to the environmental struggle, as SPPF members did not criticise industrial society as a whole, but rather its 'excesses', the 'outrageousness' of certain attacks, and 'assaults' – in short everything that went beyond common sense and moderation. To do so, and to better guide public opinion, the other objective of the congress was to strengthen the links between national associations.[38] As a result, the primary goals of SPPF leaders were to compare different legislation, suggest new legislation and create or strengthen personal links. Anselme Changeur summarised these goals at the opening of the congress: 'If this event can contribute to spreading our principles and creating a few new links between nations, we will be pleased: what better common ground … than beautiful landscapes!'[39]

The 1909 Congress

The First International Congress for the Protection of Landscapes finally took place from 17–21 October 1909. It was organised around five major topics: protection and legislation; forests; rural landscapes; urban landscapes; and landscapes, sciences and the arts.[40]

The sessions consisted of presentations by each national delegate on the actions undertaken by the respective association or country in connection with these topics. Contributions were made by Germany, Austria-Hungary, Belgium, Great Britain, Sweden and Switzerland.[41] However, there were also discussions regarding measures implemented in Greece, Norway, China, Egypt, Serbia, etc. These communications were supplemented, toward the end of the congress, by more theme-based research on 'The landscape at school' or 'The protection of flora and fauna'.

The final session provided an opportunity to formulate certain demands intended for governments, public opinion and associations participating in

38 Ibid.

39 Ibid., 24.

40 *Bulletin de la SPPF*, no. 30, 15 Apr. 1909: 12.

41 Conwentz was German, De Munck Belgian, Koechlin Austrian and Boni Italian.

the congress. It is worth noting the originality and modernity of some of these proposals. There were suggestions to create national parks, and to save certain endangered species such as beavers, ibex and flamingos; to encourage heritage education for children and adults; to slow the uniformisation of cities by preserving their local characteristics and conserve their harmony by establishing an 'aesthetic zoning law'; to combat abusive advertising, in both the country and cities…[42] Charles Beauquier gladly affirmed that 'we are all working for the same international cause, defending the shared heritage of humanity'.[43]

In ending on such a vibrant note of international communion, the congress offered its participants, the press, and SPPF members reading the meeting summary the appearance of a resounding success. In the months that followed, its success was repeatedly celebrated in the *Bulletin de la SPPF*, which suggested a new era had begun. It is precisely this faith on the part of SPPF leaders that we will explore here: was the 1909 congress truly the vector for a hitherto unknown international dynamic,[44] or did it result from a pre-existing movement?

The legacy of the 1909 Congress

A short-lived dynamic

The central question is the legacy of the congress. Despite the outpouring of compliments in the columns of the *Bulletin de la SPPF*, the results were scant. The demands put forward were highly relevant, but concrete effects were long in coming. There were of course a few positive consequences directly inspired by the congress, undoubtedly including better knowledge of national legislation, which certainly strengthened the environmental cause. For example, in 1911 SPPF members believed that the passage in Belgium of a law for the protection of natural monuments was a direct effect of the Paris congress.[45] International meetings ensued.[46] In its wake, Charles Beauquier presented a bill to the French Parliament on plans for the extension and

42 De Clermont et al., *Le Premier Congrès*.

43 *Bulletin de la SPPF*, no. 33, 15 Jan. 1910: 65.

44 While attendance at this congress was European, its ambitions surpassed the sole framework of the Old World.

45 *Bulletin de la SPPF*, Aug. 1911: 13.

46 International Congress on Public Art, Brussels, 1910; Town Planning Conference, London, 1910. See *Bulletin de la SPPF*, no. 38, Nov. 1910: 17.

embellishment of cities.[47] Conwentz continued to praise French initiatives, and wanted to take inspiration from the SPPF's moderate approach in his struggle against industrial actors, and his alliance with professionals from the tourism industry.[48] Still, this activity and its foreign impact were most certainly more a continuation of exchanges and actions that preceded the congress than the sign of a new dynamism initiated by it.

This is no doubt demonstrated by the failure of its successors. In Paris, it was decided that the various associations that met in 1909 would continue to meet regularly, and that the next congress would take place in 1911. In reality, only two other congresses were held before the First World War: one in June 1912 in Stuttgart, the other in November 1913 in Bern. From the sole perspective of international participation, the first was a failure: Richardson Evans, who received a late invitation to come to Stuttgart in April 1912, responded a month after it was held. His correspondent urged him to participate, and to provide him with the names of other English figures who could attend, affirming that he had also written to the Secretaries of Public Education and 'Public Buildings',[49] asking them to send representatives as well. Even more surprising was the absence of delegates from the SPPF: Charles Beauquier, its president, Anselme Changeur, its secretary-general, and Raoul de Clermont, the primary organiser of the congress of 1909, all excused themselves from the meeting.[50] They nevertheless sent the communications they were supposed to present,[51] which were read by other participants at a conference that appears to have chiefly brought together Central European countries. There were of course other French representatives, notably an inspector general of historic monuments, a member from Touring Club de France, and a former custodian of Fontainebleau forest, although the congress went almost unnoticed in France, not to mention the United Kingdom.[52] The Bern congress brought together delegates from seventeen European and non-European countries (the United

47 *Bulletin de la SPPF*, no. 39, Dec. 1910.

48 Wöbse, 'Les liaisons sinueuses', 110.

49 To my knowledge, no such secretary existed.

50 The current state of research does not make it possible to formulate an explanation for this absence.

51 These communications were respectively about the protection of landscapes in France, the protection of villages and constitutive elements of the landscape, and the means available to address abusive advertising.

52 It is tersely and briefly mentioned in the *Bulletin de la SPPF*, no. 56, July 1912: 5; the Touring Club's magazine completely ignores it, along with the Bern congress. When Raoul de Clermont wrote a report in 1925 about the International Congress for the Protection of Nature held two years earlier, he mentioned the 1909 congress among its forerunners, but was entirely silent about

States, Japan, etc.), and agreed upon the creation of an Advisory Commission for the International Protection of Nature, which was nipped in the bud by the Great War.[53] At this point, the initiative seems to have escaped France, and the SPPF in particular: only Raoul de Clermont remained a favoured speaker at this meeting, which incidentally broadened its field of action by taking an interest not only in landscapes, but also flora and fauna.

The European dynamic consequently did not accelerate following the meeting in Paris. This failure, which partly originated from growing tension between European countries, also demonstrates in my opinion a new configuration of the international movement, which gravitated toward the Germanic sphere of influence.

From English influence to the German model

Luigi Piccioni has suggested the existence – even within a shared conceptual framework around the patrimonialisation of nature – of three distinct and influential cultural areas in Europe: the United Kingdom; Central Europe and Scandinavia influenced by Germany and its concept of *Heimat*; and a Latin zone embodied especially by France and Italy.[54] Alan Confino has also distinguished a Germanic space from the rest of Europe, seeing it as the only place where the nation could truly be identified with a landscape.[55] The 1909 congress helps show the bridges connecting these different areas, particularly between France and Germany, and confirms the relative isolation of Britain at the time.

The tutelary figure of nature preservation associations in France, Italy, and to a lesser extent in Germany was the Briton John Ruskin. His writing marked an entire generation of activists in the United Kingdom. He was one of the only Englishmen mentioned by Hugo Conwentz in his work *The Care of Natural Monuments with Special Reference to Great Britain and Germany*, the very fact of whose publication in 1909 demonstrates the in-

that of 1912 (de Clermont et al., *Premier Congrès*). The magazine *Nature*, which had mentioned the Paris congress, did not refer to the one in Stuttgart.

53 Martin Holdgate, *The Green Web: A Union for World Conservation* (Abingdon: Earthscan, 1999), pp. 10–11.

54 Piccioni, 'L'influence de la France', 98.

55 Alan Confino, *The Nation as a Local Metaphor. Württemberg, Imperial Germany and National Memory, 1871–1918* (London: University of North Carolina Press, 1997).

fluence of these two countries.[56] In the founding article of the SPPF, Jean Lahor still mentioned Ruskin, along with William Morris, to whom he devoted a book in 1897. He also established a parallel between the activity of the future organisation, and that of the already numerous English environmental associations, which he subsequently enumerated. It is noteworthy that, throughout the article, England is the only country given such exhaustive presentation. Furthermore, by virtue of the vocabulary he used, Jean Lahor probably took inspiration from England, evoking the need for defenders of nature who want to achieve their ends – who abandon overly 'sentimental' arguments in favour of more 'practical' reasoning[57] – and by calling his opponents 'the utilitarians'.[58] These were precisely the terms that framed the environmental debate in England.[59] It is possible that this is only a convergence; however it underscores the proximity of thought between Lahor and his British counterparts. The *Bulletin de la SPPF* tried to extend this proximity by mentioning the example of England from time to time. Even during the debate for the law of 1906, Ruskin was mentioned as a major source of influence.[60] For that matter, it is impossible to deny the relatively early dynamism of the British in matters of environmental protection, with strong and effective action beginning in the 1870s; in 1907, a law regarding the regulation of advertising and the ability of the National Trust to declare lands inalienable appeared as the crowning achievements of this activism.[61]

Still, it is evident that this English influence waned, especially during the 1909 congress: while almost all European nationalities were represented in the committees that organised and composed it, the United Kingdom was absent. In addition, no British delegate travelled to Paris, with only a single presentation by Richardson Evans being read, which naturally left centre stage for SCAPA, thereby somewhat distorting a more complex reality. While contacts were maintained, it was quite obvious that the direction of the movement no longer came from England, with British influence being replaced by the German model.

56 Hugo Conwentz, *The Care of Natural Monuments with Special Reference to Great Britain and Germany* (Cambridge: Cambridge University Press, 1909).

57 Jean Lahor, 'Une société à créer', 527.

58 Ibid., 530.

59 Mathis, *In Nature We Trust*, ch. iv.

60 Maurice Faure, recorder of the law, *Journal des Débats*, 28 Mar. 1906, Sénat, p. 282.

61 Mathis, *In Nature We Trust*, pp. 398–404.

This model was clearly personified by Dr Hugo Conwentz, who as the Custodian of Natural Monuments in Prussia played an eminent role in the environmental movement in Germany. He was undoubtedly the most influential foreign figure at the SPPF, if one is to believe the association's bulletins, which mention his activities, cite him and praise his works, especially *On the Care of Natural Monuments*. He was of course the first person, just after Charles Beauquier, to head a working session during the 1909 congress. More generally, during the years preceding and following the congress, most of the articles in the *Bulletin de la SPPF* about protecting natural beauty presented German associations and legislation as a model to inspire the rest of Europe.[62] It hardly comes as a surprise then that Charles Beauquier affirmed, during one of the sessions of the congress, that 'all we have to do is imitate what's being done in Germany'.[63]

The German model was quite simply attractive, as it was the only country at the time with an administration tasked with the protection of landscapes. It was especially based on an ideology that French activists – and at least a part of Central Europe – were sympathetic toward, and that was personified by the *Heimatschutz* movement. Luigi Piccioni has quite rightly made it one of the three most important influences on the early twentieth century European environmental movement, along with the 1906 Beauquier Law and the national parks created in Sweden.[64] This association,[65] which was founded in 1904 and had offshoots in Switzerland and Austria-Hungary, protected the '*petite patrie*', as it was defined by the German delegate: 'respect for the image of the country is developed and [preserved] by this love of the local steeple, a basis for patriotic sentiment that is fully summarised in this word: *Heimatschutz*'.[66] In this framework, the conception of environmental protection was much larger than that in France or even in the United Kingdom, as it was not concerned solely with preserving urban or natural landscapes, but everything closely or remotely connected to national identity, including

62 See for example *Bulletin de la SPPF*: no. 25, 15 Jan. 190: 293–296; no. 30, 15 Apr. 1909; no. 35, 1 June 1910.

63 Charles Beauquier, in de Clermont et al., *Le Premier Congrès*, p. 12.

64 Piccioni, 'L'influence de la France', 107. Despite the importance of the movement in the UK, its international influence was actually limited, aside from the posthumous intellectual aura of Ruskin and Morris.

65 On *Heimatschutz*, see, for example, Confino, *The Nation*; William Rollins, *A Greener Vision of Home* (Ann Arbor: University of Michigan Press, 1997).

66 Dr. Fuchs, 'L'œuvre du *Heimatschutz*', in de Clermont et al., *Le Premier Congrès*, p. 39.

animals, plants and historic monuments (within a broader movement that also sought to protect traditional clothing and folklore).[67]

This approach by way of what German activists called 'natural monuments' quite evidently attracted the other delegates present in 1909. The French team, which had organised the meeting, was also very sympathetic to it, with there most likely being a passing of the baton of sorts between Lahor, who was more oriented toward the Anglo-Saxon world but who died a few months before the congress, and the team surrounding Charles Beauquier and Raoul de Clermont, who greatly admired the German model.[68] It was present during a resolution that was passed regarding the international conference on the preservation of natural resources being planned by US president Theodore Roosevelt in The Hague: the congress wanted this meeting to integrate the programme of *Heimatschutz*.[69] The British review *Nature* was not mistaken, emphasising this German influence in its report on the meeting.[70]

This coming together around the German conception of the environmental movement can of course be explained by the concurrent eclipse of the UK, the successful activities taking place in Germany and most certainly by a common regionalist influence, at least with regard to France. Caroline Ford has rightly emphasised the close links between the SPPF and the regionalist movement, often with shared members and leaders, as well as identical values in the attachment to local traditions and landscapes that must be preserved.[71] In 1909, Charles Beauquier distinguished, from the very opening of the congress, between the '*Vaterland*' and the '*petite patrie*', or 'what we could call the "*matrie*"'.[72] This intertwining of two scales of patriotism, the local and the national, was present

67 *Actes de la Recherche en Sciences Sociales*, issue on *L'Identité* **35** (1980); *Bulletin de la SPPF* No. 25 (Jan. 1908): 293–296.

68 It is possible that the biographical backgrounds of the various individuals also promoted this shift in influence, although this theory remains to be supported: Charles Beauquier and Raoul de Clermont were from Eastern France (from the Doubs); the latter was incidentally an attaché at the French embassy in Bern (See Yamina Larabi, Piotr Daszkiewicz and Patrick Blandin, 'Premier Congrès international pour la protection de la nature etc. Hommage à Raoul de Clermont', *Courrier de l'environnement de l'INRA* **52** (2004): 117–121).

69 De Clermont et al., *Le Premier Congrès*, p. 72. Theodore Roosevelt had sent invitations to 45 countries to meet at The Hague in 1909 regarding the question of natural resource preservation. This initiative was buried by his successor in the White House.

70 *Nature* **83** (2116) 19 May 1910: 345.

71 Ford, *Natural Interests*, pp. 108–109. See also Anne-Marie Thiesse, *La création des identités nationales* (Paris: Seuil, 1999).

72 De Clermont et al., *Le Premier Congrès*, p. 12.

throughout Europe according to François Walter,[73] although it was theorised and applied more vigorously in Germany, which quite naturally established itself at the head of the European movement. We are subsequently able to better grasp why – in the context of the growing international tension during the years following the Paris congress, and the increasingly sharp division of the continent – a dynamic led by Wilhelmine Germany ran out of steam. This reason is nevertheless insufficient on its own to explain the slowdown in international cooperation, particularly in Europe. Deadlock specific to the movement itself should also be taken into consideration.

Difficulty moving beyond the national scale

Goodwill was nevertheless present, as there was a desire to create an international society bringing together all the national organisations, in order to strengthen the movement. This wish had already been expressed in 1907 by Richardson Evans, in a letter to Anselme Changeur: 'My own very strong feeling is that if we are to succeed in our humanising mission, it must be by making the movement international, that is to say, bringing those in every country who feel similarly into touch with each other.'[74] If this was not an initial goal of the 1909 congress, which simply mentioned a 'moral bond between peoples', it ultimately became one in the form of an 'International Union for the Protection of the Motherland'.[75] Raoul de Clermont was subsequently entrusted with a new task, that of conducting an investigative commission for the creation of an 'International Federation of Societies for the Preservation of Natural and Regional Treasures':[76] the name simultaneously connected it to the conference sought by Roosevelt, as well as the extensive German concept of *Heimatschutz*. De Clermont was supposed to present the report on this topic at the next congress, although as we saw earlier he did not attend, and the contribution he sent was on an entirely different subject. He revived the idea

73 Walter, *Les Figures paysagères*, pp. 274–280.

74 Letter from Richardson Evans to Anselme Changeur, 18 Nov. 1907, London Metropolitan Archives, A/SCA/III/2/10 (Europe).

75 De Clermont et al., *Le Premier Congrès*, p. 43. Its ambition was clearly international, extending beyond the borders of Europe by including the United States. Yet it seems to me that there was a great deal of illusion on the part of the contributors to this congress. Raoul de Clermont's suggestion to include *petites patries* in the conference on the planet's energy resources being planned by Roosevelt a few months later was more the stuff of a pious pledge...

76 De Clermont et al., *Le Premier Congrès*, p. 72.

in September 1910 during a literary and artistic congress in Luxembourg.[77] In 1913, during the Bern Congress, he once again requested the implementation of an international commission to serve as a clearing-house for information on the protection of sites and monuments.[78]

Beyond these stated ambitions, it was ultimately the ambiguity regarding the missions of such a federation that slowed its implementation. If it was simply a matter of putting people in relation with one another, as proposed by Evans, what was the purpose? This had already been done. Furthermore, there was already a great deal of information exchange. To justify a new alliance, especially within a tense international context, it was important to go further, although this would entail national associations submitting to a new governing body that would dominate them – something that both the French and Germans fiercely refused.[79] Hugo Conwentz, for example, always advocated a non-restrictive form of international cooperation. This resistance was present during the international conference for the protection of nature held in Bern in 1913 by the naturalist Paul Sarasin:[80] it took its place instead within a tradition that was parallel to the 1909 congress, as the central issue was the protection of flora and fauna rather more than landscapes. Still, the unavoidable Conwentz was present, as were French representatives from the Muséum national d'Histoire naturelle and the Touring Club. An advisory commission tasked with gathering and diffusing information on the protection of species was established,[81] whose prerogatives were limited precisely due to openly expressed national resistance.[82]

The source of this apparent unwillingness to establish an international federation can be traced to the difficulty of internationalising what was a national and even nationalist movement. Luigi Piccioni has shown the difficulty, notably with respect to patrimonial protection of nature, of transplanting a concept from one country to another. While the notions of 'landscape' and 'natural monuments' enjoyed considerable success, the French idea of 'natural

77 De Clermont et al., *Le Premier Congrès*, p. vi.

78 Holdgate, *The Green Web*, p. 10.

79 A simple governing body *coordinating* national activities would not have been of great interest.

80 Holgate, *The Green Web*, p. 11, briefly discusses this resistance.

81 Donato Bergandi and Patrick Blandin. 'De la protection de la nature au développement durable: Genèse d'un oxymore éthique et politique', *Revue d'histoire des sciences* **65** (2012):116–117; and Anna-Katharina Wöbse, 'Separating spheres: Paul Sarasin and his global nature protection scheme', *Australian Journal of Politics and History* **61** (2015): 339–351.

82 Wöbse, 'Les liaisons sinueuses', 113.

sites of artistic interest', which was the foundation of the 1906 Beauquier Law, struggled to spread, despite being unanimously praised abroad. The same was true of the Italian concept of 'natural beauty' (*bellezze naturali*), whose marginal success led to its abandonment in Italy itself.[83] Even more fundamentally, following the German model made the defence of national identity the primary reason for protecting natural and historic monuments, sometimes to the detriment of other arguments, such as economic or aesthetic ones. This helps explain why it was so problematic to found an international federation protecting the nation… At most, as suggested by Fuchs in 1909, such a union would help combat what he referred to as 'cosmopolitism', without a doubt implying a form of devastating modernity originating from the United States – one that took aim, he pointed out, 'first and foremost [at] the conservation of national characteristics'.[84] With this in mind, we nevertheless cannot see what purpose such a federation would have served, or how it could have acted.

All in all, the 1909 congress can be considered as both an outcome and a phase. Indeed it crowned the internationalisation of movements for the patrimonialisation of nature that had begun in the late nineteenth century, attesting to increased Germanic influence, to the detriment of the UK, as well as to the strong bonds that had been established between associations since the early twentieth century. While in the short term it did not trigger a European dynamic, it nevertheless was a stage in implementing – for lack of a genuinely European space for debate – a transnational space of communication in Europe bringing together intellectuals, artists, scientists and a few politicians around the question of preserving national landscapes. A certain amount of seemingly fairly favourable publicity was given to these exchanges in the press.

As in many other areas, the First World War would profoundly transform this movement, as the scope of destruction alarmed people more than ever about the harmful power of humans and the limits of the Earth.[85] The patriotic approach to the protection of nature would begin to diminish – without

83 Piccioni, 'L'influence de la France', 99–100

84 Dr. Fuchs, 'L'œuvre du *Heimatschutz*', 43.

85 Mathis, '*La Terre vaine*. Mutations du sentiment de la nature', in Alain Corbin, Jean-François Courtine and Georges Vigarello, *Histoire des émotions* vol. III (Paris: Seuil, 2017), pp. 201–202.

disappearing, far from it – in favour of more scientific conceptions. François Walter has rightly asserted that 'after 1920, discourse on the protection of nature cleared itself of nationalist accents'.[86] This enabled an internationalisation on new bases: the congresses of 1923 and 1931 no longer focused exclusively on landscapes, but also included endangered flora and fauna, more in keeping with the foundations established in Bern in 1913 than the conclusions of Paris in 1909; naturalists gradually imposed themselves, with ecology overshadowing the aesthetic-patriotic approach. These meetings took place, in a way, amid the ruins of the 1909 congress, as they often included the same men and even the same associations. While there were new faces, along with new highlighted issues, the networks established in 1909 were not completely wiped out. Finally, the links established between environmental engagement and regionalism did not fade entirely, as they were sometimes present during the structuring of the environmental movements of the 1970s.[87] France ultimately continued to play an important role in this internationalisation of the movement, although it did so by increasingly including its colonies, as did other imperial powers.[88] It was thus a 'green' international with a different face that became established during the interwar period, whose offshoots survived the Second World War and which served as a basis for creating the International Union for the Protection of Nature in 1948, in which Europeans had an important role until the 1960s.[89]

86 Walter, *Les figures paysagères*, p. 279.

87 On this topic, see the divergences underscored by Martin Siloret between the environmental movement in Brittany, which was highly marked by regionalism, and that of Wales, which was less connected to it. Martin Siloret, 'La structurataion partisane de l'écologie politique: une comparaison Bretagne – Pays de Galles (1974-1995)', Ph.D. in history under the direction of Jacqueline Sainclivier, Université de Rennes 2, 2017, ch. 2.

88 Ford, *Natural Interests*, ch. v.

89 Holdgate, *The Green Web*.

CHAPTER 2.

THE HISTORICAL ROOTS OF THE EUROPEAN CULTURE OF CATASTROPHES

François Walter

The environment occupies a prominent place among those historical fields attracting keen interest. Beginning with Michelet in the nineteenth century, it became common to consider the relations between humans and their environment as a long-term struggle, in which the former would inevitably win out by harnessing the forces of nature and mastering adversity. More recently, with the reversal of perspective prompted by the discrediting of overly anthropocentric views on the topic, it has become clear that, on the contrary, human societies are indeed the primary predator on the planet. Of course the bacteria of 3.5 billion years ago – the very ones that began the recycling of carbon and made life possible – changed the environment much more radically than we humans of the twenty-first century. Still, humans have emerged as a 'macroparasite' that incessantly transforms the planet, to the point of endangering it altogether. From there it is just a step to reflecting on history in general as a catastrophic scenario, a step that certain historians have no fear of taking.[1]

In 1990, the philosopher Michel Serres began his book *The Natural Contract* by describing the duel with sticks (*Duelo a Garrotazos*) painted by Francisco Goya in 1820–23. Two men are fighting in a patch of quicksand, which Serres comments on in the following terms: 'With every move they make, a slimy hole swallows them up, so that they are gradually burying themselves together'. The pace at which they sink depends on their aggressiveness. Yet 'the belligerents don't notice the abyss they're rushing into; from outside, however, we see it clearly'. This blindness of human beings who are occupied with their small and large disputes, as if nothing were happening, prevents them from detecting the slimy ground in which society as a whole is floundering. The conclusion is clear:

1 This assessment received broad media attention after scientists such as James Lovelock sounded the alarm during the 1980s. Such suppositions are at the foundation of global history research on the Anthropocene. See, for example, among the classics, John R. McNeill, *Something New Under the Sun: An Environmental History of the Twentieth-century World* (London: Allen Lane, 2000).

> earth, waters, and climate, the mute world, the voiceless things once placed as a decor surrounding the usual spectacles, all those things that never interested anyone, from now on thrust themselves brutally and without warning into our schemes and maneuvers. They burst in on our culture, which had never formed anything but a local, vague, and cosmetic idea of them: nature.
> What was once local – this river, that swamp – is now global: Planet Earth.[2]

Natural hazards – events that are always unforeseeable and potentially threatening – and more or less vulnerable societies come face to face. Under certain circumstances, their interaction transforms into an extreme event that is both destructive and abrupt: this is what is commonly meant by the term 'catastrophe'. A short and oft-cited phrase by Max Frisch ably captures what may seem obvious, but must constantly be pointed out: 'Only human beings can recognize catastrophes, provided they survive them; Nature recognizes no catastrophes.'[3]

Moving beyond this observation, some authors believe that a desire for catastrophe is a constitutive element of postmodern culture, like the duty of remembrance that serves as its counterpart.[4] Affective engagement with the past and catastrophic sensibility converge in practice. Without a doubt, the flow of memory characteristic of the system of historicity in place since the 1990s is linked to the great catastrophes of the twentieth century, including wars, the Holocaust and genocides. Furthermore, remembrance on other levels still remains a working-through of deep, often suppressed, social and cultural wounds, which burst forth brutally. Memory includes a sacred dimension from which history precisely tries to free itself, in an effort to provide a detached view of the break, one that is both constructed and distant. Hence, in studying alpine society, researchers have successfully emphasised the essential memorial component of the management of natural catastrophes.[5] This includes a commemorative dimension that is codified and legitimised (monuments and scenographies). It also takes place through narratives and images, which constitute a genuine 'collective knowledge of the catastrophe'.[6]

2 Michel Serres, *The Natural Contract*, trans. Elizabeth MacArthur and William Paulson (Ann Arbor: University of Michigan Press, 1995), pp. 1 and 3. This reading is of course foreign to Goya, as the disappearance of the protagonists' legs is simply the result of a poor restoration of the work!

3 Max Frisch, *Man in the Holocene*, trans. Geoffrey Skelton (New York: Harcourt Brace, 1981), p. 79.

4 This is the argument proposed by the collection of articles edited by Peter Gray and Kendrick Oliver, *The Memory of Catastrophe* (Manchester & New York: Manchester University Press, 2004).

5 René Favier and Anne-Marie Granet-Abisset (eds), *Histoire et mémoire des risques naturels* (Grenoble: MSH-Alpes, 2000).

6 René Favier and Anne-Marie Granet-Abisset (eds), *Récits et représentations des catastrophes depuis l'Antiquité* (Grenoble: MSH-Alpes, 2005).

This arrangement can be found in other European contexts, especially in the Mediterranean.[7] This movement does not solely affect Europe, and includes a patrimonial dimension when the sites of extreme events are transformed into memorials or recognised by UNESCO's 'world heritage' label.[8]

Social sciences and catastrophes

Historical anthropology, which is primarily interested in the structuring of these experiences, has made the vulnerability of human societies a central notion since the 1980s. It is because the explanatory factors reside more in the society itself, rather than in natural conditions, that assessing the degree of vulnerability has become a central topic of research.[9] Vulnerability, which results from economic and social inequality, is distinct, varying across the societies, historical periods, and modes in which the event qualified as catastrophic is perceived and represented.[10] This concept inevitably relates to a characteristic of fragility, to something that is suffered. Yet it also opens onto resilience, either the mechanisms or technological resources that allow for confronting a catastrophe, which is to say the system's capacity to regain its previous state of balance. Catastrophes became a genuine historical subject approximately twenty years ago, when understanding an event itself was joined by new focus on the distinctive features of the social group and context that determine the capacity to anticipate, react, resist, and recover from the potential realisation of a risk. This approach includes the cultural resources that define how potentially harmful events are perceived and inscribed with meaning.[11]

Research that is already well-established has explored catastrophic phenomena within a broad social context, by emphasising their almost

7 See Domenico Cecere et al. (eds), *Disaster Narratives in Early Modern Naples: Politics, Communication and Culture* (Rome: Viella, 2018).

8 A good example is the icon used for the 'Saguenay flood', which depicts the small white house in Chicoutimi (Quebec), in the heart of the memorial park commemorating the 1996 catastrophe. Among examples of patrimonialisation, one could cite the recent addition (Nov. 2018) to the intangible cultural heritage list of 'avalanche danger management', jointly received by Austria and Switzerland.

9 Regarding the inclusion of risk in the social sciences, see Claude Gilbert, 'Quels risques pour la recherche en sciences humaines et sociales?', in Dominique Bourg, Pierre-Benoît Joly and Alain Kaufmann (eds), *Du risque à la menace: penser la catastrophe* (Paris: PUF, 2013), pp. 217–236.

10 Susanna M. Hoffman and Anthony Oliver-Smith (eds), *Catastrophe & Culture: The Anthropology of Disaster* (Santa Fe: School of American Research Press, 2002), pp. 60–62.

11 Current research generally distinguishes five characteristics: 1) the catastrophe itself, always indexed to what humans suffer from it; 2) objective natural dangers such as hurricanes, lightning or avalanches; 3) social, economic, physical and psychological vulnerability; 4) resilience; 5) cultural resources.

structural nature and revealing a society's mentality.[12] It is also impossible not to cite the foundational article by Lucien Febvre, who over sixty years ago proposed studying the need for security and its various manifestations in Western culture.[13] There is also no lack of monographs that have studied major urban fires, volcanic eruptions, and earthquakes.

In Europe, it was in the Germanic field in particular that research on risks and catastrophes (*Katastrophenforschung*) was established as a topic earlier than elsewhere.[14] The expression, which is derived from the English terms hazard research or disaster research, does not have an equivalent in other European languages, which is due not only to the inventive flexibility of the German language,[15] but also and especially to the existence of an authentic disciplinary field. This type of research is interested in the process of catastrophe management, and therefore in the conditions of action in situations of crisis (what is called disaster management, or *Katastrophenmanagement* in German).[16] The context for this was no doubt awareness of the potential technological failures and environmental dangers of industrialisation and nuclear energy. More recently, it has turned toward analysis of the risks connected to climate change.[17]

For all that, the cultural history of catastrophes owes its rise and affirmation as an independent field of research to a historian specialising in the early modern period, Manfred Jakubowski-Tiessen (1948–), who wrote a study on the storm tides of 1717 that devastated North Sea coasts from the Netherlands to Denmark.[18] Without neglecting the impact and management of the

12 I am thinking in particular of Jean Delumeau's explorations of fear and the feeling of security in the West, with natural calamities serving as one of the matrices: Jean Delumeau, *Rassurer et protéger: le sentiment de sécurité dans l'Occident d'autrefois* (Paris: Fayard, 1989).

13 Lucien Febvre, 'Pour une histoire d'un sentiment: le besoin de sécurité', *Annales E.S.C.* **11** (1956): 244–247.

14 See Walter François, 'Thinking the disaster: A historical approach', in Gabriele Duerbeck, Urte Stobbe, Hubert Zapf and Evi Zemanek (eds), *Ecological Thought in German Literature and Culture (Ecocritical Theory and Practice)* (London: Lexington Books, 2017), pp. 161–174.

15 In the early 1990s, there were no less than 82 compound words in German including risk either as a suffix or prefix! See Wolfgang Bonss, *Vom Risiko : Unsicherheit und Ungewissheit in der Moderne* (Hamburg: Hamburger Ed., 1995).

16 See Wolf R. Dombrowsky, 'Sozialwissenschaftliche Katastrophenforschung und Sicherheitsdiskurs', in Siedschlag Alexander (ed.), *Methoden der Sicherheitspolitischen Analyse* (Berlin: Springer, 2014), pp. 223–236.

17 Martin Voss (ed.), *Der Klimawandel. Sozialwissenschaftliche Perspektiven* (Wiesbaden: VS Verlag für Sozialwissenschaften, 2010).

18 Manfred Jakubowski-Tiessen, *Sturmflut 1717: die Bewältigung einer Naturkatastrophe in der Frühen Neuzeit* (Munich: Oldenbourg, 1992).

catastrophe, the author was attentive to contemporary discourse regarding the event. The task of interpreting its meaning was left to experts from the period, namely men of letters and essentially theologians. Their reading of it connected knowledge of the natural sciences with an analysis of divine intervention – nuanced by their belonging to either Lutheran orthodoxy or Pietism – and references to Enlightenment ideas. Without yet gauging the innovative impact of this decentring of the issue, Jakubowski-Tiessen had initiated the new approach of the cultural history of catastrophes.

The heuristic association of disaster research and cultural anthropology resulted from the rediscovery of the religious dimension. It was no longer possible, as had too often been the case, to preserve two major paradigms: one older and considered obsolete, interpreting the catastrophe as a supernatural and exogenous phenomenon; the other connected to modernity, presenting it as a natural and endogenous phenomenon, a subject of scientific knowledge. Henceforth, the religious and the symbolic also contributed to a comprehensive explanation over the *longue durée*. It is therefore important to avoid thinking that the disenchantment of the world, initiated by the Enlightenment, definitively relegated the validation of this type of intelligibility to the past. In fact, the rational ontological topos of modernity did not simply replace an earlier model of interpretation, but rather superimposed itself on the former, thereby increasing the number of explanatory hypotheses, which were so many resources available for societies confronted by the need to understand and explain the world. Rational and religious readings were not necessarily seen as being antagonistic to one another, and have cohabited over the *longue durée* up to the present.[19] Crises and catastrophes are first and foremost indicators of an understanding of the world. As such, there has been a gradual development of the significations ascribed to natural events. Legitimacy of interpretation became a consideration, an opportunity for rivalry among those who possess the authority to produce theological, scientific, management or simply narrative discourse.[20]

The field gradually developed from the environmentalist standpoint, which made it possible to connect information from the natural and social

19 On religious interpretation as a global explanatory model, see Manfred Jakubowski-Tiessen, 'Mythos und Erinnerung: einige kommentierende Anmerkungen über Städte aus Trümmern', in Andreas Ranft and Stephan Selzer (eds), *Städte aus Trümmern: Katastrophenbewältigung zwischen Antike und Moderne* (Gottingen: Vandenhoeck & Ruprecht, 2004), pp. 274–286.

20 The symbolic and religious as a global explanation is the central argument of our cultural history of catastrophes. See François Walter, *Catastrophes: une histoire culturelle XVIe–XXIe siècle* (Paris: Seuil, 2008).

sciences.[21] In the French-speaking world, the social history of catastrophes responded to the requests of local and regional authorities interested in the risks of avalanches and torrential flooding in mountainous environments.[22] Not satisfied with their role as purveyors of old documents, which the natural sciences expected of them, researchers in the social sciences engaged in risk prevention projects in order to give their work on the memory and culture of risk genuine legitimacy, as well as to play a uniting role in this composite and necessarily fragmented disciplinary field.

Evoking a European culture of catastrophe calls for a dual approach: a history of 'cultures of risk' as well as a history of the 'culture of risks'. The first relates to the knowledge required by practices of risk management (from protection against catastrophes up through reconstruction).[23] The second especially emphasises the perceptions and behaviours of Western society in the face of environmental risks, along with the social and territorial differentiations that characterise their realisation.[24] Beyond their heuristic finality, these two approaches complement one another, and connect in a highly pragmatic way.

The interest of the social sciences in these questions is of course closely linked to the emergence of a public sphere in which social concerns are expressed. The major shift took place during the 1970s. A series of catastrophes highlighted the vulnerability of technological systems: the sinkings of the *Torrey Canyon* (1967) and the *Amoco-Cadiz* (1978); and the chemical

21 See F. Walter, 'Paysage et environnement en histoire: échapper au brouillage', *Information géographique* **3** (2014): 26–41. See also Christof Mauch and Christian Pfister (eds), *Natural Disasters, Cultural Responses. Case Studies Toward a Global Environmental History* (Plymouth: Lexington Books, 2009). A good review of late twentieth-century European literature from the perspective of an American historian can be found in J.R. McNeill, 'Observations on the nature and culture of environmental history', *History and Theory* **42** (2003): 5–43. For recent history in the field of catastrophes, see Stefan Willer, 'Katastrophen: Natur – Kultur – Geschichte. Ein Forschungsbericht', in *H-Soz-Kult* 13 Sept. 2018, http://hsozkult.geschichte.hu-berlin.de/forum/2018-09-001

22 René Favier and Anne-Marie Granet-Abisset (eds), *Récits et représentations des catastrophes depuis l'Antiquité* (Grenoble: Maison des Sciences de l'Homme-Alpes, 2005); Favier and Granet-Abisset (eds), *Histoire et mémoire des risques naturels* (Grenoble: CNRS - Maison des Sciences de l'Homme-Alpes, 2000).

23 The cultural history of risks is notably illustrated by Emmanuelle Collas-Heddeland et al., *Pour une histoire culturelle du risque : genèse, évolution, actualité du concept dans les sociétés occidentales* (Strasbourg: Éditions Histoire et anthropologie, 2004). See also F. Walter, 'Pour une histoire culturelle des risques naturels', in Walter, Bernardino Fantini, and Pascal Delvaux (eds), *Les cultures du risque (XVIe–XXIe siècle)* (Geneva: Presses d'histoire suisse, 2006), pp. 1–28.

24 This research field was notably defined by Sandrine Glatron, 'Culture des risques', in Vincent Moriniaux (ed.), *Les risques* (Nantes: éditions du temps, 2003), pp. 71–87.

accidents of Bolsover (1968), Bitterfeld (1968), Flixborough (1974) and Seveso (1976). The economic recession of the mid-1970s helped bring new attention to topics of risk and prevention. Growing concerns initially surrounding the proliferation of nuclear weapons (Euromissile crisis), which subsequently became very real with the Chernobyl (Ukraine) reactor fire in 1986, gave a planetary dimension to these preoccupations. It was initially in Europe that critical ecology revealed the apocalyptic blindness of a society capable of self-destruction. Philosophers such as Günther Anders and Hans Jonas theorised the planned catastrophe. The dark possibilities of climate change were grafted onto these foundations in the late twentieth century.

Multiple interpretive sequences

If we now try to structure into descriptive models the diverse representations that later underpinned practices for risk and catastrophe management during the last three centuries, three sequences emerge: 1) Societies of protection up to the eighteenth century; 2) Societies of prevention during the nineteenth and twentieth centuries; 3) Risk society (since the late twentieth century).[25] The temporal divisions are purely indicative, as the duration of one attitude was not necessarily exclusive relative to the ensuing or preceding one: their temporalities are interlinked. In reality, complex practices overlapped, ceaselessly driven by new waves that did not, for all that, eliminate earlier realities. At each of these stages and crises, different social actors, based on their conditions and contexts, encoded information, reconstructed for their own use, and updated through power relations their confrontation with risks and catastrophes.

European societies and protection

When they experienced a catastrophe, whatever it might be, so-called traditional societies from the seventeenth and eighteenth centuries knew quite well how to connect the event to natural factors. The hazards or dangers connected to the climatic or geographic context were well known, and were connected to the possible event, whose probability of realisation was totally unpredictable. Protective measures were therefore taken and ceaselessly improved depending on the experience of the catastrophe. Coasts exposed to storm tides (the North Sea) were lined with networks of dykes. In the

25 Here we are returning to and expanding the sequences sketched out in our article 'Catastrophes', in Dominique Bourg and Alain Papaux (eds), *Dictionnaire de la pensée écologique* (Paris: PUF, 2015), pp. 131–135.

Alps, villages threatened by avalanches were topped by protective forests, in which it was strictly forbidden to chop down trees. Beginning with the Middle Ages, cities issued construction regulations to protect against fire. Even though earlier societies clearly developed authentic knowledge of natural limits, we should be wary of overestimating ancestral knowledge. In many domains (avalanches, floods), knowledge remained fragmentary and evolving. Wisdom and good sense were not necessarily sufficient in the face of highly unpredictable risks.

If a catastrophic event took place despite these protective systems, it was probably because it was not entirely reducible to natural causes. Exceptional events had exceptional causes. This is where the supernatural dimension intervenes, the action of Divine Providence, which exercises its power of retribution. Many old engravings illustrating burning cities inevitably represent lightning in the sky, and above the clouds a punitive God. In Judeo-Christian traditions, lightning was one of the instruments used to punish humans. As a result, the essential question was to know why God decided specifically to punish a particular city at a particular time. The suppositions that served as answers were recorded in the great number of sermons that provided the meaning of the event, which incidentally were for a long time the only vehicles for diffusing information. To take just one example, when Saint Michael's church in Hamburg caught fire by lightning strike in 1750, no fewer than twenty sermons offered clever theological craft to explain the destruction of a church, while the neighbouring homes suspected of all kinds of turpitude were spared.[26] During the eighteenth century, occasional and later periodical publications also covered these events, and following the example of scientific texts, tended to minimise or even elide completely the Providentialist dimension of the catastrophe, which nevertheless remained highly present until the following century in traditionally Protestant countries.

The insistence on evoking the punitive action of God served not only to inform, but also and especially to influence future behaviour, in an effort to better protect against the hazard. The associated lexical field contains biblical vocabulary, including terms such as scourge, calamity, and disaster,

26 Walter, *Catastrophes*, pp. 114–115. For examples in a highly different context, see Armando Alberola et al. (eds), *Desastre natural, vida cotidiana y religiosidad popular en la España moderna y contemporánea* (Alicante: Universidad de Alicante, 2009).

but very rarely 'catastrophe'.[27]

In simplifying, one could say that a moral and social disorder which contemporaries of the event strove to identify led to direct intervention by God, who momentarily changed the rules by which natural phenomena function in order to punish. In terms of managing the catastrophe, the direct consequence was the strengthening of norms, both technological (regulating construction) and moral (for instance a ban on behaviour deemed to be licentious, such as dancing, or restrictions on alcohol consumption). In addition, there was recourse to religious rites such as the invocation of patron saints, with whom the community symbolically concluded a contract. Pilgrimages and other processions could function over a very long period of time as reducers of uncertainty in the face of the unforeseeable nature of hazards. To take just one example, facing the advancing Aletsch Glacier threatening their pastureland and homes, inhabitants of the village of Fiesch (Valais Alps) committed in 1678 to mending their ways as part of a procession held each year on 31 July. The measure has worked so well that, in the early twenty-first century, with the glacier withdrawing beyond their expectations, the local community now fears water scarcity. As a result, in 2010 it took measures to obtain papal authorisation to invert the ritual's direction, so as now pray for the preservation of the glacier. It would be wrong to wax sarcastic about the convictions of these mountain dwellers.

This reactive ensemble can be referred to as 'restorative reaction'. Society reacts in an effort to return to the order preceding the transgression, often identified with an ethical fault, whether individual or collective, obvious or latent. With regard to fires, this helps us better to understand attitudes that our purely rationalist contemporary protective standards might consider absurd. When a fire occurs, the local community can hesitate over whether to first extinguish it, or to proceed with exercises in piety (collective prayer and rituals of intercession). The logic of this hesitation, along with the primacy granted to the spiritual attitude instead of the collective effort to fight the fire, is rooted in the conviction that there is no point in fighting against a material fire while divine anger remains unappeased.[28] The most pragmatic communities combine the two attitudes: while able-bodied men go about putting

27 The word was used occasionally, especially during the Lisbon earthquake (1755). On the different types of discourse, see Andrea Janku et al. (eds), *Historical Disasters in Context: Science, Religion, and Politics* (London: Routledge, 2012).

28 Maria Luisa Allemeyer, *Fewersnoth und Flammenschwert: Stadtbrände in der Frühen Neuzeit* (Gottingen: Vandenhoeck & Ruprecht, 2007).

out the fire, women, the elderly and children withdraw to one side to pray. This kind of scene became a commonplace in iconographic representations.

Earlier societies subsequently appear as societies of 'protection' against hazards, in the face of known and recurring risks. They process danger retroactively, through 'management carefully framed by uncertainty, and a method by which responsibility is shared between humans and non-humans'.[29]

European societies and prevention (nineteenth to twentieth centuries)

Between our system of intelligibility for catastrophes and that of earlier societies, there is room for notions of planning and prevention. During an initial phase, which overwhelmingly concerns the nineteenth century, there was particular talk of 'planning', a concept related to that of chance, which is inherent in the hazards of existence. On an individual level, we also try to take into consideration a future that is not just the cyclical repetition of the past. However, a mentality that is concerned with preventing new disasters by taking a whole society into consideration can truly develop only when the struggle against natural forces, along with their mastery by sciences and technology, is seen as a purpose of human history. Instead of a fundamentalist conception of Providence, the interpretive framework is surely that of the nature/society dichotomy and the conflictual relations between the two. Instead of a fatality that is suffered, natural disasters enter the domain of 'prevention'. This is another term whose meaning has been expanded. Until the early twentieth century, the word 'prevention' was used primarily in law to designate the right to exercise a prerogative before a third-party. It was only gradually that it took on its contemporary meaning of measures seeking to diminish the risk of accidents or illness, in other words to include the future within the perception of hazards.

This change in perception was not straightforward, especially because it seemed to infringe upon the laws of Nature as sought by Providence. Was defending oneself from lightning by installing a lightning rod on one's roof, or avoiding illness through preventative vaccination, a way of avoiding the possibility of divine punishment? This type of questioning greatly disturbed minds in the eighteenth century.

Two domains served as experiments for preventive practices. The oldest

29 René Favier (ed.), *Les pouvoirs publics face aux risques naturels dans l'histoire* (Grenoble: MSH-Alpes, 2002).

one is a concern for mitigating the damage caused by fire through construction norms and the first insurance establishments, which emerged in Germany and England during the second half of the seventeenth century, and in the remainder of Europe during the eighteenth century. Later, late nineteenth-century fears of health catastrophes were inscribed in a context in which prevention required the certainty of science. It therefore developed in the wake of Pasteurian discoveries regarding contagion, which clearly demonstrated that the wellbeing of an individual could depend on the behaviour of the person next to them. As a result, the fight against tuberculosis led to the establishment of a kind of medicalised establishment precisely known as a 'preventorium', before ill persons were treated in 'sanatoriums'. In an almost obsessive manner, it fell to public authorities to identify the vectors or to indicate those responsible for the risk, in order to eradicate the potential source of the accident or epidemic. This was dominant until the 1970s, when the public sphere expanded and grew more complex, and prevention chiefly took the form of information. This was the ideal of engineers and technicians, who were convinced that science and technology could master the forces of nature. It is also important to note that territorial management policies were born from this movement. From their very beginning, they constructed a legitimising discourse based on the perspective of a catastrophe to be avoided, whether in connection with forests (fight against flooding attributed to reckless clearing of high-altitude land), protection against avalanches, systematic use of dykes along rivers to guard against floods or sanitary control of water resources.

It was in this context that the word 'catastrophe' in its usual sense, as a major accident with horrendous consequences, gradually replaced within discourse (especially that of the media reporting about them) terms with stronger connotations (scourge, disaster) from the preceding period.[30] The new concept began to spread in the mid-nineteenth century, as the legitimate discourse for providing a plausible explanation distanced itself from the religious sphere of interpretation.

With regard to fires, this logic was particularly successful. The fires that devastated entire cities over the centuries became rare and even tended to disappear after 1850, which explains the media impact of major residual catastrophes, such as the Hamburg fire of 1842. However, traditional practices

30 The word 'catastrophe' originally belonged to the semantic register of dramatic theatre, denoting an ill-fated end to the plot. Its current meaning as an extreme event hardly dates back before the 1860s, although the word was sometimes used during the eighteenth century.

of a religious kind did not disappear. During major floods (which tended to become frequent with urbanisation and the trend of global warming), people made recourse to collective events of religious fervour, although these came after efforts to reinforce dykes.[31] They proved highly effective, for as we know peak flooding is always followed by a drop in water level!

If we try to reduce this period to a simple pattern, one could say that it favoured a reactive sequence in which accidents were attributed to chance instead of Providence, or connected to bad fortune or breakdowns in technological measures. This led to a certain reduction in the sense of individual responsibility, in favour of intervention by the state, social institutions and insurance. Such an evolution was obviously connected to a new industrial civilisation that experienced increasing risks due to industrial and mining activity. When a catastrophe occurred, it was crucial to re-establish the situation that existed before the event. The reconstruction phase was an important part of the post-traumatic phase, with the primary goal of consolidating anything that could 'prevent' the probability of a hazard.

The situation is stabilised through reaction, with the catastrophe emerging more than ever as a social construction ascribing value to technological assets, engineering knowledge, civic solidarity, the competence of public authorities, the effectiveness of health systems and coverage by insurance establishments. This reactive model, which was used by those in power, seeks to minimise the responsibility of state authorities and economic leaders by attributing the errors to the inherent risks of technological development. Accidents are consequently part of the natural order of things, an inevitable aspect of the system that must be addressed by the insurance-based society, and eventually solved by technology.

The model seems to have functioned well, at least until a new type of catastrophe – sparked by what was called 'major technological risks' during the 1970s – challenged certain established certainties or methods for applying prevention. These methods would quickly spread across the globe, although experimentation with them initially took place in Europe.

31 See, for example, the 1866 floods of the Loire River in Guillaume Cuchet, 'Trois aspects de la crise des représentations de l'action de Dieu dans l'histoire au XIXe siècle', *Transversalités* **128** (2013): 13–25.

The 'Risk Society' (1990 to the present)

Since the nineteenth century, collective perception has slowly shifted from a vision based on protection, and later planning, to one based on risk, which is to say a notion oriented even more directly toward the future. While earlier societies were confronted with recurring risks, which were highly localised or regional, contemporary societies now also face new and emerging risks. The former only considered known risks whose danger was established by experience, and was confined to the uncertainty of the hazard. By way of precaution, the latter also took into consideration hypothetical and potential risks before their realisation, as soon as an intellectual process confirmed their plausibility. For that matter, the scale of extreme phenomena expanded, taking on not only a national but also an increasingly transnational and even global scale.

This distinction between objective probabilities, whose distribution we know, and probabilities that are simply constructed, is a recent one. It provides reassurance and support for the convictions of those who believe that contemporary society is capable of managing uncertainty. The very use of the word 'risk' to designate the condition of contemporary humans became common during the late 1970s. In Europe this concept was still unclear in its usage before the crisis of 1973–74 (called the first oil crisis), but later became pervasive, as if the word played a role as a rational substitute for the concerns sparked by the announced scarcity of energy resources. However, on a more general level, increased uncertainty and the awareness of vulnerabilities took their place within the context of societal changes: instability of labour markets, the dreaded effects of neoliberalism and globalisation of the economy and the environmental turning point. There was now the conviction that the dangers created by humans were infinitely more serious than any natural catastrophe!

Risk is always an intellectual construction. It is calculable, which means that it is grasped only when we measure its random character, in which we assess the chances of realisation of an encounter between a hazard and a vulnerability. Concretely, this means that it is possible to express the frequency of floods of a certain scope, but obviously not to specify the date. There is a certain confusion in public opinion on this topic, hence the surprise of residents near the rivers of Central Europe, who experienced the 'flood of the century' in 2002, before contending with the 'flood of the millennium' in 2013! The risk's probability of realisation in a way thwarted the formidable mechanism for distancing represented by the very notion of risk itself. There is also a misleading perception in the Alps, when various nivological services daily evaluate the danger of an avalanche based on a scale of seriousness. It is clearly not a matter of specifying the level of risk, although among the public there is a tendency to equate danger and risk. Aren't experts there to calculate and minimise exposure to risk? In other words, the job of experts is to bring risks into existence using quantification and cartography. When the avalanche report announces maximum danger, the intended audience (off-trail hikers) understands maximum 'risk', which is clearly an extrapolation, for this risk – in the event it can be calculated (which is impossible in terrain as differentiated as that in mountainous environments) – varies from one slope to another depending on criteria such as gradient and exposure. This is why, when a terrible event occurs, the discourse of survivors vacillates between a number of equally irrational attitudes: some return to traditional considerations regarding the inscrutable intentions of Divine Providence, even if it means seeing the event as a warning. Others speak of the need to respect the mountain, whose natural and supernatural power punishes those who violate unwritten laws (especially city-dwellers with a penchant to see the snowy peaks as no more than a playground). Finally, at a push, Nature itself is suspected of striking in totally immoral fashion, for any type of catastrophe is out of place in a leisure or vacation setting. This attitude is glaringly present after the tsunamis that devastate tropical islands.

Some analysts go so far as to normalise our society's global perception as being that of a 'risk society'. This expression was proposed by the German sociologist Ulrich Beck, who believed that the major difference between the past and the present is that the latter is 'characterized essentially by a *lack*: the impossibility of an *external* attribution of hazards ... While all earlier cultures and phases of social development confronted threats in various way, society today is *confronted by itself*...' As a result, there is no longer anything external to the social world. Even nature is integrated, to the point that there is no longer 'any reserve to which we can reject the "collateral damage" of our actions'.[32] Risks are produced by society itself. The immediate consequence of such an epistemological choice is to strip the natural catastrophe of its nature as an unforeseeable event, and to categorise it as an accident provoked by human incompetence. Yet are they still true dangers, or simply a modification of thresholds of tolerance? It is a question that bears asking, despite the lack of a simple answer.

32 Ulrich Beck, *Risk Society: Towards a New Modernity*, trans. Mark Ritter (London: SAGE Publications, 1992), p. 183. Unlike the two preceding sequences, the model refers to society in the singular, as this form of risk management has a transnational and global dimension.

Formerly, in history, only urban fires belonged to this category, in which it is difficult to separate the effects of natural conditions from those linked to technological choices, lack of precaution or simple recklessness. Today, the tragic reality of major industrial accidents (such as black tides, chemical catastrophes, events with a global impact such as Chernobyl or Fukushima) has confirmed the relevance of this way of conceiving risks. 'Man-made hazards' henceforth dominate, inseparable from a society's degree of development. Previously, natural dangers and extreme events prompted specific social responses. In this new configuration, social practices themselves – by heightening dependence on technological systems that are highly vulnerable due to their interconnectedness – are helping to transform hazard into catastrophe!

What are the paradigms of this new way of thinking? Very briefly stated, it is based on references from the physical sciences that are diverted by the social sciences, and then transformed into social phenomena. An example of this is the Second Law of Thermodynamics (Law of Entropy), which underscores a system's tendency toward disorder and the irreversibility of change. Next, it is important to note the propensity to quantify risks, which is so pronounced that no risk whatsoever can exist until it has been duly mapped and quantified in its probabilities of occurrence. Risk management is based on statistics. This is how we protect ourselves against the possibility of a flood, which is based on a threshold that must not be passed within a given range of time. Nuclear reactors must be shielded from high-water levels, which are statistically measured based on their occurrences and a limit that on average can be surpassed only once every 10,000 years! And yet... Due to the rapid mediatisation of potentially catastrophic events, contemporary societies are increasingly aware of their extreme vulnerability. The more sophisticated the technological systems developed during this second period, the more their exposure to collapse proves evident. Extreme dependence on interconnected networks (energy, information, transportation) actually increases vulnerability.

To put it plainly, this means that the average citizen, who is continually reassured by the discourse of experts, is less and less prepared to suffer the hazards of everyday life, whatever they may be. For that matter, new activities (athletic ones in particular) realise threats that had hitherto remained potential. Finally, how does one form an opinion regarding phenomena as complex as climate change, in which real and observable signs (the shrinking of glaciers for example), which are sometimes counter-intuitive (recurring cold spells), must be integrated within hypothetical and virtual scenarios (the progression of average temperatures). Overall, contemporary society has been unable to narrow the margin separating natural hazards from social vulnerability. Civil society, which is hostage to the media attention that transforms models into realities, seems to be increasingly deprived of critical distance.

An illustration of this is the 2010 eruption of a volcano in Iceland with an unpronounceable name (Eyjafjallajökull). In itself, this natural phenomenon was not a catastrophe, as it took place in an uninhabited area. However, it became one in the ensuing hours and days, as it covered a large part of Northern Europe with ash, paralysing air traffic on a global scale.

In the face of rising uncertainty, a proposal has served as a panacea since the 1990s and guides management by public authorities, namely the 'principle of precaution' (already institutionalised in 1987): taking proportionate measures to prevent irreversible risks at an acceptable economic cost. The limits of this new dogma have already been tested by terrorism, new pandemics, and Fukushima (12 March 2011). The latter catastrophe, which will over time undoubtedly mark a shift in our relation to risks, personifies the essence of post-industrial risk: it is global, simultaneously invisible and furtive, organic and mutating, and incessantly adapting.

At the same time, the new concept of resilience has tended to supplant that of vulnerability. Originating from the field of psychology approximately twenty years ago, it designates the capacity to overcome post-traumatic stress. While the notion of vulnerability has a connotation of passiveness, resilience places greater emphasis on the confrontation of reality and capacity for action, with the goal being to re-establish the balance of the social system.[33]

There remains the sensitive question of the particularity of Europe, which has been continually underscored in the preceding pages. Of course, the shifts discussed appear to be inseparable from the pace of an increasingly industrialised and urbanised society. From this point of view, the public sphere in which the culture of catastrophe developed was initially European and Western. It was in the European space that public management of the consequences of catastrophic events was tested beginning in the eighteenth

33 On this subject, see Gilbert Claude, 'De l'affrontement des risques à la résilience. Une approche politique de la prévention', *Communication & langages* **176** (2013): 65–78.

century.[34] Destructive earthquakes (Lisbon in 1755, Messina in 1908), epidemics (cholera in 1831–32) and the slaughter of the two world wars were so many events that forged expertise (for better and for worse). It is hardly possible to disregard the weight of this history when confronted by today's planetary risk society. In this sense it is legitimate to speak of a European culture of catastrophe.

In looking more closely, and not simply being content with listing the advances of cultures of risk, history can help teach us to live with danger despite our knowledge and efforts to protect ourselves from it. It helps make a reflexive and proactive risk management more credible by mobilising civic responsibility to a much greater degree than before. This attitude – potentially a new practice model for risks – appears increasingly necessary in a world confronted more than ever by uncertainty, one that is not yet sufficiently accustomed to functioning in just-in-time mode across all domains. It is not a lack of technology that generates insecurity, but rather the difficulty in admitting that risk is henceforth inherent to the way of life. Managing external hazards is no longer sufficient, as what matters is realising how much the unthinkable and the uncertain are part of normal life, something that is illustrated by today's outward signs of climate change, which was long deemed to be improbable. The world of possibilities has transformed into the fulfilment of the probable. It is important for us to take this into account if we want to continue to endure.

34 For the beginnings of a comparative history, see Gerrit Jasper Schenk, *Historical Disaster Experiences: Towards a Comparative and Transcultural History of Disasters across Asia and Europe* (New York: Springer Verlag, 2017).

CHAPTER 3.

EUROPE AND CHERNOBYL: CONTESTED LOCALISATIONS OF THE ACCIDENT'S ENVIRONMENTAL, POLITICAL, SOCIAL AND CULTURAL IMPACT

Karena Kalmbach

Considering the global dimension of the Chernobyl debate, does it actually make sense to inquire into a connection between Chernobyl and a European public sphere – as suggested by the editors by inviting me to contribute an article to this compilation? It does indeed. Because the question of Chernobyl's 'Europeanness' has been debated in a very particular public sphere – creating a discourse which claimed itself to be in the position to define what Europe actually is.

Was Chernobyl an accident in a European nuclear power plant? As simple as this question might seem, every trivial answer of 'yes' or 'no' implies far-reaching statements: on the geo-political boundaries of Europe as well as on nuclear issues. First, this question touches upon the issue of Ukraine's integration into the European Union – a highly politicised issue that triggered a civil war in this country in 2013. Second, an answer to the question of Chernobyl's 'Europeanness' touches upon crucial nuclear political debates and includes statements on reactor safety, nuclear emergency plans and science diplomacy. The changing discursive localisation of the accident's environmental, political, social and cultural impact as inside or outside Europe thus points to two historical transformations that took place over the last 30 years: the changes within Europe and the changes within nuclear politics. In bringing together these two discourses and pointing out their partial interconnectedness, this chapter sheds light on how changing definitions of 'Europeanness' and changing boundaries of 'nuclearity'[1] have co-shaped the coming to terms with the Chernobyl accident, in particular regarding the allocation of responsibilities and the formulations of 'lessons learned'.

If we consider the institutional level of the European Communities and the European Union, Chernobyl has definitely played an important role in the

1 Gabrielle Hecht, 'Nuclear ontologies', *Constellations* **3** (2006): 320–331.

legal and technological integration processes. In 1986, Chernobyl triggered the establishment of uniform dose limits for radioactive contamination of foodstuff within the European Communities; and, in the 2000s, its legacy paved the way for implementing Western European security standards and thus technologies in Eastern Europe, as conditions for Eastern European countries joining the European Union. In this way, Chernobyl nuclearised the European trade in foodstuff and Europeanised the nuclear techno-political system of the Western part of the former Soviet Union.

But this chapter is not primarily concerned with these legal, institutional and technical histories; it is interested in the question of how *Europe* as geographical, political and cultural entity was defined and negotiated through Chernobyl narratives. Therefore, it investigates Western European Chernobyl narratives that have been present in public discourse over the last thirty years. It starts from a consideration of the various aspects that are negotiated within Chernobyl narratives: from health effects of low-level radiation to risk-taking in modernity. Understanding these multiple layers of Chernobyl narratives and the discursive fields they are interlinked with is crucial for understanding the significance of a narrative localisation of Chernobyl as *inside or outside Europe*. After laying this groundwork, the chapter will delve into concrete Chernobyl narratives brought forward by specific actors at specific moments in time and shed light on the political implications of these narratives – political implications that reach far beyond the field of nuclear politics. In a last step, the chapter will expand on the question of the politics of the Chernobyl discourse itself and turn to sociological concepts which build upon a certain interpretation of Chernobyl. Looking at Chernobyl from this angle allows us to enlarge the question of how Europe is geographically and politically defined and negotiated *through* the Chernobyl discourse, to a contemplation of the question of how a certain interpretation of Chernobyl has created a specific communication sphere on modern risks.

This focus on 'Europe and Chernobyl' should, however, not overshadow the fact that European Chernobyl debates have remained very much contained in their national frameworks. Not only have the varying regional and local agricultural problems shaped the specific national Chernobyl debates.[2] The

2 Comparing the cases of Corsica, northern Sweden and the British Lake District sheds light on the variety of problems the Chernobyl fallout triggered in Western Europe: in Corsica, we still have a vivid debate about the question of whether the Chernobyl fallout actually caused thyroid cancer in children on this island. Through this public debate, many people in Corsica have become familiar with the debate about health effects of low-level radiation. It has probably become common

perception and interpretation of Chernobyl has also depended on various national political and cultural specifics such as, for instance, in Western Europe, the formation, role and status of nuclear 'experts' and 'counter experts'; the shape, political role and protest culture of the anti-nuclear movement; or the media system. But a comparative history of Chernobyl narratives not only tells a story of divergences. As we will see, it is precisely in the question of the 'Europeanness' of this accident that joint narratives cut across local, regional, and national particularities.

Chernobyl: Where, when, and what?

Chernobyl is not just the punctual event that took place in the form of an explosion in a nuclear power plant on the night of 25 to 26 April 1986. It is an ongoing disaster, both from environmental and social perspectives. Neither its environmental impact nor its social impact are limited to the geographic location of the power plant.[3] Chernobyl caused fallout of radionuclides across the northern hemisphere, and severely contributed to the breakup of the Soviet Union. And it directly affected the life and health of countless people, in particular in Ukraine, Belarus and Russia.[4] What is more, the history of the punctual event of the explosion does not start in 1986. The measures taken to mitigate and control the impact of the explosion and the release of radioactive particles have a long prehistory. This prehistory is shaped by the Cold War arms race and the Atoms for Peace programme. But the history of diverse

knowledge among them that radioactive iodine accumulates in sheep milk, which is pillar of the diet of many Corsican farmers. For reindeer farmers in northern Sweden, the knowledge about environmental effects of Chernobyl rather concerns the accumulation of caesium in lichen and moss. For them, Chernobyl as an event meant the mass slaughtering of their flocks. And sheep farmers in the British highlands had to learn about soil specifics that enabled the formation of radioactive hotspots on their pastures. 'Chernobyl' in these terms meant restrictions on the sale and movement of one fifth of the British sheep population.

3 Karena Kalmbach, *The Meanings of a Disaster: Chernobyl and Its Afterlives in Britain and France* (New York: Berghahn Books, forthcoming 2021); Susanne Bauer, Karena Kalmbach and Tatiana Kasperski, 'From Pripyat to Paris, from grassroots memories to globalized knowledge production: the politics of Chernobyl fallout', in Laurel MacDowell (ed.), *Nuclear Portraits: Communities, the Environment, and Public Policy* (Toronto: University of Toronto Press, 2017), pp. 149–189.

4 Adriana Petryna, *Life Exposed. Biological Citizens after Chernobyl* (Princeton: Princeton University Press, 2002); Olga Kuchinskaya, *The Politics of Invisibility. Public Knowledge about Radiation Health Effects after Chernobyl* (Cambridge: MIT Press, 2014); David Marples, *The Social Impact of the Chernobyl Disaster* (Basingstoke: Macmillan, 1988).

technologies such as satellite systems,[5] and in particular the history of medical investigations into the health effects in Hiroshima and Nagasaki[6] also play a prominent role. Furthermore, the actions undertaken in 1986 cannot be understood without knowledge of the history of the emergence of a specific safety culture within the Soviet nuclear programme;[7] the history of radiation protection;[8] and the history of the central actor within the international nuclear-political system – the IAE.[9] If we apply this *longue durée* perspective, Chernobyl transforms itself from a punctual event in the nuclear power plant *Lenin* located a hundred kilometres north of Kiev, into a network of related geographies, events and actors that have been woven together into a wider story about risk-taking in modernity, in particular by Ulrich Beck.[10]

Historical studies which specifically put their focus on nuclear risk-taking normally stop in 1986, such as the recent book by Christoph Wehner.[11] Practical reasons, like the inaccessibility of archival material, can of course justify stopping an historical investigation at a certain point in time. But it is no coincidence that these works stop in 1986: the underlying assumption is that something fundamentally changed with Chernobyl. Numerous political scientists and sociologists have applied the same assumption, and in this way justified only considering the period after 1986. Social scientists' strong interest in 'focusing events'[12] and the impact of such events on mobilisation, agenda-setting and public opinion have shaped many works discussing Chernobyl's political impact. So we have come to think of Cher-

5 Johan Gärdebo, *Environing Technology: Swedish Satellite Remote Sensing in the Making of Environment 1969–2001* (KTH Stockholm: Ph.D. thesis, 2019).

6 Susan Lindee, *Suffering Made Real: American Science and the Survivors at Hiroshima* (Chicago: University of Chicago Press, 1994).

7 Sonja D. Schmid, *Producing Power. The Pre-Chernobyl History of the Soviet Nuclear Industry* (Cambridge: MIT Press, 2015).

8 Samuel J. Walker, *Permissible Dose. A History of Radiation Protection in the Twentieth Century*, (Berkeley: University of California Press, 2000); Cyrille Foasso, *L'Histoire de la sûreté de l'énergie nucléaire civile en France, 1945–2000* (Université Lumière Lyon II: Ph.D. thesis, 2003).

9 Elisabeth Röhrlich, 'The Cold War, the developing world, and the creation of the International Atomic Energy Agency (IAEA), 1953–1957', *Cold War History* **16** (2016): 195–212.

10 Ulrich Beck, *Weltrisikogesellschaft. Auf der Suche nach der verlorenen Sicherheit* (Frankfurt am Main: Suhrkamp, 2008); Ulrich Beck, *Risikogesellschaft. Auf dem Weg in eine andere Moderne* (Frankfurt am Main: Suhrkamp, 1986).

11 Christoph Wehner, *Die Versicherung der Atomgefahr. Risikopolitik, Sicherheitsproduktion und Expertise in der Bundesrepublik Deutschland und den USA 1945–1986* (Göttingen: Wallstein, 2017).

12 Thomas B. Birkland, 'Focusing events, mobilization and agenda setting', *Journal of Public Policy* **18** (1998): 53–74.

nobyl as a breaking point and have divided academic research into historical 'before Chernobyl' and political 'after Chernobyl' stories. This periodisation, however, obscures the many continuities, in particular with regard to the aspects of the important *longue durée* perspectives which I pointed out above. We can't take for granted that Chernobyl was a 'turning point or catalyst in European environmental policy and politics'.[13] And we can't take for granted that it triggered the formation of a European public sphere focused on environmental problems.

The problems caused by Chernobyl, the reactions that these problems triggered and the memories that these problems and reactions created are different all across Europe. They do not just differ between the two sides of the former Iron Curtain. They also differ all across Eastern and Western Europe and even within one and the same country.[14] But the same memories can also be found in different geographical locations. Within these various communities of memory, Chernobyl has come to legitimise the most diverse forms of action. Depending on the discursive context, Chernobyl works as argument for anti-nuclear manifestations (anti-nuclear groups); for charity activities for Ukrainian, Belarusian and Russian children (solidarity movement); or for closer collaboration amongst emergency response forces (radiation protection institutions). These three communities of memory allocate a high historical importance to Chernobyl and consider the accident an event worth commemorating. However, there exists also the interpretation of Chernobyl as an event that is not particularly worthy of commemoration. Pro-nuclear activists and the nuclear industry have framed Chernobyl as one amongst many other industrial accidents. These actors have stressed the low number of immediate human casualties. In this narrative, Chernobyl has

13 In this regard, it makes complete sense to put a question mark after the statement: 'Chernobyl – Turning Point or Catalyst?' as did the organisers of an international conference at the Heinrich-Böll-Stiftung in Berlin in November 2016. Focusing on 'Changing Practices, Structures and Perceptions in Environmental Policy and Politics (1970s–1990s)' the conference aimed at thinking the decades 'before Chernobyl' and 'after Chernobyl' together and embedded them in their broader social-political context. The fact that almost all the presenters did not halt their narrative in 1999 but included present day developments shows that Chernobyl is not only an ongoing disaster, but also an enduring reference point in environmental and energy politics. The presentations were filmed and are available online at https://www.youtube.com/watch?v=KpW5n9GVOtg (accessed 12 May 2020).

14 Karena Kalmbach, 'Radiation and borders. Chernobyl as a national and transnational site of memory', *Global Environment* **11** (2013): 130–159; id., 'Tchernobyl – angle mort', in Étienne François and Thomas Serrier (eds), *Europa notre histoire – L'Héritage européen depuis Homère* (Paris: Les Arènes, 2017), pp. 316–318; Melanie Arndt (ed.), *Politik und Gesellschaft nach Tschernobyl: (Ost-) Europäische Perspektiven* (Berlin: Christoph Links Verlag, 2016); id., 'Memories, commemorations, and representations of Chernobyl', *Anthropology of East Europe Review* **30** (2012): 1–12.

come to legitimise the further intensification of the nuclear programme.

Chernobyl's 'death toll' has thus become a battlefield on which pro- and anti-nuclear activists fight each other. The dispute amongst scientists about the health effects of low-level radiation – which started long before 1986 – allows for the wide range of numbers that each claim to be the 'true' Chernobyl death toll. This debate is not only important for the present and future of the evacuated people and areas around Chernobyl. It gains importance on a global scale, as it influences the evaluation of the health impact of reprocessing plants, nuclear power plants, uranium mines – and, of course, other nuclear accidents. Claims about the Chernobyl death toll directly imply statements on the health impact of the 2011 Fukushima accident: the assumption of a certain number of Chernobyl victims caused by the released levels of radionuclides in 1986 indeed directly implies assumptions of a certain number of Fukushima victims caused by the released levels of radionuclides 25 years later.

Entangled in these global dimensions of the Chernobyl debate lies a discourse that is very much concerned with Europeanness, a discourse in which Chernobyl narratives became a tool of identity politics. It is this particular dimension of the Chernobyl debate to which we turn now our attention.

Negotiating Chernobyl's Europeanness

Ever since the first news reports about Chernobyl, *Europe* formed an inherent part of Chernobyl narratives. Countless accounts published in 1986 stated that large parts of *Europe* were affected by the radioactive fallout and that there was a lack of coordination amongst *European* governments, resulting in very different counter measures taken by each and every country. Statements defining the nuclear-political lessons to be learned from Chernobyl and the sanitary consequences, however, did not refer to *Europe*. These statements instead applied a Cold War mapping, stressing that *the West* did not need to worry about either severe health effects or challenges to its nuclear enterprise.

The evaluations brought forward by politicians, nuclear industry representatives and other nuclear state and industry experts from Western Europe stressed the East-West divide in particular: it was a Soviet nuclear reactor design (the RBMK reactor) that had caused the accident. For example in an interview published in the newspaper *Le Parisien* on 30 April, a representative of the French national radiation protection agency SCPRI stated:

> a major accident like the one in Chernobyl just cannot take place in France because of the difference in design that exists between the plant concerned and the type of

> plants which we build ... Our quality, safety and maintenance controls are a lot more rigorous than those in the USSR.[15]

Thus, the Western European public sphere that inquired into the accident, worried about its consequences and explained why it happened, located Chernobyl outside of its own geographic territory. The event happened in the Soviet Union, not in Europe. In the problem allocation, the political entity was given priority over the geographical entity. Given the context of the Cold War, this geographic alienation is in no way surprising. But it is interesting to see that this narrative of the 'Soviet accident' has remained powerful ever since, despite the fact that the transnationality of the accident's environmental consequences was a key element in gaining knowledge about the event. It was indeed at the Swedish nuclear power plant Forsmark that the release of radioactive particles was detected before anybody in Western Europe had heard about the accident. This material proximity might also, even as early as 1986, have let journalists and politicians locate Chernobyl in their narratives within Europe. After all, the RBMK design was also implemented in Lithuania, a country whose 'Europeanness' has hardly ever been questioned. But Chernobyl was not framed as a European nuclear accident. The fact that fallout from an accident in a Ukrainian nuclear power plant could be detected in a Swedish nuclear power plant was narrated as a sign of the accident's severity (which led to intense rumours about very high numbers of immediate radiation deaths), and not as a representation of the geographical closeness and entanglement of Eastern and Western Europe.

These alienation politics were a common feature across Western Europe. They aimed at stripping any hint of Europeanness from the accident, in order to keep Chernobyl's political, social and economic consequences on the Eastern side of the Iron Curtain. The way nuclear officials, both from the industry and public institutions, explained Chernobyl and its consequences to the wider public thus also had common features across Western Europe. Official statements stressed the safety of national nuclear power programmes and the limited harm caused by the accident outside the immediate surroundings of the plant. For instance, in the UK, 'MPs were assured by both the Prime Minister and the Environment Secretary, Mr Kenneth Baker' that 'Britain has escaped the effects of the nuclear plant disaster in the Soviet Union'.[16]

Stories about careless, drunken Soviet plant operators who were neither

15 'Le nucléaire en France: la sécurité avant tout', *Le Parisien*, 30 Apr. 1986.

16 Alain Travis, 'No radiation threat to the UK, Commons told', *The Guardian*, 30 Apr. 1986.

aware of the dangerous material they were handling nor able to properly control it became a very popular way of explaining the origins of the accident. A popular science book published in 1988 – whose author claimed to have delivered 'an historical account of what happened before, during and after the accident'[17] – went as far as including a photo of a bottle of Ukrainian vodka amongst its illustrative pictures.[18] This narrative of 'Soviet nuclear carelessness' gained central political and economic importance in 2004 and in 2007. When the Eastern European countries joined the EU, they had to apply Western European security standards to their nuclear power plants[19] – and Lithuania had to dismantle its RBMK reactors in Ignalina. There was no way that a reactor design like the one that had caused Chernobyl could be tolerated within this newly defined geo-political *Europe*.

So, while the former Western part of the Soviet Union was discursively transferred into *Eastern Europe*, the nuclear political discourse continued to locate Chernobyl outside Europe. However, this narrative – which was foremost shaped by state and industry actors – had already become severely challenged by civil society actors, mainly anti-nuclear groups and charity organisations. These groups have stressed the European dimension of the accident, particularly in their memory work around the Chernobyl anniversaries. In this regard, in November 2010, the German Association for International Education and Exchange (Internationales Bildungs- und Begegnungswerk, IBB) initiated the foundation of the European Chernobyl Network. This network was intended to become the forum of exchange of the various solidarity groups and the basis for the preparation of joint commemorative activities around the 25th anniversary of Chernobyl across Europe. This is in no way self-evident. Chernobyl memories and the meanings that are implied in the commemorations of the event differ profoundly among European countries. In France, for instance, the commemoration of Chernobyl implies a radical criticism of the state elite system,[20] while in Germany, where anti-nuclear convictions have become mainstream, commemorating Chernobyl rather serves the purpose of keeping the anti-nuclear fight alive. But despite these different connotations and implications of Chernobyl remembrance,

17 Richard F. Mould, *Chernobyl: the Real Story* (Oxford: Pergamon Press, 1988), p. ix.

18 Ibid., p. 48.

19 Thomas R. Wellock, 'The children of Chernobyl: Engineers and the campaign for safety in Soviet-designed reactors in Central and Eastern Europe', *History and Technology* **29** (2013): 3–32.

20 Karena Kalmbach, *Tschernobyl und Frankreich. Die Debatte um die Auswirkungen des Reaktorunfalls im Kontext der französischen Atompolitik und Elitenkultur* (Frankfurt a. M.: Peter Lang, 2011).

anti-nuclear groups and charity organisations across Eastern and Western Europe together have increasingly stressed the European dimension of the accident. This European dimension consists of shared responsibilities: both that to help the victims of the accident and that to prevent another nuclear accident to happen. And these shared responsibilities stem not least from the fact that Chernobyl is a place *within Europe*.

With Fukushima, anti-nuclear groups expanded on this European level of shared responsibilities and have come to stress the wider international level of nuclear responsibilities. The International Chernobyl Day is a telling example in this regard. Immediately in 2011, the International Chernobyl Day incorporated the Fukushima victims into the events that this loose network of anti-nuclear initiatives has organised for the public remembrances of the Chernobyl victims every year all across Europe.

But highlighting the Europeanness of Chernobyl also still plays a major role in anti-nuclear campaigning. The Greens in the European Parliament, in particular Rebecca Harms, have continuously put the topic of Chernobyl on the European institutional political agenda. By commissioning the TORCH report (short for: The Other Report on Chernobyl)[21] and thus critically challenging the Chernobyl narrative provided by the IAEA and WHO, the Greens in the European Parliament have paid particular attention to the long-time health effects of the Chernobyl fallout across Europe. Stressing the European dimension of the accident's environmental and sanitary effects has become increasingly important since the early 2000s, when pro-nuclear actors started to proclaim a 'nuclear renaissance' and called for public subsidies for this 'low-carbon electricity supply'. For European energy politics, this 'greening' of nuclear energy has very practical implications: if nuclear energy is considered a renewable energy, the new build of nuclear power plants qualifies for the relevant EU subsidies. It is thus in the context of the renegotiation of the EU energy politics in the framework of climate change mitigation actions that the question of the Europeanness of Chernobyl has gained major political importance. Anti-nuclear Chernobyl narratives have thus discursively moved Chernobyl from the past Soviet Union into the geopolitical present of Eastern Europe.

But it is not only the anti-nuclear movement that has argued against the dominant state and industry alienation politics that locate Chernobyl outside

21 Ian Fairlie and David Sumner, *The Other Report on Chernobyl* (Berlin/Brussels/Kiev: The Greens in the European Parliament, 2006), http://www.chernobylreport.org/torch.pdf (accessed 12 May 2020).

Europe. The Chernobyl solidarity movement also stressed the Europeanness of the accident. The term Chernobyl solidarity movement is self-coined and means the collective of nongovernmental groups that provide humanitarian aid to the regions in Belarus, Ukraine and western Russia which have been most affected by the radioactive fallout. These groups are mainly known to a wider public through their organisation of recreational holidays abroad for the 'Chernobyl children' and the collection of clothes, medicine and presents for these children. Furthermore, many of these initiatives collect money that is invested in the infrastructure of hospitals and orphanages. Through bringing hundreds of thousands of children to Western Europe for recreational stays, and motivating thousands of people to travel to the affected regions to help and meet the people there, the solidarity movement has built many individual bridges across the former East-West-divide.[22] Stressing the European dimension in this shared responsibility to help the victims has been considered as an integrative factor.

As we have seen, the localisation of Chernobyl as inside or outside Europe has gained particular importance in the framework of Western European nuclear politics. But the implications of this discursive localisation reach far beyond this specific techno-political field. Hand in hand with the question of the accident's Europeanness went the question of which parts of the former Soviet Union should actually be considered part of the cultural or political entity 'Europe'. Every answer to this question implied a concrete statement on present and future responsibilities in overcoming the accident's sanitary and environmental impact – and in preventing future nuclear accidents from happening. Furthermore, the question of Chernobyl's Europeanness also implied statements on past responsibilities: if Western European nuclear experts were so quick in indicating all the shortcomings of the RBMK plant

22 IBB (ed.), *Tschernobyl und die europäische Solidaritätsbewegung* (Dortmund: IBB Dortmund, 2011); Astrid Sahm, 'Auf dem Weg in eine transnationale Gesellschaft? Belarus und die internationale Tschernobyl-Hilfe' *Osteuropa* **56** (2006): 105–116; Melanie Arndt, 'Verunsicherung vor und nach der Katastrophe: Von der Anti-AKW-Bewegung zum Engagement für die "Tschernobyl-Kinder"', *Zeithistorische Forschungen* **7** (2010): 240–258. With regard to the environmental effects of Chernobyl, these recreational stays, however, might have rather alienated Chernobyl from Western Europe by providing the image that these children would only need to stay for a couple of weeks per year in healthy Western European environments to improve their health conditions. It would be interesting to know if the presence of 'Chernobyl children' in, for instance, the British Lake District reminded the people there that their environment, too, had experienced severe consequences of the Chernobyl fallout in 1986 / 1987; or if the presence of these children rather had the effect of allocating Chernobyl's consequences to a region far away – a region to which these children would return after having experienced the uncontaminated and healthy environment of the Lake District.

design, why did they not raise their voices earlier? Within the framework of science diplomacy, wasn't it indeed these experts who had moved relatively freely across the Iron Curtain? Why was the blame solely put on the Soviet plant operators and the Soviet nuclear techno-political system when the international community of nuclear experts was well aware of what was going on? After all, within this international community of experts, Western Europeans played a central role – so what was their responsibility? Sure, there were just a few voices in the Chernobyl debate that stressed the complicity of Western Europe in the causes of the accident. But the fact that these voices exist shows how multi-layered the question of Chernobyl's Europeanness actually is.

Chernobyl and the emergence of a (European?) communication sphere on modern risks

In a last step, this chapter will expand on the European politics implied in the Chernobyl discourse and inquire into sociological concepts that build upon a certain interpretation of Chernobyl – and inquire into the communication sphere in which these sociological concepts became powerful.

Although the Chernobyl experience differed profoundly for people across Western European local, regional and national settings, there is one feature that many of these experiences have in common: the open disagreement amongst experts. In the days and weeks following the first news of the accident, Western European mass media offered a forum to nuclear experts in which they contested each other's evaluations of the accident's impact. Possible health effects caused by the fallout, the very level of the fallout rates, the way how fallout measurement should be taken: all was up for debate. If we believe in the claim that loss of trust in expertise is a characteristic of postmodern society, Chernobyl might well have worked as catalyst in this process. But while we should be cautious in creating such universal narratives – for instance, in the UK, Chernobyl didn't challenge at all the credibility of nuclear experts – it is interesting to see that the Chernobyl experience is the cornerstone of one of today's most popular theories of the postmodern society: the risk society, developed by Ulrich Beck, later extended to the global risk society. In the (global) risk society, it is no longer a specific group defined by location, class, gender or race that is threatened by a particular risk. In the (global) risk society, risks become universal – and they can turn into a concrete threat for literally everybody.

Beck's theory is so closely linked to Chernobyl that his notation of 'anthropological shock' is sometimes used as a metaphor for the accident itself. The reason for this is not only that Beck provided one of the first sociological analyses of Chernobyl[23] but also that the book that made him world-famous was published in the accident's direct aftermath.[24] Beck finished the writing of *Risikogesellschaft* (Risk Society) shortly before the Chernobyl accident, so it did not influence the text itself. However, it did lead him to write in May 1986 a pre-preface with the title 'Aus gegebenem Anlaß' (Due to Recent Events)[25] which was added to the publication, though not included in the English translation. So while the theory developed in *Risk Society* was not framed by Chernobyl, Beck's perception of Chernobyl was fully framed by his theory, as can clearly be understood from his pre-preface. As he himself declared, Chernobyl unfortunately proved his theory right.[26] Chernobyl was for Beck 'das Ende der "anderen"' (the end of "the others").[27] From the moment of the accident on, due to the 'Allbetroffenheit'[28] the distinction between us, the non-infected, and them, the infected, no longer existed. In this perspective, Chernobyl represents a turning point in history – the moment in which the era of the risk society established itself beyond any doubt.

If we have a closer look at the argument Beck developed in *Risk Society*, we see how his theoretical framework allowed him a certain reading of Chernobyl's immediate aftermath as he could identify in the public discourse some of the dynamics he had just generalised in his writing. According to Beck, the risks faced by people of the risk society are, in the first place, due to toxic threats to their health. These threats are caused by industry and high technology and appear as pollution in the air, water, soil, foodstuff, clothes and furniture, as well as in the unknown consequences of genetic manipulation or irradiation. In this regard, Beck provided a theoretical framework to the concerns raised by environmental movements since the 1970s. Ra-

23 Ulrich Beck, 'Der anthropologische Schock. Tschernobyl und die Konturen der Risikogesellschaft', *Merkur: Deutsche Zeitschrift für europäisches Denken* **8** (1986): 653–663.

24 Beck, *Risikogesellschaft*.

25 Ibid., p. 7.

26 Ibid., p. 10: 'Die Rede von der (industriellen) Risikogesellschaft ... hat einen bitteren Beigeschmack von Wahrheit erhalten. Vieles, das im Schreiben noch argumentativ erkämpft wurde – die Nichtwahrnehmbarkeit der Gefahren, ihre Wissensabhängigkeit, ihre Übernationalität, die "ökologische Enteignung", der Umschlag von Normalität in Absurdität usw. – liest sich nach Tschernobyl wie eine platte Beschreibung der Gegenwart. Ach, wäre es die Beschreibung einer Zukunft geblieben, die es zu verhindern gilt!'

27 Beck, *Risikogesellschaft*, p. 7.

28 Ibid.

dioactivity, which Beck himself in his book puts foremost into the context of its civilian use in terms of the production of electricity, rather than its military use, plays a special role in Beck's concept of risk: 'By risks, I mean above all radioactivity, which completely evades human perceptive abilities'[29]. The risks of the risk society are 'particularly open to social definition and construction' as they 'initially only exist in terms of the (scientific or anti-scientific) knowledge about them and thus they can be changed, magnified, dramatised or minimised within knowledge'. In this context, we also find Beck's observation of an 'Opening up of the Political' as he claims that, due to this open definition and construction, 'the mass media and the scientific and legal professions in charge of defining risks become key social and political positions'. Despite the social definition and construction of knowledge about the risks, this 'knowledge gains a new political significance' as it can be the key to survival: 'in risk positions consciousness determines being'.[30] It could be argued that this emerging academic field of risk sociology reflects the emergence of a wider public discourse on risk-taking in modernity. It would be worth investigating to what degree this emerging public discourse on risk took on a specific European dimension. In any case, Chernobyl came to occupy a central role in risk theory, and thus it is important to shed light on the kinds of Chernobyl narratives present in these works.

Like Ulrich Beck, Wolfgang Bonß in his book *Vom Risiko* (On Risk)[31] strongly referred to Chernobyl in order to underpin and exemplify his arguments. According to him, Chernobyl illustrated the devastating consequences that were possible in the event of a failure of tightly interlinked high technologies and he integrated his reflections on the discursive reactions toward Chernobyl in his theorisation of coping with risks and uncertainties in modernised modernity.

But there was also disagreement with making the Chernobyl experience the lynchpin of modernisation theories. Niklas Luhmann wrote his book *Soziologie des Risikos* (Risk: A Sociological Theory)[32] as an answer to, or rather a criticism of, Ulrich Beck. In the first chapter, Luhmann stated: 'Sociology has finally also turned its attention to the problem of risk; or at least laid claim to the term of risk. Following the ebbing of anti-capitalist

29 Ulrich Beck, *Risk Society* (London: Sage Publications, 1992), p. 22.

30 Ibid., p. 23.

31 Wolfgang Bonß, *Vom Risiko. Ungewißheit und Unsicherheit in der Moderne* (Hamburg: Hamburger Edition, 1995).

32 Niklas Luhmann, *Risk: A Sociological Theory* (Berlin: de Gruyter, 1993).

prejudice, it now finds a new opportunity to fill its old role with new content, namely to warn society.'[33] The footnote at the end of this sentence referred to Beck's *Risk Society*. According to Luhmann, Beck's attempt to theorise modern society through the category of risk fails completely. The problem that Luhmann basically had with Beck, but also with the majority of theoreticians in the field of sociology, was that they did not apply his theory of systems and, more concretely, that there was no definition of risk one could work with profitably. Therefore, Luhmann saw it as his main task to develop a definition of risk himself. He came to the conclusion that risk is basically a 'highly hierarchical contingency arrangement' that can best be approached – as is the case in general within his system theory – through a distinction. For Luhmann, the distinction that must be applied in this case is the one between risk and danger.[34] Risk is for Luhmann something that cannot increase. There are no more risks in the world now than before. All that has changed is the perception that people have to take more and more risky decisions. From Luhmann's point of view, this is due to the fact that in the past people just did not have a fully developed decision awareness as they had greater trust in divine forces.[35] So, according to Luhmann, a risk is nothing that can be avoided, as it is not there as such; it means only that we cannot predict the outcome of our decisions.

This concept of risk provides Luhmann with a view on public debates in post-modern society that is opposed to Beck's. Where Beck calls for more participatory rights in order to decrease risks, Luhmann asks for an application of his definition of risk in order to

> cool down considerably the unnecessarily heated public discussion on risk-related topics, and allow a more moderate tone to prevail ... There is no risk-free behaviour [and] ... no absolute safety or security ... One cannot avoid risks if one makes any decision at all ... And in the modern world not deciding is, of course, also a decision.[36]

Looking from Luhmann's point of view at the immediate response to

33 Ibid., p. 5.

34 Ibid., p. 17; in Luhmann's words: 'The distinction presupposes (thus differing from other distinctions) that uncertainty exists in relation to future loss. There are then two possibilities. The potential loss is either regarded as a consequence of the decision, that is to say, it is attributed to the decision. We then speak of risk – to be more exact of the risk of decision. Or the possible loss is considered to have been caused externally, that is to say, it is attributed to the environment. In this case we speak of danger.'

35 Ibid., p. 7.

36 Ibid., p. 28.

Chernobyl, as well as at the long-term debate about the health impact of the radiation released by the accident, one obtains a very different image from the one Beck describes. Contrary to Beck, neither does Luhmann specifically integrate Chernobyl in his argumentation. He sees nuclear power as a risk, but as a risk can never in itself be a problem, an argument against nuclear power, from Luhmann's point of view, cannot be based on the statement that it is too risky, as this risk perception is in any case only a social construction.[37]

The reasons why I am elaborating so extensively on these two different conceptualisations of risk are threefold. First, they mark the two extremes of the debate about the role of risk in the (post-) modern society. Second, internationally, Beck and Luhmann have become main reference points for theoretical approaches to risk sociology. Third, and most importantly, these two theories reflect the fact that theories in the field of risk sociology are profoundly shaped by individual risk perception of the author him- or herself. Over the last decades, social science research has turned Chernobyl into a central basis for theories on public risk perception, public understanding of science, expert lay person interaction and agenda setting.[38] But it is here, at the very foundation of the conceptual thinking about the constitution of modern societies, that theorisations of Chernobyl have become the most powerful in terms of their fundamental impact on social science research.

Chernobyl has thus created a specific academic communication sphere on modern risks. It is a striking fact that this communication sphere on modern risks has been severely imprinted by the Chernobyl experiences of two distinct West-German, middle-aged, middle-class, white, male sociology professors. In this regard, Chernobyl's contested Europeanness gains central importance in relation to the emergence of a public communication sphere on modern risks, urging us to further reflect on the question to what

37 Ibid., p. 101; in Luhmann's words: 'Nuclear power generation is a risk, even if we may be certain that a serious accident will occur only once every thousand years – although we do not know when. In this question it is a matter of the degree of sensitivity of probabilities and the extent of loss – that is to say to social constructs subject to temporal influences.'

38 Karena Kalmbach, 'Revisiting the nuclear age. State of the art research in nuclear history', *Neue Politische Literatur* **62** (2017): 49–69.

degree this public communication sphere on modern risk actually defines itself as inherently shaped by *European* experiences and to what degree it forms a cornerstone of twentieth century European identity.

CHAPTER 4.

THE WESTERN EUROPEAN PUBLIC SPHERE AND THE ENVIRONMENT IN EASTERN EUROPE DURING THE COLD WAR: BETWEEN MODEL, UTILISATION AND DENUNCIATION

Michel Dupuy

After the fall of the Berlin wall, the West seemingly discovered for the first time, apart from the Chernobyl disaster, the environmental damage in the East, with the drying of the Aral Sea serving as an archetypal symbol. Environmental damage in Eastern countries had nevertheless been noted at the level of senior administration in the West, although without becoming a part of East-West relations for lack of genuine political considerations. Only the USSR during the 1970s strove, during the Helsinki Conference, to make the environment a part of diplomacy, for it knew that the West was divided on the question, especially with regard to acid rain. The conference's final act on security and cooperation in Europe included the topic in the second basket.[1]

However, recent research in environmental history on the topic of communist Europe has demonstrated that environmental issues emerged at the same time on both sides of the Iron Curtain, with the East being on par with the West, including in matters of scientific ecology, with the Soviet geochemist Vladimir Vernadsky's formulation of the concept of the biosphere in 1926.[2] In addition, the environment became institutionalised at the turn of the 1970s, with the creation in both the West and the East of dedicated ministries, along with a series of laws seeking not only to protect nature but also to combat pollution, including in the GDR (1970), Czechoslovakia (1973), France (1976), the FRG (1976), etc.

In fact, until the late 1960s, environmental damage in Eastern countries

1 Eugeny Chosudovsky, *'East-West' Diplomacy for Environment in the United Nations* (New York: UNITAR, 1988); Michel Dupuy, 'Science, pouvoir et pluies acides en RDA', in Laurent Coumel, Raphaël Morera and Alexis Vrignon (eds), *Pouvoirs et Environnement. Entre confiance et défiance, XVe-XXIe siècle* (Rennes: Presses Universitaires de Rennes, 2017), pp. 159–173.

2 Coumel and Dupuy, 'Les trois écologies à l'Est. Quel tournant environnemental en RDA et en URSS?', in Anahita Grisoni and Rosa Sierra (eds), *Nachhaltigkeit und Transition: Politik und Akteure. Transition écologique et durabilité: Politiques et acteurs* (Frankfurt: Campus Verlag, 2018), pp. 229–252.

was not ignored by national media in the West, whether it involved the pollution caused by the use of lignite in the GDR, or Lake Baikal or the Volga in the USSR. At the time, there were particular fears of radioactivity following nuclear tests in the atmosphere, which the USSR and the US renounced in 1963. These fears were transmitted, among others, by scientists from the US, notably within the *St. Louis Citizens' Committee for Nuclear Information.*[3] The notions of 'environmental crisis' (1967), 'ecocide' (1970) and 'environmental catastrophe' invented in the West initially referred to the damage caused in capitalist countries, for instance through the use of napalm on the rainforests of Vietnam, or the sinking of the *Torrey Canyon* oil tanker (18 March 1967), which affected French and British coasts. The expressions 'environmental catastrophe' and 'environmental crisis' appeared in the East German press beginning in 1973, but were used to designate environmental damage in the West.[4]

The West focused on environmental damage in the East if it was directly concerned, as in pollution of the Baltic Sea, the paper mill in Blankenstein (GDR) that polluted the city of Hof (FRG), the Elbe laden with industrial waste from Czechoslovakia and especially the GDR, and Chernobyl (1986), among others. Other instances of environmental damage gradually appeared in the media during the early 1970s with the Stockholm Conference on the Human Environment (1972), where the West questioned its development model. For all that, in the late 1970s Eastern Europe was increasingly affected by an environmental crisis (water and air pollution), which called the communist system into question. Movements with an environmental sensibility developed in most countries in the East despite dictatorship, and established public spheres on the topic, subsequently opening the way for protest.

When the environmental issues of Eastern European countries were mentioned in the Western European public sphere, it was of course the sign of a shared concern for environmental matters, but also a challenge to the model of growth. This presence within a Western European public sphere nevertheless provided an opportunity for environmental movements in communist countries to expose the damage they had suffered, by counting on a space of communication beyond the Iron Curtain via Western media including the BBC, Radio Liberty, Radio Free Europe and West German

3 William Krasner, 'Baby tooth survey – first results', *Environment* **55** (2013): 18–24.

4 Fjodor Krotkow, 'Der Menschheit droht keine Katastrophe', *Berliner Zeitung*, 29 Apr. 1973; 'Leben nur im Gleichgewicht mit der natürlichen Umwelt', *Berliner Zeitung*, 12 Aug. 1973.

stations such as ARD and ZDF, whose shows were listened to and viewed in the East.

The construction of a field of shared concerns

While there were shared concerns in environmental matters between Western and Eastern Europe, along with equal enthusiasm for leisure activities in 'natural' spaces (mountain, sea, forest), the East raised questions for the West with respect to its environmental protection policy. The true division, however, was connected to civilian nuclear power.

Shared concern between the West and the East

Regarding the protection of nature, during the 1950s and 1960s there was a European market of images for animal shows on television, including in the USSR. For instance, Frédéric Rossif regularly sought them out for his show *La Vie des Animaux* [*The Life of the Animals*] (1952–1966) on TF1 (French television channel 1). The West German zoologist Bernhard Grzimek, who produced the show *Ein Platz für Tiere [A Place for the Animals]* (1956–1987), began filming in the USSR in 1963.[5] The book by the French naturalist Jean Dorst, *Avant que nature ne meure [Before Nature Dies],* was published in 1965, and translated into Russian in 1966.[6] Nature protection was indeed a common concern.[7]

Still, the environment was not given its own column in the press, or its own section on television. The topic began to emerge in the late 1960s among Western European and North American editorial boards, all while remaining marginal in comparison to other subjects (political, economical, cultural), and in competition with journalists reporting on scientific and technological developments.[8]

In both the West and the East, scientists nevertheless tried to use the press to share their concerns. For instance, on 24 October 1967, the Swedish daily *Dagens Nyheter* published an article by Svante Odén, an agronomist

5 Claudia Sewig, *Der Mann, der die Tiere liebte: Bernhard Grzimek. Biografie* (Cologne: Bastei, 2009).

6 This book was translated into English in 1970. Jean Dorst, *Before Nature Dies* (Boston: Houghton Mifflin, 1970).

7 Douglas R. Weiner, *A Little Corner of Freedom. Russian Nature Protection from Stalin to Gorbatchev* (Berkeley: University of California Press, 1999).

8 Dupuy, 'Scientifiques, télévision et écologie: entre vulgarisateur et lanceur d'alerte', *Temps des Médias* **2** (2015): 182–199.

tasked by the Swedish government with drafting a report on the increasing acidification of rainwater due to sulphur dioxide emissions from Central Europe:[9] the discussions surrounding acid rain had begun in Europe.

Still, the case that made it past the Iron Curtain was Baikal. This lake was praised by Jules Verne in *Michel Strogoff*, while from 1958 onward industrial projects had been proposed and criticised in the Soviet press by defenders of nature. In the West, the issue appeared in an AFP news dispatch from 12 May 1966, following a collective letter that appeared in *Komsomolskaya Pravda* on 11 May, signed by academics and scientists against the creation of a cellulose factory.[10] It was reprinted by the *New York Times* and *Le Monde* the following day.[11] The news item highlighted the media warning issued by scientists, as well as its reception by the accused company, the cellulose factory, which had to implement a technical solution to remedy the pollution.

In the run-up to the Stockholm Conference in 1972, references to Baikal increased, as witnessed by the digitised press archives of both Switzerland and the FRG.[12] It became a symbol of the struggle against water pollution for both the budding environmental movement and sympathetic communists. On 29 April 1973, a documentary on Siberia presented in the collection *Lettres d'un bout du monde*, directed by Jean-Emile Jeannesson with the participation of the State Committee for Soviet Television on Siberia, raised the issue of Baikal, prompting the journalist to say: 'In Irkutsk people told me that "The Americans pollute their natural water reserves. We prevented that!" Scientists prevented the chemical industry from setting up on the lake's shore'.[13] The transfer of the Baikal affair toward Western Europe was promoted by Soviet authorities, who used it to display their effectiveness in combatting pollution, and with it the superiority of the communist model.

Environmental damage in both the USSR and other countries in the East was available to the French, West German and Swiss press, in the event that it focused on the subject, which it hardly did. In fact, this was not a topic

9 Svante Odén, 'The acidification of precipitation', *Dagens Nyheter*, 24 Oct. 1967.

10 Slava Lubomudrov, 'Environmental politics in the Soviet Union: The Baikal controversy', *Canadian Slavonic Papers / Revue canadienne des slavistes* **20** (1978): 529–543.

11 'Soviet conservationists see peril to Lake Baikal', *The New York Times*, 13 May 1966; 'Des savants soviétiques protestent contre l'installation d'une usine sur les rives du lac Baïkal', *Le Monde*, 13 May 1966.

12 O. Lainé, 'Au bord du Baïkal: une usine qui ne pollue pas la nature', *Journal de Genève*, 25 May 1972; 'Sowjet-Union: "Wir töten die Erde"', *Spiegel* **4** (1972).

13 USSR-Siberia: Part 2, *Lettres d'un bout du monde* 2, 29 Apr. 1973.

of interest for correspondents in Eastern Europe, who were specialised in more distinguished subjects such as diplomacy and politics. Pollution was absent even in personal accounts of stays in the East, even though nature was present. Still, there were articles in both the Soviet press and that of the GDR, in accordance with the concerns of civil society. Publications within French documentation mentioned it, but without outlets in the press, nothing transformed into a sustained affair or societal issue. The media attention given to whale hunting by the USSR is symptomatic in this respect. The Soviet state was of course accused alongside Japan in the mid-1970s, although most articles devoted to the subject discuss Japan.

Between model and condemning industrial society

The emergence of environmental protection as a political issue in both the West and the East naturally raised the question of political model, and also led to questions regarding industrial society in the wake of the 1972 report for the Club of Rome. In a 1970 report on the environment in the USSR, the geographer Alain Giroux provided a list of the damage: dropping water level in the Caspian and Aral Seas, soil degradation, pollution of the Volga, Lake Baikal, etc. He especially underscored both the ineffectiveness and absence of sanctions, and concluded that industrial society was endangering the environment.[14]

With the politicising of environmental questions and their presence on the agenda, there emerged a genuine questioning of economic model. In the USSR, the topic of the planet's limits was debated in November 1972 within the pages of the journal *Voprossi Filosofi*, among others.[15] This looming lack of resources called for the development of civil nuclear power, placing the economy within a closed process without waste, and spreading socialism on a global scale. Previously censored works on environmental issues in the USSR and the GDR also appeared, but without resonating in the West.[16]

This same socialist model was highlighted in the West. For instance, a

14 Alain Giroux, 'Mise en valeur et protection de la nature', *Problèmes politiques et sociaux* **13** (1970): 26–38.

15 Dupuy, 'Shortage of resources and political model in the GDR: 1971–1989', (paper presented at the *The Right Use of the Earth* conference, Paris – Ecole Normale Supérieure, 29 May–1 June 2018).

16 Dupuy, 'Justifying air pollution in the GDR 1949–1989', in Eli Rubin et al. (eds), *Ecologies of German Socialism* (Oxford: Peter Lang, 2019), pp. 115–145; Jürgen Kuczynski, *Das Gleichgewicht der Null. Zu den Theorien des Null-Wachstum* (Berlin: Akad.-Verl, 1973); Evgenij K. Fjodorow, *Die Wechselwirkung zwischen Natur und Gesellschaft* (Berlin: Deutscher Verlag der Wissenschaften, 1974).

report on air pollution in Poland, broadcast on TF1 on 9 October 1970, took Katowice as an example. Jean-Pierre Alessandri condemned the capitalist system: 'Humans must live in harmony with nature, and must be capable of understanding it and subordinating it to their needs; capitalist industrial activity has lasted 150 years. It has shaken the invigorating balance of the forces of nature.'[17] The discourse is voluntaristic, seeking to control nature by making better use of its power and wealth. It shows the Polish state in action, with the application of laws requiring new factories to equip themselves with filters or to risk fines, along with the education and mobilisation of citizens within workers' councils. The tone is optimistic. France was invited by the journalist to strengthen its legislative arsenal, fines and citizen action.

On 22 November 1975, the French environmental journal *Combat nature* published an article on Warsaw and its architecture. 'In Warsaw the quest for individual profit does not exist; one can see and appreciate a more human environment thanks to abundant green spaces that compensate for concrete', a remark that the author, Alain de Swarte, extended to Polish cities by boasting about the absence of cars in city centres.[18] This was incidentally one of the very rare articles devoted to Eastern Europe by the environmental press in France. The populations of Eastern Europe also wanted cars – the symbol of the capitalist world par excellence – although they were condemned by Marxist East German philosophers sensitive to environmental thinking, such as Wolfgang Harich and Robert Havemann, who saw them as promoting private property, a sign of capitalism. Industrial development was also called into question in the GDR in connection with the Protestant church and international ecumenical conferences – albeit without explicitly condemning socialism – especially in a booklet produced in Wittenberg in 1982, entitled '*Die Erde ist zu retten': Umweltkrise, Christlicher Glaube, Handlungsmöglichkeiten [The Earth is to Save: Environmental Crisis, Christian Faith, Possible Actions]*.

On 8 June 1974, the Bulgarian journalist Stéphane Groueff was invited on the *Homo Sapiens* show airing on FR3 (French television channel 3) to discuss his book *L'homme et la terre* [*Man and the Earth*], in which he condemned Soviet dams and river diversions in Siberia, especially the diversion of the Pechora river toward the Volga in an attempt to solve dropping

17 'En Pologne: quatrième exemple la planification', *XXème siècle* **9** (Oct. 1970).

18 Alain de Swarte, 'Varsovie: sauvegarde architecturale et déclin de l'urbanisme', *Combat nature* (Nov. 1975): 15–18.

water levels in the Caspian and provide irrigation water for farmland.[19] In his remarks he denounced this Promethean vision of humans facing nature, but not communism.

As a result, a shift slowly began to unfold: it was no longer an ideological or economic system that was being questioned, but actually a form of growth based on the exploitation of natural resources, directly in line with the Club of Rome's 1972 report. However, the discourse against industrial society was based on examples taken from the West rather than the East.

Civil nuclear power: A division

The question of civil nuclear power in Western Europe played a key role in the development of environmental movements.[20] However, in the East, scientists and environmental organisations saw nuclear energy as an alternative to pollution, while citizens did not express any true opposition.

The handling of the Kyshtym disaster is particularly revelatory in this regard. On 29 September 1957, the explosion of nuclear waste in the Urals affected 20,000 km^2 of land. The accident, which had been kept secret for a long time, began to generate attention after the English popular science journal *New Scientist* published an article on 4 November 1976 by Zhores Medvedev, a Soviet radiobiologist who defected to the West in 1973.[21] His demonstration was based on articles that had appeared in the Russian radiobiology press about the Kyshtym area.

In reaction, on 8 November 1976 the *Times* of London published an interview with Sir John Hill, president of the United Kingdom Atomic Energy Authority. He believed that it was impossible that the 'Russians' had not followed security norms, and cast doubt on whether the explosion took place, qualifying it as 'pure fiction, rubbish and [a] figment of [the] imagination'. One day later, following an AFP news dispatch, UPI, Reuters, AP and *le Monde* published an article on the subject, repeating the conclusions of Sir John Hill and French nuclear authorities.[22]

19 Book of the month, *Homo Sapiens* 3, 8 June 1974.

20 Franz-Josef Brüggemeier and Jens Ivo Engels, *Natur- und Umweltschutz nach 1945* (Frankfurt: Campus Verlag, 2005).

21 Zhores Medvedev, 'Two decades of dissidence', *New Scientist* (Nov. 1976): 264–267.

22 'However, British and French nuclear circles are much more reserved regarding the nuclear accident discussed by M. Medvedev today. They especially emphasize that it was in fact waste, and that a nuclear explosion is impossible', 'Un accident nucléaire s'est effectivement produit en Union Soviétique en 1957 ou 1958', *Le Monde*, 12 Nov. 1976.

On 13 May 1978, Medvedev was invited by the environmental organisation *Amis de la Terre* [Friends of the Earth] to hold a press conference at the Collège de France, in the laboratory of Marcel Froissart, a member of the Groupe de Scientifiques pour l'information sur l'Energie Nucléaire [Group of Scientists for the Information on Nuclear Energy]. The conservative newspaper *Le Figaro* did not send a journalist, for they were not 'interested in Russian dissidents who were not pro-nuclear', such as Andreï Sakharov.[23] This defiance with regard to antinuclear remarks from the East was also present on 18 January 1978, when TF1 planned to begin its evening news with an interview with Medvedev, which was cancelled at the last moment.[24] Medvedev published a book on this topic in German in April 1979, and then in the US in July. The author was interviewed on the channel ARD on 9 July for the show *Bilder aus der Wissenschaft*, which aired at 21:50. His book was only translated from English to French in 1988, despite being announced for 1979 in an interview with *Paris Match*.[25]

When news of the disaster spread publicly in Western Europe, it was met with opposition by public authorities, who knew it would be impossible for Western journalists to visit the site, as the area was kept behind a wall of secrecy. Furthermore, in the West European public sphere, dissidents engaged in discourses on human rights and freedom of expression, but not on ecology. The Kyshtym disaster remained in the memory of the environmental movement, as did Windscale (10 October 1957) in Great Britain, but without reaching the symbolic impact of the Chernobyl disaster. French and British authorities broadly developed a policy of casting doubt on this event, first by denying the reality of the event, and then by communicating very little on the subject.

In the West the subject of energy was central to building and mobilising the environmental movement, whether it involved nuclear energy or

23 Interview between P. Erskine and Jaurès Medvedev, 'Marx aurait-il été antinucléaire?' *La Gueule ouverte* **216** (June 1978): 14.

24 'Int savant russe Medvedev', *IT1 20H*, 1, 18 Jan. 1978. The news anchor nevertheless preserved the following summary: 'The USSR began to construct nuclear reactors in 1945. But the problem of managing radioactive waste was made without sufficient precautions. The accumulation of waste caused an explosion that dispersed it in the air above industrial areas. This waste was of course radioactive.'

25 Zhores A. Medvedev, *Bericht und Analyse der bisher geheimgehaltenen Atomkatastrophe in der UdSSR* (Hamburg: Hoffmann u. Campe, 1979); *Zhores A. Medvedev, Nuclear Disaster in the Urals* (New York: W.W. Norton. 1979); Zhores A. Medvedev, *Désastre nucléaire en Oural* (Cherbourg: Isoète, 1988); 'La grande catastrophe atomique soviétique', *Paris Match* 1577 (1979): 3–17.

the two oil crises.[26] However, in the East, the attention of populations and environmental movements focused firstly on water and air pollution, with nuclear power appearing as a healthy alternative.

The environmental crisis extends to countries in the east

With the Prague Spring and the publication of books by dissidents such as Alexander Solzhenitsyn, the communist world experienced a moral and economic crisis, which was compounded by an environmental crisis. It was economically and materially incapable of addressing the causes behind pollution, which became a sensitive and even secret subject, despite making it past borders via numerous outlets.

The environmental crisis

In 1978, the West German publisher Possev (Sowing) published a work in Russian by Ze'ev Wolfson (alias Boris Komarov) on the destruction of nature in the USSR. The book was written in the USSR, and secretly sent to the West. From 1970 to 1977, Wolfson had worked on educational television programmes on the environment, as well as in the Soviet government's department of biology. His work was translated into German in 1979, English in 1980 and French in 1981, the year in which the author emigrated to Israel; then into Italian in 1983.[27] For the first time a Soviet author had published a book on the destruction of nature in a socialist country. The terms 'environmental crisis' and 'environmental catastrophe' henceforth applied to the East. In 1979, the German weekly *Der Spiegel* spoke of an ecological crisis in connection with the destruction of nature in the USSR.[28] On 26 October 1983, taking up an AFP dispatch from the previous day, the *Gazette de Lausanne* spoke of an 'environmental catastrophe' in the Ukraine in connection to the pollution of a river by ammonia following an explosion in a factory.[29] In the late 1970s, Radio Free Europe also adopted a critical tone toward environmental management in the GDR,

26 Mark Landsman, *Dictatorship and Demand: The Politics of Consumerism in East Germany* (Cambridge, MA: Harvard University Press, 2005).

27 Boris Komarov, *Le rouge et le vert. La destruction de la nature en. U.R.S.S.* (Paris: Seuil, 1981); Marshall I. Goldmann, 'The identity of Boris Komarov – at least', *Environmental Conservation* **12** (1985): 180.

28 'Soll doch die Taiga eingeächert werden', *Der Spiegel*, 12 Nov. 1979.

29 'AFP, Pollution d'ammoniac dans le Dniestr', *La gazette de Lausanne*, 26 Oct. 1983.

in programmes such as *Spiegel* (beginning in 1978), using data produced by East German scientists.[30]

This crisis now affecting countries in the East took its place within the globalisation of environmental crises: the hole in the ozone layer, destruction of the Amazon, and especially acid rain with the disappearance of forests due to forest dieback (*Waldsterben*).[31] In this context, the forests of the Ore Mountains of Bohemia became a stand-in for the German, Swiss and French press, as their fate prefigured that of German forests. The damage was caused in large part by Czechoslovakian industry, and drew the attention of environmental groups, along with neighbouring populations in the GDR and Czechoslovakia.

This environmental crisis was being felt just as environmental groups were emerging in Eastern Europe, especially in 1978 in the GDR under the authority of the Protestant church following debates sparked by the Club of Rome. In January 1986, the Stasi counted 42 environmental groups in the GDR, including 28 active ones forming an alternative public sphere.[32] In Lithuania the count reached fifty.[33] These groups began to emerge in Poland from 1979 onward, with the Polish Ecological Club, as well as during the 1980s in Czechoslovakia, Hungary, etc. East German groups found outlets, notably in the FRG with the journalist Peter Wensierski, who in 1981 published *Beton ist Beton*, in which each chapter was written either by a member of the Protestant Church in the GDR, such as Peter Gensichen, or collectively, as with the Working Groups of Wittenberg or Greifswald.[34]

This environmental crisis in the East was, as in the West, connected to an energy crisis, which was compounded by a crisis of planned economies. The rise in oil prices implemented by the USSR (except for Poland), along with the growing indebtedness of countries in the East in comparison to the West, weakened their economic system.[35] The environment became a national issue in this context, transforming into eco-nationalism. For example, one month

30 'Umwelt: lenkt die DDR ein?' *Der Spiegel* **39** (1978): 18–19.

31 Dupuy, *Histoire de la pollution atmosphérique en Europe et en RDA* (Paris: L'Harmattan, 2003).

32 Hans-Peter Gensichen, 'Christen und Kirchen in der DDR', in Peter Bohley (ed.), *Erlebte DDR-Geschichte: Zeitzeugen Berichte* (Berlin: Links, 2014), pp. 57–75.

33 Jane I. Dawson, *Eco-nationalism. Anti-nuclear Activism and National Identity in Russia, Lithuania, and Ukraine* (Durham: Duke University Press, 1996).

34 Peter Wensierski and Büscher Wolfgang, *Beton ist Beton. Zivilisationskritik aus der DDR* (Hattingen: Edition Transit, 1981).

35 André Steiner, 'From Soviet occupation zone to "new Eastern states". A survey', in Hartmut Berghoff and Uta Andrea Balbier (eds), *The East German Economy, 1945–2010, Falling Behind or Catching Up?* (Cambridge: Cambridge University Press, 2013), pp. 17–51.

before Chernobyl, 350 Armenians called for the closing of a nuclear power plant located in a seismic zone.[36] The Chernobyl disaster challenged the nuclear option in Eastern European countries, Poland in particular, where the building of a nuclear power plant at Zarnowiec (near Gdansk) was now the subject of protest.[37] The enlargement of the Ignalina nuclear power plant (Lithuania) also sparked contestation, and led to strong popular opposition in 1988.[38]

These different manifestations drew increased attention in Western Europe to environmental damage in countries in the East, especially in France, where the journal *La nouvelle alternative, revue pour les droits et les libertés démocratiques en Europe de l'Est* devoted an entire column to it beginning in 1986.

Information crossing borders

In the face of censorship and control over Western journalists, access to sources of information on Eastern countries took place through non-public channels. In April 1984, an Alsatian naturalist association organised a tour through Eastern Europe to observe the damage caused to forests by acid rain. The group included engineers, scientists, and journalists. The GDR and Poland refused to grant the group access to their territory; only Czechoslovakia accepted, but without journalists. *Le Monde* ultimately published an article on 25 February 1985, as it had commissioned one of its engineers to provide a report on his observations.[39]

Information circulated through multiple networks, notably those of the church. On 26–27 April 1986, *Le Figaro* published an article entitled 'Pologne: grave menace écologique' [Poland: serious ecological threat], whose source was a Paris-based journal close to the Polish episcopate, *Znaki Czasu* (Signs of the Times). In Czechoslovakia, a report from the academy of sciences on the country's ecological situation made its way into the hands of a Charter 77 member, and later passed to the West via the network surrounding the Christian association Entr'aide et action [Help and Action]. It was ultimately the subject of an article in *Le Monde* on 7 January 1984.[40]

36 Letter dated 31 Mar. 1986 to M. Gorbachev / Radio Free Europe / Radio Liberty Archive AS 5822.

37 Agnieszka Laddach, 'History and present. The Żarnowiec nuclear power plant. Proposition of research project', paper presented at the *Not Just Chernobyl* conference, Poznan, 21–22 April 2016.

38 Dawson, *Eco-nationalism*.

39 'Le massacre a commencé en Tchécoslovaquie', *Le Monde*, 25 Feb. 1985.

40 The association published a bulletin that included Jean-Marie Domenach in its liaison committee. 'L'Académie des sciences a établi un rapport alarmant sur la situation écologique', *Le Monde*, 7 Jan. 1984.

Of course the development of green movements that established links with the West facilitated the exchange of information, while during periods of relative liberalisation, information on environmental damage became more free in a few Eastern European countries. This was the case in Poland from August 1980 until the declaration of martial law on 13 December 1981, as demonstrated by the article regarding pollution in the voivod of Katowice that appeared in the *New Scientist* on 22 October 1981, which was translated in the ecological journal *Le Courrier de la Baleine*.[41] The data presented in this article came from a report drafted by scientists from the Polish Ecological Club. In the USSR, perestroika also freed up speech in the media on these matters.[42]

Tensions surrounding environmental issues between Eastern European countries even appeared in Western European media starting in 1985, revealing that the communist bloc was also not united, and that the environment was a source of tension. On 8 April 1985, *le Monde* exposed the dispute between Hungary and Czechoslovakia surrounding construction of a dam on the Hungarian side of the Danube, and returned to the topic on 25 September 1985. The weekly *l'Express* covered it on 10 April 1987. This emergence of a cross-border public sphere also resulted from the creation of The Danube Circle-Duna Kör, a movement consisting of intellectuals opposed to the dam, knowing that Hungarian authorities were no longer inclined to proceed due to the cost of the works, and therefore allowed information to filter through.[43]

On 20 December 1986, in an AFP news dispatch, Poland denounced Prague's failures in connection with the effects on the Oder River of fuel oil pollution originating in Czechoslovakia. On 14 November 1987, an AFP dispatch reported on a chlorine cloud over the city of Ruse in Bulgaria, a cloud that had originated from a chemical factory in Romania, the site of a caustic soda production factory since 1984.[44]

However, none of this environmental news transformed into a closely

41 Lloyd Timberlake, 'Pologne: Le pays le plus pollué du monde', *Le Courrier de la Baleine* **60** (1982): 14–17.

42 Serhiy Choliy, 'People had voice: Individual initiative and population relocation process in late Soviet Union (1986-1991)', paper presented at the Not just Chernobyl conference, Poznan, 21–22 April 2016.

43 See the contribution by Daniela Neubauer in this volume.

44 AFP, 'Un nuage de chlore sur la ville de Rousse à la frontière roumaine', 14 Nov. 1987; Vladimir Socor, 'A row with Bulgaria over Pollution: the Giurgiu Chemical Plant', *Radio Free Europe Research*, 25 Nov. 1987.

monitored case or major societal issue. Some of them reappeared regularly in the media, but at intervals stretching over multiple months or years, such as the Baikal affair or the construction of a dam between Hungary and Czechoslovakia, sometimes ultimately becoming a part of environmental memory or symbols. This reflected a gradual rise in environmental concerns within editorial boards, although they remained marginal during the 1980s.[45]

Environmental matters appeared in the media, notably in France and the FRG, but continued to be minor, with the exception of the cross-border pollution affecting Western Europe (Elbe River, Baltic Sea, etc.). Moreover, the protests against environmental damage were not necessarily passed on to the West, whether the activities of environmental groups in the Protestant church within the GDR, or in the USSR during perestroika.

Being present in the Western European public sphere to be heard in the East

Aside from Radio Free Europe, Radio Liberty, and the BBC, Eastern European countries also picked up Western television channels, for instance in Estonia (Finnish television). The GDR was beneath the waves of RIAS, Radio in the American Sector, whose broadcasts reached Western Poland and Northern Czechoslovakia. East Germans could also watch West German channels (ARD, ZDF). The influence of West German media could be gauged in the aftermath of the Chernobyl disaster, when East German parks were emptied of children for a number of weeks, as worried parents followed the instructions of West German television.[46] Svetlana Alexievitch has gathered a number of accounts from Belorussia that mention listening to Radio Free Europe following the Chernobyl disaster.[47] The ability to pick up media from the West encouraged environmental groups in the East to pass on information to the West, knowing that they would enjoy media coverage, with West German outlets being particularly important.

45 Dupuy, *Traitement et représentations du concept de biodiversité à la télévision* (Paris: INRA, 2018).

46 Melanie Arndt, *Tchernobyl. Auswirkungen des Reaktorunfalls auf die Bundesrepublik Deutschland und die DDR* (Erfurt: Landeszentrale für politische Bildung Thüringen, 2011).

47 Svetlana Alexievitch, *La supplication: Tchernobyl, chroniques du monde après l'apocalypse* (Paris: J'ai lu, 2005).

Taking advantage of the West German public sphere

West Germany offered East German intellectuals and dissidents a space for publication. In 1981, the East German novelist Monika Maron published *Flugasche* [*Fly Ash*] in the FRG, in which she directly tackled pollution in the city of Bitterfeld. She had actually been negotiating its publication in the GDR since 1976, under the title *Und morgen komme ich wieder* [*And Tomorrow I'll Be Back*].[48] However, literary authorities wanted the author to make corrections that described, according to East German minister of culture Klaus Höpcke: 'not only the destructive, but also the positive consequences of work for human beings'.[49] After protracted negotiations Maron refused to accept new corrections, and passed her manuscript on to the West, where it was published by S. Fischer.

Political exiles in the West, such as Roland Jahn, tried to help environmental groups in the GDR from their side of West Berlin by sending equipment to print their booklets via diplomats, members of the Bundestag and journalists, who were not searched at the border. Jahn even sought to contact major companies in the West such as Siemens, Sony, Xerox and Toshiba to obtain printers and video cameras, in order to pass them on to environmental groups in the GDR, often in vain. In the summer of 1986, he nevertheless succeeded in sending a VHS camera to the GDR, an uncommon and costly piece of equipment at the time. He was also in contact with Peter Wensierski, who worked for the *Kontraste* programme airing on ARD (West German channel 1), and who was banned from visiting the GDR in 1984 following his publications on the state of the environment in that country.[50] The first film produced, entitled *Uns Stinkt's* [*It Stinks to Us*], was aired on 3 March 1987 on ARD. Two others followed, the first directed by Michael Beleites on the exploitation of uranium ore in the GDR, broadcast on 3 November 1987; and the second on the city of Bitterfeld, scheduled for 27 September 1988, filmed by a West German journalist and an ecologist who was a member of the Arche group (East Berlin).[51]

48 Büro für Urheberrechte, 'Aktennotiz', 19 Jan. 1981, BArch: DR 1 16910.

49 Klaus Höpcke made this declaration to West German journalists during the Leipzig book festival after the publication of Monika Maron's book. Uwe Wittsstock, 'Verordnetes Schweigen', *Frankfurter Allgemeine Zeitung*, 14 Apr. 1981.

50 Stefan Wolle et al., *Operation Fernsehen. Die Stasi und die Medien in Ost und West* (Göttingen: Vandenhoeck & Ruprecht, 2008).

51 *Michal Beleites*, '*Uranbergbau* in der DDR', ARD Reihe Kontraste, 3 Nov. *1987*; *Arche*, '*Bitteres aus Bitterfeld*', ARD Reihe Kontraste, 27 Sept. 1988.

The show on Bitterfeld was viewed by both local inhabitants and others living across the GDR, for instance a man from Freiberg who sent a petition (*Eingabe*) to the East German minister for the environment on 2 October 1988: 'We know that numerous measurements have been carried out in the Freiberg area. Why have the results and conclusions not been communicated to the public?' He added:

> Furthermore, we find it very regrettable that such information only reaches people in the GDR through Western mass media, and never through our own press. From this I can only conclude that you are powerless in the face of these things or, as I have already suggested, that you have no knowledge of them.[52]

On 5 October 1988, the authorities reacted by drafting a report from the industry and raw material department of the Central Committee. It was written in collaboration with local authorities from the area surrounding the industrial chemical combine of Bitterfeld. Counterarguments were provided to local administrations, data were challenged and special emphasis was placed on the fact that pollution was far worse in Western European countries.[53]

Again with respect to pollution, the German Institute for Economic Research, located in West Berlin, conducted an investigation beginning in 1984 on sulphur dioxide (SO_2) emissions in the GDR, which had committed that same year to a thirty per cent reduction in emissions by 1993 as part of the *Convention on Long-distance Transboundary Air Pollution*. The emissions data were intended for the United Nations Economic Commission for Europe in Geneva. The GDR sent falsified data, reporting 4 kt of annual SO_2 emissions since 1980 (in reality 4.4 kt). The study by the Berlin institute was published on 25 July 1985, with emissions estimated for 1982 at 4.9 kt (in reality 4.5 kt). These estimates were made based on East German publications and by crosschecking information. The institute demonstrated in particular that given its energy needs, the GDR could not meet the thirty per cent reduction in SO_2 emissions by 1993. These conclusions were reprinted by the daily newspapers *Die Welt*, on 25 July 1985, and *Frankfurter Rundschau*, on 31 July 1985.[54] The issue became political, forcing the GDR to make the real data public.

52 https://landesarchiv.sachsen-anhalt.de/fileadmin/Bibliothek/ Politik_und_Verwaltung/MI/LHA/externa_alt/89_06/8990_Juni_Protest_9.htm (accessed 22 Dec. 2018).

53 https://landesarchiv.sachsenanhalt.de/fileadmin/Bibliothek/Politik_und_Verwaltung/MI/LHA/externa_alt/89_06/8990_Juni_Umwelt_3.htm (accessed 8 Aug. 2017).

54 Jochen Bethkenhagen et al., 'Luftverunreinigung in der DDR: die Emission von Schwefeldioxid und Stickoxiden', *DIW Wochenbericht* (1985): 337–346.

Becoming rooted in the West

Among intellectuals from the East who published their writings on political ecology in the West, it is important to cite Wolfgang Harich and Rudolf Bahro. Harich was sentenced in 1957 to ten years in prison, as he was an advocate for democratic socialism. At the time he was already aware of matters of global ecology through the work of the limnologist August Thienemann. In 1971, following the debate surrounding the Club of Rome report, he engaged in favour of ecology. In 1975 he published *Kommunismus ohne Wachstum? Babeuf und der 'Club of Rome'* [*Communism without Growth: Babeuf and the 'Club of Rome'*] with the tacit agreement of the East German regime. His work was translated into Spanish in 1978 and into Swedish in 1979. From 1979 to 1981, he was authorised to travel to the West after recovering his title as doctor, of which he had been stripped in 1957. He took part in conferences in Austria, the FRG, Switzerland and Spain.[55]

Rudolf Bahro was a journalist deeply marked by the Prague Spring in 1968. He subsequently began a reflection on the environment. In, 1977 he published his work *Die Alternative* in the FRG.[56] Its publication was accompanied by an interview with a journalist from *Spiegel* on 22 August, along with a self-interview for Rias. On 23 August, Bahro was arrested. Support committees immediately sprang up in the FRG, France (François Maspero), Italy, Great Britain, etc.[57] His book was translated into English in 1978, and into French in 1979. He was freed on 11 October 1979, and extradited to the FRG one week later. In the meantime, François Maspero had founded the journal *Alternative,* in order to defend human rights in countries in the East.[58]

The two authors developed a Marxist critique of the ecological crisis, condemning the communist system, which for them had set out on the capitalist path. Yet in the West, journalists saw them as dissidents, and expected a discourse against communism. In his press conferences Harich attacked the roots of consumerism, which he situated in the West, and also spoke about the environment, whereas the journalists wanted to hear about dissidence. Faced with this discourse, the German magazine *Stern* refused to publish an

55 Siegfried Prokop, *Ich bin zu früh geboren* (Berlin: Dietz Verlag, 1997).

56 Rudolf Bahro, *Die Alternative: Zur Kritik des real existierenden Sozialismus* (Cologne: Europäische Verlagsanstalt, 1977).

57 Liselotte Julius, 'Le combat pour Rudolf Bahro', in Bahro (ed.), *Je continuerai mon chemin* (Paris: Maspero, 1977), pp. 19–38.

58 Luc Pinhas, 'François Maspero, le passeur engagé', *Documentation et bibliothèques* **56** (2010): 187–194.

interview with him on 9 November 1979. He was convinced that the longer the solution to the environmental crisis – connected to the unreasonable exploitation of raw materials – was delayed, the greater the need for an authoritarian regime. He believed that communist society was better equipped to resolve this crisis, as it referred to use value rather than exchange value.[59] Bahro had to contend with the same questions as Harich at the outset. Just a few months after his extradition, he criticised a media system that wanted to make him into a dissident and critic of the GDR.[60] Still, in order to address fellow citizens on ecological matters, dissidents from the GDR had to use the West German public sphere, such as *Radio Glasnost*, which had been broadcasting shows on these topics from West Berlin since August 1987.[61]

Despite his critiques, Bahro engaged, on the advice of Carl Amery and Rudi Dutschke, in the FRG's green movement, in which he was the representative for this radical ecology, alongside socialist, realist and eco-libertarian ecologies.[62] For Bahro it was essential in particular to deindustrialise and not to collaborate with political authorities. From 1982 to 1984, he was even a member of the leadership of the Greens, and resigned in 1985 after refusing the realist path that had taken them over. His Marxist critique resonated in France with Pierre Juquin, who was one of the founders of eco-socialism along with André Gorz, Murray Bookchin, etc. This branch within the Eurocommunist movement was quite marked in Spain, where Bahro and Harich were invited. In addition, the translation of Harich's book into Spanish owed much to Manuel Sacristán, a Marxist philosopher and member of the PSUC (Unified Socialist Party of Catalonia), which had also been affected by the Prague Spring.[63]

Environmental matters in Eastern European countries interested the West on the political level when the communist system was called into question.

59 Prokop et al., *Ein Streiter für Deutschland. Auseinandersetzung mit Wolfgang Harich* (Berlin: Edition Ost. 1996).

60 Rudolf Bahro, *Elemente einer neuen Politik* (Berlin: VielFalt bei Olle & Wolter, 1980).

61 Paul Hockenos, *Berlin Calling* (New York: The New Press, 2017).

62 Udo Baron, *Kalter Krieg und heisser Frieden* (Hamburg: Lit Verlag Münster, 2003).

63 José M. Faraldo, 'Entangled Eurocommunism: Santiago Carrillo, the Spanish Communist Party and the Eastern Bloc during the Spanish transition to democracy, 1968–1982', *Contemporary European History* **26** (2017): 647–668.

After the Soviet system's collapse, journalists could freely travel to these countries, and published an increasing number of reports presenting living conditions amid a polluted universe, truly exposing the regime. While the ecological movement in Western Europe did not at all develop in relation to what was happening in the East, the environmental movement in the East was able to extend beyond borders and create a European public sphere to raise awareness of the true state of its environment (Komarov), and to inform the populations of these countries through different media. After the collapse of popular democracies, the transition to a post-Fordist economy, and adherence to environmental norms following EU membership, Eastern Europe practically disappeared from Western media, with the exception of Chernobyl and a few symbolic sites that were already familiar to naturalists before 1989, such as Baikal or Bialowieza Forest on the border between Poland and Belorussia.

This disappearance can partly be explained by the coming of democracy, as the space of regulation was henceforth national. The eco-nationalism that had served to affirm an identity against the USSR in Baltic countries, Armenia and Poland gave way to global environmental issues, such as the protection of biodiversity following the Rio Summit, the struggle against global warming and promotion of sustainable development. This was joined by Western eco-colonialism that raised concerns regarding GMOs, industrial waste pouring into Eastern Europe due to its laxer norms in comparison with Western Europe and the promotion of nuclear power in the name of energy independence,[64] which was rehabilitated in Lithuania and Armenia after having been despised. This took place in countries feeling the full force of the political and economic transition, with subsequent mass unemployment. Eco-tourism emerged as a possible path; in Latvia, with support from the WWF and local ecologists, the natural reserve of Slitere was transformed into a national park in the name of preserving biodiversity, with no regard for the local population, whose activities did not mesh with the WWF's pre-agrarian vision.[65]

While environmental groups in the East used the West European public sphere before 1989, environmental organisations from the West were present in the Eastern European public sphere during the 1990s, bringing with them

64 Krista Harper, *Wild Capitalism. Environmental Activists and Post-Socialist Political Ecology in Hungary* (New York: Columbia University Press, 2006).

65 Katrina Z.S. Schwartz, *Nature and National Identity after Communism* (Pittsburgh: University of Pittsburgh Press, 2006).

global concerns, including ones that were not yet shared by environmental actors in the East.

PART II

THE SHAPING AND USE OF THE EUROPEAN PUBLIC SPHERE ON ENVIRONMENTAL ISSUES: ABOUT THE INFLUENCE OF TRANSNATIONAL ACTIVISTS AND MOVEMENTS

CHAPTER 5.

THE IMPACT OF EAST GERMAN NATURE CONSERVATIONISTS ON THE EUROPEAN ENVIRONMENTAL CONSCIOUSNESS IN THE TWENTIETH CENTURY.

Astrid Mignon Kirchhof

When the German Democratic Republic (GDR) became part of the Federal Republic of Germany (FRG) on 3 October 1990, the unified Germany acquired five new national parks (Jasmund, Vorpommern, Müritz, Harz and Sächsische Schweiz), six biosphere reserves and three nature reserves. Codified in the Unification Treaty, the protected tracts of land are often referred to as the 'crown jewels of German unity'.[1] The legal codification of these crown jewels was the result of persistent political commitment during what were literally the last days of the foundering GDR. The person responsible for designating the fourteen large-scale protected areas was agricultural scientist and biologist Michael Succow, Deputy Minister for Nature Conservation, Ecology and Water Management in the period January to May 1990.[2] Supported by a committed team of colleagues, he enacted East Germany's national park programme, placing 4.5 per cent of the country's surface area under protection[3] while the state was crumbling all around. Designated tracts of land along the former German-German border are now integrated into the European Green Belt. Over dozens of years, a stretch of valuable biotopes had developed along the Iron Curtain, untouched by human interventions. Now, the death strip

1 Ulrich Messner, 'Nur einmal im Leben: Der Kampf um die Müritz und die Entstehung des Nationalparkprogramms', in Nationalpark, *Wo Mensch und Wildnis sich begegnen,* No 149, 03/2010, pp. 21–24.

2 With no more than two nature conservation officials employed at the central level at the GDR Ministry for Agriculture, the establishment of a separate Ministry of Nature Conservation was considered an urgent necessity. Cf. Michael Succow, 'Persönliche Erinnerungen an eine bewegte Zeit', in Michael Succow, Lebrecht Jeschke and Hans Dieter Knapp (eds), *Naturschutz in Deutschland: Rückblicke - Einblicke - Ausblicke* (Berlin: Ch. Links Verlag, 2012), pp. 63-70, here p. 63.

3 Originally, the protected reserves were to cover 10.8% of the total surface area. However, with the People's Chamber deciding in August upon the GDR's swift unification with the Federal Republic on 3 October 1990, the national park programme was reduced to the extent achievable in the remaining period of time. As a result, only 4.5% was placed under protection: see mail from Uwe Wegener to the author dated 5 Oct. 2017.

between Eastern and Western Europe was turning into a unifying lifeline. Driven by a joint commitment to protect nature, cooperative action between European states ensued.[4]

The following article proposes that the ideas and concepts of pioneering GDR nature conservationists were instrumental in the emergence of environmental consciousness in Europe. Among the trail-blazers of the concept of European nature conservation were the two pioneers of environmental protection in the GDR, Erna and Kurt Kretschmann.[5] Their belief in a good life in harmony with nature found an echo in the ideas of the GDR's ecological movement in the 1980s and even survived the end of the East German state: Not only were the convictions of the couple and their fellow campaigners codified in the Unification Treaty, they also manifested themselves in the establishment of a Europe-wide exchange network for nature conservationists. Hence, the protected areas along the Iron Curtain, which made history as the European Green Belt, also have their roots in the visions held by East German nature conservationists and ecologists.

A life dedicated to nature conservation

Born in Stettin in 1912, Erna Jahnke was two years the senior of Kurt Kretschmann, who was born in Berlin in 1914. Before dedicating their lives to nature conservation, they both practised different professions: Erna was a nursery teacher and Kurt was a qualified tailor. In an interview held in late 1990, they explained how they came to be nature conservationists. While Kurt Kretschmann referred to his critique of life in big cities, which awoke his longing for nature, Erna declared that her life was changed when she met Kurt. At the time, Kurt had already turned his back on Berlin, sharing a simple life with a friend in a forest hut in Brandenburg. Both men fell in love with Erna who lived in the neighbouring village of Rüdnitz.[6] Dur-

4 'Das grüne Band: Vom Todesstreifen zur Lebenslinie' ('The green belt: From death strip to lifeline') is the title of a brochure on the German and European Green Belt published by the Bundesamt für Naturschutz at https://www.bund.net/gruenes-band/(retrieved 13 Jan. 2020).

5 Erna Kretschmann: b. Stettin, 12 Nov. 1912; d. Bad Freienwalde, 6 Jan. 2001. Kurt Kretschmann: b. Berlin, 2 March 1914; d. Bad Freienwalde 20 Jan. 2007. In recognition of their services, the Kretschmanns received numerous awards during their lifetime. Cf. 'Nabu-Ehrenpräsident Kurt Kretschmann deutscher Gesamtsieger beim Europäischen Umweltpreis', in *Nabu Pressedienst, 101 (Nov. 1999)*, Erna-und-Kurt-Kretschmann-Archiv (KreA), 280.

6 Marion Schulz, *Ein Leben in Harmonie. Kurt und Erna Kretschmann – für den Schutz und die Bewahrung der Natur* (Neuenhagen: Findling, 1999), p. 13.

ing this era, Kurt Kretschmann's philosophy of life was moving towards reform-based beliefs and he was striving for a life in nature, far from the big city. The reform movement was a melting pot of various ethic beliefs and reform-based approaches which were all connected and cannot be separated one from another: vegetarianism, critique of capitalism, anti-alcoholism, rejection of technology and big cities, pacifism and affinity with nature.[7]

Kurt was drafted into the army in 1936 and deserted his post eight years later. Looked after by Erna, who had become his wife, he hid for a number of weeks in an underground shed in Bad Freienwalde (Brandenburg) where the couple lived most of their lives. Since the end of the National Socialist dictatorship came as a great relief to the Kretschmanns, they wanted to give something back to the new state and be actively involved in its establishment and preservation:

> Free at last. An incredible feeling. A miracle beyond words. We had survived and wanted to express our gratitude. At once, we joined the Communist Party. My wife became a member of the district council and I became one of the 4 local party leaders. In charge of the agitation and propaganda sector, I was also responsible for political education which I oversaw for a period of four years. After that, we launched our nature conservation work.[8]

Introduced directly after the war, the GDR government's nature conservation activities were based on three pillars – politics/administration, science and civic engagement – and were also implemented at three different levels: the Ministry for Agriculture and Forestry (Ministerium für Land- und Forstwirtschaft), the Institute for Landscape Planning and Nature Conservation (Institut für Landschaftsforschung und Naturschutz) and the GDR's Cultural League for the Democratic Renewal of Germany (Kulturbund zur demokratischen Erneuerung der DDR), a central mass organisation.[9] In the early days, the Kretschmanns focused predominantly on the Cultural League. Having become members of the local Freienwalde group of the Communist Party of Germany (KPD) in 1945 and the So-

7 Kurt Kretschmann, *Erinnerungen an meinen im Hitler-Krieg gefallenen Freund Herbert Marquardt* (Biesenthal: Hoffnungstaler Werkstätten, 2002); Kurt Kretschmann, *Gedichte gegen den Krieg – 'Gewalt ist die Waffe des Schwachen, Gewaltlosigkeit die des Starken' – Mahatma Gandhi*, with the assistance of Daniel Fischer [n.p., n.d.]; Kurt Kretschmann, *Unsere Eß- und Trinkgewohnheiten unter die Lupe genommen – Erfahrungen aus 70-jähriger fleischloser Ernährung* (Bad Freienwalde: [n.d]).

8 Ibid., p. 30.

9 Hermann Behrens, 'Naturschutz in der DDR', in Stiftung Naturschutzgeschichte (ed.), *Wegmarken, Beiträge zur Geschichte des Naturschutzes, Festschrift für Wolfgang Pflug* (Essen: Klartext, 2000), pp. 189–258, here p. 206.

cialist Unity Party of Germany (SED) in 1946, the couple now joined the Cultural League's Central Committee of the Friends of Nature and Heimat. Erna Kretschmann was also active in the Cultural League's Central Expert Committee for Landscape Planning and Nature Conservation and held the position of District Secretary for Nature and Heimat at the Frankfurt/Oder[10] branch of the Cultural League.[11]

Eventually, the Kretschmanns extended their commitment beyond the voluntary level and started working within the context of the government's nature conservation programme. Up until 1964, Erna Kretschmann held various positions, among them District Councillor for National Education and Consultant for Landscape Planning and Nature Conservation on the Council of the District of Oberbarnim.[12] Kurt Kretschmann, in turn, was employed as Special Representative for Environmental Protection Matters by the Brandenburg Ministry for Agriculture and Forestry as of 1951 and took over the State Office of Nature Conservation in Brandenburg. The Kretschmanns launched two large-scale projects which were to become pioneering, unique enterprises. One project involved setting up ecological training courses at *Müritzhof* – the first of their kind worldwide; the other entailed their lifestyle and resource management at their self-designed *Haus der Naturpflege* (house of nature care), which Erna Kretschmann referred to as the 'crucible of nature conservation in the GDR'.[13]

Müritzhof, a farm in Mecklenburg, in what was then the GDR's largest protected area in today's county of Mecklenburgische Seenplatte, was set up as a 'central training establishment for nature conservation' in 1954.[14] The Kretschmanns ran the institution for six years, until 1960. Taking on a total of 1,200 trainees who worked at the lowest level in villages and communes, they instructed them in effective, hands-on nature conservation. Among other aspects, the Kretschmanns taught their charges how to approach the authorities and deal with their objections. Both during this period and before they had taken over the management of *Müritzhof*, the couple had repeatedly

10 From 1952–1990 Bad Freienwalde was a city in the district of Frankfurt/Oder.

11 Cf. CV of Erna Kretschmann, in Haus der Naturpflege e.V. (ed.) *Erinnerungen an Erna Kretschmann* (Bad Freienwalde: 2012), pp. 18–19.

12 Ibid.

13 Schulz, 'Ein Leben in Harmonie', 26.

14 Ludwig Bauer, 'Naturschutzarbeit der 1950er und 1960er Jahre in der ehemaligen DDR', in Stiftung Naturschutzgeschichte (ed.), *Natur im Sinn. Zeitzeugen im Naturschutz* (Essen: Klartext, 2001), pp. 47–61, here p. 53.

locked horns with official institutions. Their rule violations and unyielding, militant conduct represented a challenge to the SED state and its need to exert control, leading to friction with superiors and administrative offices. Their unconventional actions and the unconditionality with which they sought to change prevailing circumstances according to their own philosophy transgressed the narrow world of bureaucrats and functionaries. As well as discontinuing their work at *Müritzhof* in 1960, Kurt and Erna Kretschmann also took the fundamental decision to move back to Bad Freienwalde. Living largely independent of any official structures, they set up a centre of nature and culture, their *Haus der Naturpflege*, which they managed until 1982.[15]

Impact of the *fin de siècle* reform movement on the Kretschmanns' intellectual roots

In terms of the history of thoughts, the philosophy adopted by Kurt and Erna Kretschmann is rooted in the reform movement of the late nineteenth and early twentieth centuries. The holistic approach adopted by the life reformers awakened the Kretschmann's interest in non-European, specifically Asian traditions, such as Buddhism or Confucianism, in which holism plays an eminent role. At an event in Berlin in the 1930s, Kurt Kretschmann had met the Swiss life reformer Werner Zimmermann, whose teaching was to have the most profound effect on the couple.[16] Zimmermann advocated anarchy, anti-capitalism and socialism in freedom as well as life reform in the sense of a transformation of human life, organic farming and free sexuality.[17] Zimmermann's specific philosophy not only promoted the replacement of the prevailing way of life with life in its natural state but also called for an anarchic economic system – the so-called free economy. In 1934 Zimmermann founded the *Wirtschaftsring* (WIR), a self-help initiative which simultaneously represented 'a practical form of free socialism'.[18] Zimmermann combined his model of life and economic system with nutritional reform and advocated a new form of sexuality where sexual union was independent of lust.

15 See Behrens, 'Wende-Wege', 106.

16 See Florentine Fritzen, *Gesünder leben. Die Lebensreformbewegung im 20. Jahrhundert* (Stuttgart: Franz Steiner, 2006), p. 295.

17 Meike Sophie Baader, *Erziehung als Erlösung: Transformationen des Religiösen in der Reformpädagogik* (Weinheim: Juventa, 2005), pp. 230–234. Also: Kretschmann, *Gedichte gegen den Krieg*, p. 64.

18 Günter Bartsch, *Die NWO-Bewegung Silvio Gesells. Geschichtlicher Grundriß 1891 – 1992/93* (Lütjenburg: Gauke, 1994) at http://userpage.fu-berlin.de/~roehrigw/bartsch/geschichte/1_II_29.htm (retrieved 15 Mar. 2018).

The reform movement emerged in reaction to industrialisation and urbanisation. Its proponents no longer relied on government reforms and instead believed in self reform. Their idea was that, although a lifestyle in harmony with nature should be adopted by the individual, the aggregation of these individual lifestyles should lead to the desired social reforms. Rather than production, they focused on the reform of consumption which they hoped would result in the creation of a harmonious and conflict-free society. For the reform movement of the late nineteenth century, the social question was predominantly one of morality, which meant that their critique was directed exclusively at the moral consequences of civilisation, as opposed to social inequity. This reduction of social ills to moral circumstances allowed reformers to individualise causes and possible solutions and distanced them from a view where socio-structural causes or economic conditions are at the root of the problem.[19]

The propagated materialism and centrally-guided economy of the GDR were diametrically opposed to this belief, not only because they focused first and foremost on the manner of production as opposed to consumption, but also because, under Marxism, reforms are the exclusive remit of the state instead of the individual. Official nature conservation and environmental protection policy also followed this dictum. The 1970 Socialist Land Cultivation Act (Gesetz über die planmäßige Gestaltung der sozialistischen Landeskultur) once more sanctioned the fundamental reference to Karl Marx who maintained that it is the being that determines the consciousness.[20] With this theorem, Marx distanced himself from his teacher Georg Hegel who had posited that consciousness determines being, a philosophy which was taken up again by new social movements, among them the environmentalists, in the Federal Republic in the 1970s, and a few years later in the GDR. The socialist idea, which still underpinned the West German student movement of the 1960s and its focus on production rather than consumption and unequal economic conditions, thus lost some of its cohesive force amongst opposition members in East and West.

Kurt and Erna Kretschmann were certain that the GDR was morally superior to the Federal Republic, and they never lost their belief in a just and

19 Cf. Eva Barlösius, *Naturgemäße Lebensführung. Zur Geschichte der Lebensreform um die Jahrhundertwende* (Frankfurt/Main: Campus, 1997), pp. 170 et seq. and 198 et seq.

20 Hugo Weinitschke, *Ein Jahr Landeskulturgesetz in der Deutschen Demokratischen Republik – Probleme und Erfahrungen aus der Tätigkeit des Deutschen Kulturbundes*, p. 10, Freilicht- und Volkskundemuseum Schwerin (FVS), KB, Landeskultur/Umweltschutz/Naturschutz, 27, 1.

fair world. The communist philosophy of a classless society and the absence of private ownership, the historical legitimacy of which they took as a forgone conclusion, did not conflict with their reform-based ideals: quoting Bettina von Arnim, Erna Kretschmann put it like this: 'there will, however, come a time when, despite all deception and force, truth will imbue the hearts of the people, of the poor, and the reign of the wealthy will come to an end'.[21]

The fact that ideas and beliefs are transformed upon their conveyance and reception, adjusted to new circumstances and modified in the process explains why, for a long time, the Kretschmanns managed to live in harmony with themselves, undisturbed by the state: adapting their concepts to the prevailing circumstances in the GDR, they merged their reform-based ideals with those aspects of the socialist vision that were in tune with their basic outlook. By contrast, although the members of the GDR environmental movement of the 1980s adopted the concept of the good life promulgated by the older nature conservationists as well as referring back to the individualisation theory, they were subject to much greater repression due to their clear critique of the communist doctrine and the SED party.

By linking their reform-based model with the Communist vision, the couple created its own counterworld[22] but avoided any perilous opposition to the system. As the historian Thomas Lindenberger expounds, it was indeed possible to articulate interests and needs in the GDR and, one might add, to live one's own life, as long as one avoided any serious conflicts with the powers that be.[23] That Erna and Kurt Kretschmann's activities did not go unnoticed by the Ministry for State Security (MfS) has been corroborated

21 Quote by Gisela Heller, 'Geliebt und verstanden werden ist das höchste Glück', in *Erinnerungen an Erna Kretschmann* (Bad Freienwalde: Haus der Naturpflege, 2012), pp. 5–8, here p. 7.

22 See the introduction to a Special issue on this concept: Nina Leonhard and Astrid Mignon Kirchhof, 'Gegenwelten', in *Geschichte und Gesellschaft* **41** (1) (2015): 5-16.

23 See Thomas Lindenberger, 'Herrschaft und Eigen-Sinn in der Diktatur. Das Alltagsleben der DDR und sein Platz in der Erinnerungskultur des vereinten Deutschlands', in Bundeszentrale für politische Bildung (ed.), *Aus Politik und Zeitgeschichte. B 40/2000*, http://www.bpb.de/apuz/25409/herrschaft-und-eigen-sinn-in-der-diktatur (retrieved 1 Oct. 2018). Although the *Eigensinn* (stubbornness) concept introduced by Thomas Lindenberger may provide some explanation for the lifestyle of the Kretschmanns, it does not sufficiently account for the ambivalent relationship many nature conservationists had with the GDR. I therefore speak of counterworlds which allowed hosts of environmental activists to create niches where they could escape the clutches of the state although they were neither hostile nor opposed to the socialist state, and in some cases even declared themselves to be apolitical. In the examples I have chosen, this counterworld lifestyle (which was certainly more than just a behavioural variety) was often founded on nature- and reform-based beliefs which, on the one hand, compellingly suggested a 'free' lifestyle and, on the other, advocated ideological proximity with socialism. Cf. Leonhard and Kirchhof, 'Gegenwelten', 71 –106.

by the unofficial collaborators who watched them over many decades. While in the 1950s, Kurt was still seen as a 'crank'[24] whose pronouncements the District Control Committee considered implausible,[25] assessments of the Kretschmanns became significantly more moderate over the coming decades, and they were described as an apolitical couple whose primary objective was nature conservation.[26] The state's tacit conclusion that the Kretschmanns had no political clout was, however, very much mistaken.

The missionaries of the house of nature care

Upon their return to Bad Freienwalde in 1960, Kurt and Erna Kretschmann set up a new home and became predominantly self-sufficient, emulating a way of life they had already tried and tested in the 1930s. Over the years, they also established a public 'centre of nature and culture' which regularly attracted visitors from Germany and the rest of Europe. People called on the Kretschmanns to discuss ideas on nature conservation and link up with fellow campaigners. In the 1970s, informers reported 70,000 to 80,000 visitors[27] over a period of fifteen years, approximately 5,000 a year, who came to 'talk about nature conversation with Erna and Kurt'.[28] Among those who flocked to the *Haus der Naturpflege* were veterinarians, farmers, professors, musicians, authors, nature conservationists, teachers, politicians, students and ordinary people.[29] At the same time, the couple was in contact with

24 Letter by the Bad Freienwalde District Office to the State Security Regional Office regarding Kurt Kretschmann, 21 Feb. 1958, p. 15, Behörde des Bundesbeauftragten der Stasi Unterlagenbehörde (BstU), Bezirksverwaltung (BV) Frankfurt/Oder, Kreisdienststelle (KD) Bad Freienwalde, ZMA 4884.

25 Letter from the Bad Freienwalde District Office to the State Security Regional Office regarding Kurt Kretschmann, 28 Feb. 1958, p. 23, BStU, BV Frankfurt Oder, KD Bad Freienwalde, ZMA 4884.

26 Report by the informal collaborator in special operations (Inoffizieller Mitarbeiter für einen besonderen Einsatz, IME) 'H. Hockum' on Kurt and Erna Kretschmann, 1988, p. 94, BStU, BV Frankfurt Oder, KD Bad Freienwalde, ZMA 4884.

27 For the figures, see *Operativ-Information* 3/77, 28 Apr. 1977, BStU, Bezirksverwaltung für Staatssicherheit (BVfS) Frankfurt Oder, dept. X 300, folio 38. 7,000 visitors per year are mentioned by the writer Gisela Heller in her contribution to the brochure in memory of Erna Kretschmann. Cf. Heller, 'Geliebt und verstanden werden ist das höchste Glück...', p. 6..

28 Cf. Sybille Knospe, 'Kurt Kretschmann - Naturschutz kennt keine Mauern'. Presentation at the conference *Über die Mauer. Deutsch-deutsche Kontakte im Naturschutz*, Potsdam – Haus der Natur, 5 Dec. 2014, http://www.haus-der-naturpflege.de/uploads/PDF/TagungPotsdam20141205_Rede_SKnospe.pdf, pp. 3, 4, and 8 (retrieved 8 Jan. 2020).

29 Cf. articles in the commemorative brochure entitled *Erinnerungen an Erna Kretschmann* (Bad Freienwalde: Haus der Naturpflege, 2012).

hundreds of others,[30] establishing the basis of their Europe-wide nature conservation network. Their exchanges were, however, not limited to written correspondence but also led to personal meetings. It was mostly Erna who maintained the written correspondence: 'Rather an impressive achievement, given that the Kretschmanns' circle of friends and acquaintances was constantly growing, both privately and through their nature conservation work, and Erna never wrote any run-of-the-mill letters',[31] remembers Rolf Göpel from Bremen. Having been invited to *Müritzhof* by the Kretschmanns as a student in the 1950s, he returned for numerous visits over the following decades and later acted as a multiplier. He delivered nature conservation material from the Federal Republic of Germany to the Kretschmanns and established contacts with West Germany, for example with Hartmut Heckenroth, then head of the State Ornithological Institute of Lower Saxony, who visited the Kretschmanns several times.[32]

The Kretschmanns also maintained close written and personal contact with other West German nature conservationists, such as Carl Duve, the Head of the Nature Conservation Authority of Hamburg (Naturschutzamt Hamburg). In the 1950s, Duve, who also headed the Hamburg branch of the Verein Naturschutzpark (nature reserve association), talked with the Kretschmanns about one of their lifelong ambitions, the creation of national parks, suggesting the integration of the Eastern bank of the Müritz into a transnational European nature reserve. Moreover, on Carl Duve's recommendation, Kurt Kretschmann was to become the GDR's representative at the International Union for Conservation of Nature and Natural Resources (IUCN).[33] To Kretschmann's great disappointment, this plan did not come to fruition, presumably because it was unclear in what form East Germany could join this international body. Based on its recently adopted Hallstein Doctrine, West Germany had filed an

30 As pointed out by the archivist of the Erna-und-Kurt-Kretschmann archive, Gebhard Schultz, *Erna-und-Kurt-Kretschmann-Archive – Online-Findbuch Schriftgutbestand* at http://www.haus-der-naturpflege.de/uploads/PDF/KreA_Online-Findbuch_2011.pdf, introduction to the search index, p. 13 (retrieved 12 Jan. 2020).

31 Rolf Göpel, 'Kontakte zu Kurt Kretschmann – Praktische Erfahrungen seit 1956', presentation at the conference *Über die Mauer. Deutsch-Deutsche Kontakte im Naturschutz*, Potsdam – Haus der Natur, 5 Dec. 2014, http://www.haus-der-naturpflege.de/uploads/PDF/TagungPotsdam20141205_Rede_RGoepel.pdf, (retrieved 8 Jan. 2020).

32 Wilhelm Breuer, 'Wahrscheinlich ist Liebe im Spiel. Hartmut Heckenroth im Porträt', in *Nationalpark – Wo Mensch und Wildnis sich begegnen*, Nr. 167, 03/2015, pp. 34–35, http://www.egeeulen.de/files/nationalpark_169_15_heckenroth.pdf (retrieved 15 Mar. 2018).

33 Knospe, 'Kurt Kretschmann', 2–3.

objection against the GDR's admission when the East German Institute for Landscape Planning and Nature Conservation applied for membership of the IUCN in 1956. Espoused by the West German government the preceding year, the Doctrine claimed an exclusive mandate to represent Germany at the international level and opposed the recognition of the GDR as a separate state. In the end, although the GDR could not join the Union as a state, the IUCN decided that membership of individuals and organisations was permissible. After years of being relegated to guest status, the East German Institute for Landscape Planning and Nature Conservation finally became an official member of the IUCN in 1965 – ten years after Carl Duve and the Kretschmanns first discussed their ambition to create a national park in East Germany.[34]

In the 1970s, the Kretschmanns' European network expanded further thanks to their establishment of the White Stork Working Group (Arbeitskreis Weißstorch) in Bad Freienwalde. The Kretschmanns had become aware that there were very few storks left in the region, so Erna set about investigating the matter: 'After weeks of writing letters and talking on the phone, my wife found out that only 5 out of an original 34 stork couples were left in our district'.[35] Kurt and Erna decided to do something about this. Taking their first steps to save the white stork in their immediate environment, they soon focused on the rest of the GDR and eventually on other countries, stimulating a debate at the international level. With the help of her correspondents, Erna Kretschmann swiftly set up a European interest group for the protection of the white stork.[36] Thanks to a local teacher who spoke six languages and provided the necessary translation services, the Kretschmanns were also able to communicate with activists abroad. Nevertheless, many of their fellow campaigners also spoke German, for instance Andrej Stollmann, a frequent visitor and host of the Kretschmanns from Czechoslovakia.[37] The Erna-and-Kurt-Kretschmann Archive contains a list from the 1980s which specifies the home countries of their contacts

34 Hans-Werner Frohn, Jürgen Rosebrock, 'Naturschutz im geteilten Deutschland. Deutsch-deutsche Naturschutzkontakte 1945-1969', *Natur und Landschaft. Zeitschrift für Naturschutz und Landschaftspflege* **83** (7) (2008): 325–328.

35 The founding date, 1978, can be found in the following brochure: Kurt Kretschmann, *Dem Weißstorch zuliebe – Ein Gruß aus dem Storchmuseum im Storchenturm von Rathsdorf-Altgaul*, published by Naturschutzbund Deutschland (NABU) e.V. (Bonn: *[n.d.*, 1990s]); *Kurt Kretschmann, Erna Kretschmann – ein Nachruf*, n.d., KreA, 014.

36 Kretschmann, *Dem Weißstorch zuliebe*, p. 10, own count.

37 See, for instance, Letter from Erna and Kurt Kretschmann to Andrej Stollmann, 15 Jan. 1981, KreA, 279.

in Eastern and Western Europe: Austria, Bulgaria, Czechoslovakia, Denmark, France, FRG including West Berlin, Hungary, Netherlands, Poland, Romania, Soviet Union and Spain.[38] With individuals and organisations in these countries, among them members of the Ornithological Union in Copenhagen, the Natural Science Faculty in Cordoba and the Natural History Museum in Vienna, Erna and Kurt Kretschmann swapped slide shows, films, books, magazines and display boards from exhibitions on the subject of white storks in Europe. The couple also initiated an international white stork count in 1984 and organised visits to Bad Freienwalde.[39]

Wishing to set an example for their host of visitors, guests and followers, the Kretschmanns demonstrated how to maintain, protect and sensibly manage nature. Not only did they encourage their visitors to get involved, they also practised what they preached, for instance experimenting with organic fertilisers and solar energy at their *Haus der Naturpflege*.[40] The supporters of the GDR's future environmental movement and numerous individuals who were indirectly involved saw them as an oasis of calm, as representatives and conveyors of a nature-centred concept of a better life. Throughout their lifelong 'educational work',[41] the Kretschmanns perfectly complemented each other, generated public attention and consistently played a missionary role, even beyond their death. Individuals remained loyal to the concept of nature conservation and reflected the beliefs of the Kretschmanns in their work for influential social institutions, for instance *Nabu*, the Nature and Biodiversity Conservation Union. Among these individuals are the couple Mechthild and Christoph Kaatz who set up Storchenhof, a centre for injured storks and other big birds in Loburg, Saxony-Anhalt, in 1979. According to Mechthild Kaatz, their contact with the Kretschmanns still benefits the Storchenhof centre today: 'We are honouring the legacy of the two Kretschmanns by carrying on their work in NABU's national white stork working group'.[42]

38 List of white stork protection groups with which Erna and Kurt Kretschmann were in contact, KreA, 316. See also Correspondence with the Danish Ornithological Union, KreA, 316.

39 See Letter from Kurt and Erna Kretschmann to Mr Shifter of the Natural History Museum, Vienna, 2 Nov. 1983; Letter of the Natural Science Faculty of the University of Cordoba, 22. May 1983; Letter from Tommy Dybbro of the Danish Ornithological Union to Kurt Kretschmann, 25 Aug. 1980, KreA, 316.

40 Schulz, 'Ein Leben in Harmonie', 25-30.

41 Anita Tack, 'Beeindruckendes Lebenswerk', in Haus der Naturpflege e.V. (ed.), *Erinnerungen an Erna Kretschmann*, Supplement

42 Mechtild Kaatz, 'Ihr schaute die Menschlichkeit aus dem Gesicht', in Haus der Naturpflege e.V. (ed.), *Erinnerung an Erna Kretschmann*, pp. 11–12, here p. 12.

When it came to their lifelong 'educational work'[43] Erna and Kurt complemented each other perfectly despite, or perhaps precisely because of their rather different personalities:

> Much of what he [Kurt, AMK] preached with the tongues of angels back then actually became law nationwide. Some people who heard him at that time will object here that Kurt appeared to him rather as an archangel who interferes with the flaming sword. Yes, he was a zealot, an inconvenient one.[44]

Erna Kretschmann, by contrast, was a source of calm and stability, both within their marriage and in contact with the hosts of visitors who referred to her as a 'ray of sunshine who lifted people's spirits, a quiet revolutionary'[45] with 'diplomatic skills'[46] and a 'conciliatory yet energetic and purposeful manner'.[47] According to Anita Tack, former Brandenburg Minister of the Environment, Health and Consumer Protection, Erna Kretschmann 'found kindred spirits, brought them together and united them in the pursuit of shared values and objectives'.[48]

The couple's pervasive influence is also reflected in the enduring relationship that developed between the Kretschmanns and the above-mentioned Michael Succow. Having visited the Kretschmanns for the first time at age twelve, Succow remained a lifelong friend. Thinking back on their relationship, Succow repeatedly states that they acted both as his spiritual parents and as the pioneers of nature conservation and environmental protection in the GDR.[49] Not only did they teach him ecological concepts in his youth, they also advised him in later years, with Kurt Kretschmann admonishing him at some point: 'Censorious bystanders will not advance the cause of nature conservation'.[50] Succow took this advice to heart and endeavoured to further the cause from within the system in his capacity as biologist and

43 Tack , 'Beeindruckendes Lebenswerk', Supplement.

44 Heller, 'Geliebt und verstanden werden ist das höchste Glück', 6.

45 Ibid., 7.

46 Tack, 'Beeindruckendes Lebenswerk', Supplement.

47 Kaatz, 'Ihr schaute die Menschlichkeit aus dem Gesicht', 12.

48 Tack, 'Beeindruckendes Lebenswerk', Supplement.

49 Michael Succow, Lebrecht Jeschke and Hans Dieter Knapp, 'Unsere Ostdeutschen Vordenker', in Michael Succow, Lebrecht Jeschke and Hans Dieter Knapp (eds), *Naturschutz in Deutschland* (Berlin: Ch. Links, 2012) pp. 35-44, here pp. 39-40.

50 Quote by Michael Succow, see Uta Andresen, 'Succows Programm', in *Berlin Tageszeitung*, 19 Mar. 2005, http://www.taz.de/ArchivSuche/!633967&s=Uta%2BAndresen&SuchRahmen=Print/ (retrieved 5 Feb. 2019).

university professor.[51] In 1970, the year he gained his doctorate, Succow took up one of the Kretschmann's key concerns, sending a petition to the GDR's parliament which suggested the establishment of national parks. The petition was unsuccessful. Tracing back to nineteenth century ideas, the national park concept also had other proponents in the GDR aside from Succow.[52] The Kretschmanns had submitted three proposals for various national parks since the 1950s.[53] They were supported by Reimar Gilsenbach, author and human rights activist from the Lower Rhine region, who had moved to Saxony in 1947. As editor of the Cultural League's *Friends of Nature and Heimat*, Gilsenbach wielded a sharp pen, fighting eloquently for the establishment of national parks.[54] At the time, the Kretschmanns were unaware that forty years later, as the GDR was entering its last throes, this dream would actually come true. In contrast to Kurt and Erna Kretschmann, Gilsenbach, who was ten years their junior, was actively involved in the GDR's nascent environmental and peace movement. Michael Succow was one of the participants of the 'Brodowin talks' launched by Gilsenbach, which provided authors, environmentalists, state and cultural officials, employees of large companies, scientists and artists with an opportunity to discuss nature and environmental conservation issues as well as the national park scheme.[55] Succow remembers it as 'a movement that brought together bright minds who set out to reform the GDR system'.[56] In his capacity of East Germany's Deputy Minister for the Environment, Succow eventually proposed a national park scheme as pursued by the Kretschmanns and other GDR conservationists and implemented the idea in the form of a national park programme.

51 Typewritten CV of Michael Succow, 1991, KreA, 152.

52 Introductory: Friedemann Schmoll, *Erinnerungen an die Natur. Die Geschichte des Naturschutzes im deutschen Kaiserreich* (Frankfurt/Main: Campus, 2004) pp. 113 et seqq.

53 Succow, Jeschke and Knapp, 'Naturschutz in Deutschland', 39-40. Letter from Prof. Dr. Otto Rühle to Kurt Kretschmann, 25 Feb. 1959, Studienarchiv Umweltgeschichte (StUG), 027-32.

54 Reimar Gilsenbach, 'Die größte DDR der Welt – ein Staat ohne Nationalparke. Des Merkens Würdiges aus meiner grünen Donquichotterie', in Institut für Umweltgeschichte und Regionalentwicklung (ed.), *Naturschutz in den neuen Bundesländern – Ein Rückblick* (Marburg: BdWi, 1998), pp. 533–546.

55 Ernst Paul Dörfler, eco-chemist and co-founder of the Green Party in the GDR, confirms that the national park concept was discussed at the Brodowin talks. See mail to author of 24 Oct. 2017.

56 Cf. Reimar Gilsenbach, Hannelore Gilsenbach and Harro Hess (eds), *Wer im Gleichschritt marschiert, geht in die falsche Richtung* (Bad Münstereifel: Westkreuz, 2004), p. 282.

Europe's Green Belt

Thanks to the Bavarian branch of the German Federation for the Environment and Nature Conservation (Bund für Umwelt und Naturschutz Deutschland), which first set out to map the birdlife along the German-German border in 1979/80, knowledge about the diversity of species and habitats in the border region was already available before the Fall of the Wall. On 9 December 1989, just one month after the fall of the Wall, *Naturschutzbund* took initial steps to safeguard the areas flanking the border.[57] Initially, its focus was on integrating protected areas into the German Green Belt to safeguard and develop the core area and bordering tracts as a habitat, especially the extensive surviving semi-natural pieces of land. In addition to the national parks and other nature reserves situated along the former border, the project comprised numerous other areas, including 789 nature reserves, 402 landscape conservation areas and 9,100 natural monuments[58]. It was against this background that the national park concept pursued by Gilsenbach and the Kretschmanns since the 1950s was revisited and a proposal was made for the establishment of parks stretching from the Baltic Sea to the Bavarian Forest. At close to 1,400 kilometres from top to bottom, the conservation areas along the former border represent the longest continuous system of biotopes in Germany, linking seventeen natural landscapes from the Baltic Sea in the north to the Saxonian Vogtland in the south.[59] The creation of this network in the 1990s was not a smooth process, with leftover mines requiring detection and defusing and a number of municipalities refusing to give up their natural landscapes without a fight. The district town of Oelsnitz in the Saxonian Vogtland, for instance, wanted to use the former border strip for agricultural and forestry purposes, while the district town of Plauen, just ten kilometres down the road, did not raise any objections.[60]

57 See Rolf Weber, 'Vom "Todesstreifen" zum "Grünen Band"– dargestellt am Beispiel der sächsischen Grenze zu Bayern', in Institut für Umweltgeschichte und Regionalentwicklung e.V. (ed.), *Naturschutz in den neuen Bundesländern – ein Rückblick*, Halbband II (Marburg: BdWi, 1998), pp. 659-69; here p. 659.

58 The national park in the Harz mountains is part of the Green Belt, see Das Grüne Band – vom Todesstreifen zur Lebenslinie, https://www.nationalpark-harz.de/de/veranstaltungen/vk-20-03-2018-Das-Gr%C3%BCne-Band-%E2%80%93-vom-Todesstreifen-zur-Lebenslinie (retrieved 27 Dec. 2019).

59 Franz August Emde, 'Naturathlon wirbt für Naturlandschaften im Osten', in *Informationsdienst Wissenschaft* (ed.) (1.10.2004), https://idw-online.de/de/news86538 (retrieved 5 Feb. 2019).

60 Weber, "Vom Todesstreifen" zum "Grünen Band"', 662.

After the turn of the millennium, in 2003, the European Natural Heritage Foundation (Stiftung Europäisches Naturerbe), the Federal Agency for Nature Conservation (Bundesamt für Naturschutz), the German Federation for the Environment and Nature Conservation (Bund für Umwelt und Naturschutz Deutschland) and the IUCN joined forces to launch the cross-border European Green Belt initiative with the aim of preserving the diversity of the European natural heritage along the former Iron Curtain for future generations. Crossing 24 different countries, the belt stretches over more than 12,500 kilometres from the borders of Norway, Finland and the former Soviet Union across Europe all the way to the Black Sea coast between Bulgaria and Turkey. Analogous to the former German-German border, the erstwhile no-man's land along the European Iron Curtain had allowed nature to develop largely undisturbed by intensive human use.[61]

To ensure smooth coordination, Europe's Green Belt has been divided into three main sections, the first comprising the Scandinavian countries, Russia, Estonia, Latvia and Lithuania, the second central Europe and the third the Balkan region. Section coordinators are responsible for promoting the exchange of information, effective coordination and the implementation of projects. Each of the countries involved has appointed a representative of the responsible national ministry who is in charge of coordinating national activities. These so-called National Focal Points are tasked with promoting cooperation and communication between the work of the national ministries and the Green Belt activities. On top of this, the project also involves a maximum number of national and international non-governmental organisations.[62]

Although brought to fruition in the national park programme and absorbed into the German and European Green Belts, the Kretschmanns' vision did not remain unchallenged, even after the demise of the GDR. Both during the GDR era and after, the project attracted juxtaposing ideas, each claiming to provide the panacea for a better life in harmony with nature: while one school of thought strictly opposes any interference with nature, another believes in managing nature and placing conservation areas into the service of infrastructure, tourism or agriculture. In years past, the creation and preservation of the German network of biotopes was 'threatened by arable

61 Katharina Grund, *Linie des Lebens statt Eiserner Vorhang. Grünes Band soll grüner werden*, in Euronatur, Vol 3/2016, p. 20.

62 Cf. website of the Green Belt Organisation, http://www.europeangreenbelt.org/ (retrieved 12 Jan. 2020).

land conversion, assignment of residential and commercial areas, harbour and road construction, depletion of materials and tourism'.[63] Moreover, as early as the 1990s, individual areas whose definitive protection status was still outstanding saw their temporary protection orders expire due to a shortage of staff and financial resources at the level of the responsible state authorities, as well as the absence of an overriding concept that received general agreement.[64] Many a time, this uncertainty brought concerned nature conservationists onto the scene. They drew attention to the uniqueness of the protected areas and supported their preservation by acquiring Green Belt share certificates from ecological associations. According to BUND, the purchase of these securities is an effective response to renewed interference with nature: 'It's a straightforward deal: nature for cash'.[65]

Throughout their lives, Erna and Kurt Kretschmann had made it their mission to promote a good life in harmony with nature, an objective they pursued unconditionally. They managed to establish a European network and were engaged in a lively written and personal exchange with nature conservationists in both Eastern and Western Europe over many decades. One of their main objectives, the creation of national parks, was adopted by the GDR environmental movement which gained momentum in the 1980s. I have chosen a concrete case, that of the GDR's last Deputy Minister of the Environment Michael Succow, to show the long-term impact of these spiritual and mental roots. It is against this background that projects were kick-started even before the demise of the GDR, for instance the designation of natural reserves in East Germany, which resulted in the German Green Belt after the fall of the wall and, subsequently, integration into the European Green Belt. Up until now, research investigating the spiritual roots of the GDR environmental movement regularly singled out the international and West German debates on civilisation critique and their impact on the

63 Weber, "Vom Todesstreifen" zum "Grünen Band"', 668.

64 Ibid.

65 Sebastian Knauer, 'Grünes Band - Die Endlos-Debatte über den Todesstreifen', in *Spiegel* online 4 Nov. 2005, http://www.spiegel.de/wissenschaft/natur/gruenes-band-die-endlos-debatte-ueber-den-todesstreifen-a-383228.html (retrieved 23 Dec. 2019).

East German environmental movement.[66] With reference to these critical voices, my article has advanced the thesis that the rootedness in East German nature conservation and recourse to the ideas of its pioneers and guides had a decisive influence on the contents and actions of the GDR environmental movement as well as on individual protagonists and subsequent projects. Given that ideas neither endure nor spread if there are no actors who can support and convey them, this study presents the Kretschmanns in this mediating role, because they functioned both as a source of inspiration and as ideal promoters. The article establishes the couple's roots in the history of ideas and argues that recourse to these roots has had a significant impact on the direction and content of subsequent nature conservation projects. Accordingly, the ideas and concepts pursued by GDR nature conservationists, among them Kurt und Erna Kretschmann, played a key role in shaping the emergence of a European environmental conscience.

66 Michael Beleites argued that the roots of the GDR peace, human rights and independent environmental movement tied in with Western debates. According to Beleites, important impulses had come from the green movement in the West and the debates about the 'limits of growth' (Club of Rome report) as well as from various international church conferences, see Michael Beleites, Die unabhängige Umweltbewegung in der DDR' in Hermann Behrens, Jens Hoffmann, Institut f. Umweltgeschichte u. Regionalentwicklung e.V (eds) *Umweltschutz in der DDR. Analysen und Zeitzeugenberichte. Vol. 3 Beruflicher, ehrenamtlicher und freiwilliger Umweltschutz* (Munich: Oekom Verlag, 2007), pp. 129–224; here pp. 184 et seqq. The sociologist Detlef Pollack is also convinced that 'reception of the Western critique of capitalism and civilisation [had been] decisive' for alternative political groupings. He believes that the critique of capitalism had delivered an argument for the inclusion of the GDR as a modern industrial society in the critique of modernism. Cf. Detlef Pollack, *Politischer Protest. Politische alternative Gruppen in der DDR.* (Opladen: Leske + Budrich, 2000), p. 257.

CHAPTER 6.

WETLANDS OF PROTEST. SEEKING TRANSNATIONAL TRAJECTORIES IN HUNGARY'S ENVIRONMENTAL MOVEMENT

Daniela Neubacher

Arguing that the project would risk the water supply of hundreds of thousands of Hungarians along with the flora and fauna of a 200-kilometre-long-strip, the activists of the so called *Duna Kör* (Danube Circle) mobilised masses at the dawn of system change in Hungary. As early as November 1984 the protest movement collected around 10,000 signatures[1]. Four years later they were able to increase the number to 150,000 for a petition requesting a referendum about the Gabčíkovo-Nagymaros[2] dam project.[3] Meanwhile, on the other side of the Iron Curtain, about two hundred kilometres upstream, environmentalists were protesting against a dam project of the Austrian government. The successful protest in the Danubian wetlands of Hainburg in 1984 has been called birth date of the Green Movement in Austria and laid the foundations of a strong political force henceforth. This paper deals with the social actors of the Hungarian environmental movement in the 1980s, concentrating mostly on the activities on the leading group *Duna Kör*. Taking the concrete campaign of *Duna Kör* and its foreign partners, we will outline the transnational aspects by raising the following typological questions. Who were the main actors of the movement and how did they interact? How did they prepare and plan their actions? Which means and protest forms did they use and what common goals can be identified? In addition, the self-images and perceptions of the activists are aspects that can be analysed. Which similarities and differences do they identify among themselves? On what ideological foundations did they build up a collective identity? The focus of these research questions lies on cross-border activ-

1. Estimate provided by the activists.
2 Hereafter the dam project (Hungarian) will be referred as GNV (Gabčíkovo –Nagymarosi Vízlépcső / Gabčíkovo –Nagymaros Dams/Waterworks).
3 See Miklós Haraszti, 'The beginnings of civil society. The independent peace movement and the Danube movement in Hungary', in Vladimir Tismaneau (ed.), *In Search of Civil Society* (New York: Routledge, 1990), pp.71–87, here p. 80.

ism between opponents of the GNV in Hungary and environmentalists in Austria. The research intends to interpret this cross-border cooperation as a form of 'Europeanization from below'[4] that is based on processes of identity-building, solidarity and shared interests. Based on archive research and interviews the study reconstructs a fluid but strong network of activists that includes politicians and scientists as well as journalists and dissidents.

The ethnical, bilateral and juridical aspects of the conflict about GNV have been discussed in several papers and research works.[5] Among these studies, the development of protest and transnational activism is, however, underrepresented. Only a few researchers, such as Hubertus Knabe, Barbara Jancar-Webster and John Fitzmaurice, have raised related questions. Besides the study of Maté Szabó and Szabina Kerényi, which dealt with 'transnational influences on patterns of mobilisation with environmental movements in Hungary',[6] this specific part of Europe's transnational history has not been covered by historians. The Hainburg case has been extensively studied in Austria as a breaking point of Austria's environmental politics and as the factor of success for the Green party, which entered the parliament in 1986 with eight mandates.[7] Seeking trajectories of transnationalism means creating a novel typology. Understanding transnationalism as a process of social as well as institutional networking, based on common ideas and a collective identity, requires environmental actors to be analysed in both their campaigns and in their self-image. This paper illuminates a part of transnational history that presents an Iron Curtain with eyes, ears and even openings to both sides. Before we take a closer look at the transnational activities of environmentalists, it is necessary to follow back the different trajectories of our actors: Where did they come from? How did they find each other, what motifs kept them together and which organisational character did they establish? For this, we need to outline the major discursive developments and context of environmentalism on the eve of Hungary's transformation. In the next step, we seek to find common ground on analysing tendencies

4 Donatella Della Porta and Manuela Caiani, *Social Movements and Europeanization* (Oxford: Oxford University Press, 2009)

5 John Fitzmaurice, *Damming the Danube. Gabcikovo and Post-Communist Politics in Europe* (Oxford: Westview Press, 1998).

6 Szabina Kerényi and Maté Szabó, 'Transnational influences on patterns of mobilisation within environmental movements in Hungary', *Environmental Politics* **15** (2006): 803–20.

7 Ortrun Veichtlbauer, ETA: Environmental History Timeline Austria. Zeittafel zur Umweltgeschichte Österreichs seit 1945 (2007) Online: http://www.umweltgeschichte.aau.at/index,3191,Links.html (accessed 29 Sept. 2017).

of transnationalism, in order to identify the cross-border dimensions of environmentalists in Hungary and their fellows in neighbouring countries. By focusing on the Hainburg case in Austria and its approximate equivalent in Hungary (GNV), we then get a more differentiated picture of their gradually growing cross-border activities beyond the Iron Curtain.

A discursive history of dissidence

There is no way of speaking of opposition to the communist regime in Hungary without mentioning the uprising of 1956. The experience of success after joining forces against the regime but meeting Soviet tanks and merciless repression afterwards has deeply influenced the memory of Hungarian dissidents. In the following years Hungarian society and the regime were living under restricted circumstances.[8] According to the author and historian György Dalos, most of the activists of the uprising in 1956 who stayed in Hungary preferred to stay calm afterwards. Dalos was charged in 1968 due to 'subversive activities' and was banned from publishing his works.[9] What was left from the once strong and strategically working movement of dissidents and reformers 'was of a symbolic, cultural, discursive, and communicative character', as Maté Szabó puts it.[10] Open protests like marches or blockades were avoided and only used by small radical groups. 'The main bulk of the protest was dissident intellectuals produced critical essays, poems, and analysis, which were then censored and publicly criticised by partisans of the regime'.[11] Some of them were canalising their thoughts in philosophical discourses, like the students and colleagues of György Lukács, who formed the Marxist critical, so-called 'Budapest School'.[12] Broad criticism rather

8 György Dalos, 'Ungarn: Die intellektuelle Formierung der Opposition seit den 1970er Jahren', in Hans-Joachim Veen, Ulrich Mählert and Peter März (eds), *Wechselwirkungen Ost-West. Dissidenz, Opposition und Zivilgesellschaft 1975–1989* (Köln/Weimar/Wien: Böhlau Verlag, 2007), pp. 61–64, here p. 61.

9 Alfrun Kliems, 'Der Dissens und seine Literatur. Die kulturelle Resistenz im Inland', in Eva Behring, Alfrun Kliems and Hans-Christian Trepte (eds), *Grundbegriffe und Autoren ostmitteleuropäischer Exilliteraturen 1945–1989. Ein Beitrag zur Systematisierung und Typologisierung*. Vol. 20 'Forschungen zur Geschichte und Kultur des Östlichen Mitteleuropa' (Stuttgart: Franz Steiner Verlag, 2004), pp. 203–84, here p. 265.

10 Maté Szabó, 'Hungary', in Martin Klimke and / Joachim Scharloth (eds), *1968 in Europe. A History of Protest and Activism, 1956–1977* (New York: Palgrave Macmillan, 2008), pp. 209–18, here p. 214.

11 Szabó, 'Hungary', p. 214.

12 Dalos, 'Ungarn', p. 61.

found new ways in culture, with a growing youth subculture following a Western lifestyle and cultural orientation.[13] To give two examples: one of the dissident art groups in Budapest was called *Inconnu*. With exhibitions such as 'The Fighting Cities' (1986) it commemorated the uprising in 1956. The censorship system confiscated the collected paintings of the group and observed their activities.[14] In the field of music, the members of the punk-rock group *Coitus* were sentenced to two years in jail. Szabó states that, in comparison to other Warsaw Pact countries, Hungary was 'relatively open to outside influences'[15]. One explanation for this is the high number of ethnic Hungarians living in Czechoslovakia, Romania and Yugoslavia, producing and consuming media and books in Hungarian. Furthermore, it was easier to travel to these countries than to travel to the West. The BBC, Radio Free Europe or the Voice of America, which were broadcasting in Hungarian, fostered access to dissent and protest against the communist regime and strongly influenced the transnational relations of Hungarians. According to Szabó, the student protest of 1968 in Eastern Europe influenced the Hungarian public much less than conflicts and protest within the Eastern Bloc.[16] 'All in all, people in Hungary could reach a wide range of official and unofficial sources of information', Szabó concludes.[17] 'By around 1985, the regime had begun to lose the support, not only of old "dissidents" but also of previously loyal intellectuals and technocrats',[18] John Fitzmaurice points out:

> Much of this evolution was subterranean. It was not easily visible behind the unchanging façade of communist power. Indeed, this low-key, patchy barely visible development was inherent in the new post-1968 generation of activists all over Central Europe and certainly in Hungary and Czechoslovakia.[19]

On the diplomatic stage Austria and Hungary have been considered a role-model for the friendly relationship between two neighbour countries with different societal and political systems.[20] Due to the visa stop Austrians

13 Ibid.

14 Derek Jones (ed.), *Censorship: A World Encyclopedia*, Vol 1–4 (New York: Routledge, 2001).

15 Szabó, 'Hungary', p. 214.

16 Ibid., p. 215.

17 Ibid.

18 Fitzmaurice, *Damming the Danube*, p. 49.

19 Ibid., p. 44.

20 Maximilian Graf, 'Eine neue Geschichte des "Falls" des Eisernen Vorhanges', in Andrássy Universität Budapest (ed.), *Jahrbuch für Mitteleuropäische Studien* 2014/2015 (Vienna: new academic press, 2016), pp. 347–72, here p. 354.

have been able to travel to Hungary since 1979.[21] This liberalisation offered activists and politicians the chance to visit the neighbour country's protest events. 'In fact, Vienna and Budapest like to describe their relations as a "model" for East-West cooperation', Radio Free Europe reported in 1986. 'Ironically enough, it is precisely this context of a good atmosphere, visa-free travel, and the rest that has facilitated the concerted action of Austrian and Hungarian environmentalists ...'[22] Increasing contacts and networks beyond the Iron Curtain also favoured solidarity between the countries. During these years *Samizdat* and *Tamizdat* established a small and elite European public with a strong belief in the need of a transnational solidarity and civil society. One main discourse of intellectuals in both Eastern and Western Europe was the concept of Central Europe.[23] Emil Brix called Central Europe an 'identity container'.[24] 'The idea of a distinct Central European region had from the very beginning much to do with the wish to create a political perspective that was neither distinctly East nor West (Isván Bibo, Jenő Szűcs).'[25] By re-imagining an 'imaginary cultural landscape'[26] the intellectuals questioned the bloc-system. Their dialogues established a counter public space which tried to overcome the 'mental map of Cold War'. [27] Civil society actors such as the activists of *Duna Kör* contributed to this by organising transnational campaigns, knowledge transfer and the establishment of a collective 'green' identity.

Being 'Green' in Hungary: Between co-existence and subversion

Based on this short contextualisation, which seeks to strengthen the understanding of external influences and the reality of civil society actors on the

21 Ibid., p. 353.

22 Herbert Reed, 'Hungarian "Greens" petition Austrian parliament', in Radio Free Europe, *RAD Background Report* 96, 11 July 1986, pp.1–4.

23 Ibid.

24 Emil Brix, 'Austria and Central Europe', in Günter Bischof, Fritz Plasser, Anton Pelinka and Alexander Smith (eds), *Global Austria. Austria's Place in Europe and the World.* Contemporary Austrian Studies Vol. 20 (New Orleans: University of New Orleans Press, 2011), pp. 201–11, here p. 202.

25 Ibid., p. 203.

26 Ibid., p. 210.

27 Jan C. Berends and Frederike Kind 'Vom Untergrund in den Westen. Samizdat, Tamizdat und die Neuerfindung Mitteleuropas in den Achtzigerjahren', *Archiv für Sozialgeschichte* **45** (2005): 427–78, here 437.

eve of Hungary's system transformation, we will particularly focus on the evolution of 'green' ideas. The development of a 'green' or environmental movement in the first half of the 1980s in Hungary was surprising not only for the Hungarian regime, but also for international observers.[28] Hungarian environmentalists were considered as pioneers of public disagreement against the authoritarian regime of János Kádár. During the 1980s they developed to be a powerful single-issue movement, which consisted of scientists, engineers and intellectuals in the beginning. By protesting against GNV, they established a gravity centre for different oppositional powers like no other social movement in the Soviet bloc had done before. 'It was this issue that brought oppositional ideas and strategies together with citizens' involvement for the first time on a scale that demanded political reaction from the Party leadership', Kerényi and Szabó state.[29] To understand the movement's special role in the course of protests against the communist regime it is crucial to outline the preliminary activities for nature conservation as well as the possibilities and boundaries of civil engagement. The Kádár regime tolerated civil engagement in nature conservation to a certain extent. Some groups and initiatives were supported by the regime as long as they would submit regular reports and schemes of activities. Since 1972 the Hungarian People's Front (HNF) has officially called the 'Conservation of Environment' a part of its tasks. Approximately 600 people have worked in the respective boards throughout Hungary for environmental affairs.[30] In the late 1970s the Communist Youth Association (KISZ) decided to get active in the conservation of environment. The members created activities such as tree planting, summer camps and courses trying to raise awareness among the youth. With the Youth Council of Environment (IKT) KISZ built up a board of young experts, addressing people via small awareness campaigns in the media. However, when the critics against the GNV got louder, IKT was not able to make an official statement against the dam project.[31] Today's still existing Clean Air Action Group (CAAG) has its roots in the early 1980s as well. Before the non-profit environmental or-

28 See, among others, N.N., 'Klippen und Schwellen', *Der Spiegel* **51** (1984): 122–23.

29 Kerényi and Szabó, 'Transnational influences on patterns of mobilisation', p. 806.

30 Hubertus Knabe, *Umweltkonflikte im Sozialismus. Möglichkeiten und Grenzen gesellschaftlicher Problemartikulation in sozialistischen Systemen. Eine vergleichende Analyse der Umweltdiskussion in der DDR und Ungarn* Vol. 49, Bibliothek Wissenschaft und Politik (Köln: Verlag Wissenschaft und Politik, 1993), p. 159.

31 Knabe *Umweltkonflikte im Sozialismus*, pp. 166–67.

ganisation was founded in 1988 its members were active mainly in three different groups: the Club of Conservationists of the ELTE University, the Green Circle of the Budapest Technical University (BME) and the Group of Esperantists for Nature Protection. András Lukács, the founder and leader of CAAG, came from the latter group. Lukács and the group participated in the demonstrations of *Duna Kör*, but 'decided not to get involved' at the very beginning, Lukács says in an interview.[32] Till 1988 it was almost impossible to establish an association for environmental issues.[33] It was easier to found a '*Klub*' or '*Kör*'. These groups were not allowed to create a bank account and had to have ties to an organisation or institution, such as a student dormitory, university or cultural organisation. The rising number of clubs and circles showed an increasing interest in the environment in the first half of the 1980s.[34] Although there were different green groups to cooperate with, 'all wanted to keep their absolute independence from everybody and everything, even from one another, allying only on specific issues when necessity demanded it', Barbara Jancar-Webster reported in her study. The ELTE Club of Conservationists, which was one of the oldest unofficial organisations, served as a centre and information point in this fluid network of 'Greens'. Not only was cooperation within the local network difficult, but also that with Greens in other Eastern European countries. With 'Greenway' the ELTE Club started an English-language newsletter which reported on activities of activists. Later on, the *Danube Movement* tried to set up a stronger network – without success.[35]

Seeking trajectories of transnationalism

When analysing actor's common ideas and self-image, campaigns offer a useful subject of interest. In 2002 Christian Lahusen tried to create a typology for analysing transnational protest forms. He recommended using the characteristics of campaigns. According to him campaigns are planned, and prepared sets of communication activities aiming to achieve or prevent a change of attitudes, behaviour or decisions. Furthermore, with campaigns

32 Interview with András Lukacs, July 2017. Archive of the author.

33 Knabe, *Umweltkonflikte im Sozialismus*, pp. 168–69.

34 Ibid.

35 Barbara Jancar-Webster, 'The East European environmental movement and the transformation of East European society', in Jancar-Webster (ed.), *Environmental Action in Eastern Europe. Responses to Crisis* (New York: Routledge, 1993), pp. 192–219.

one aims to convince a certain group of people. Summing up, campaigns can be analysed by the scale, the planning, the defined goals and addressees.[36] To analyse the cross-border cooperation of *Duna Kör* and its foreign partners, we will focus on typological questions that seek to identify the main actors and their interactions, their preparations and planning as well as the means and goals of their campaigns including dimensions, perception, ideology and identification in regard to Della Porta and Caiani's idea of 'Europeanization from below'. Della Porta and Caiani, who studied social movements and their influences on creating a European public, see an enormous potential in transnational activism:

> During transnational campaigns activists begin to identify themselves as part of a European or even a global subject. Action in transnational networks also enables the construction of transnational identities through the recognition of similarities across countries.[37]

Based on the hypothesis that the Austrian and Hungarian environmentalists contributed to the establishment of a common public sphere by their cross-border cooperation, the following section seeks to answer some of the questions posed earlier.

Wetlands of protest: The cases of Hainburg and Gabčíkovo-Nagymaros

Europe's second-longest river flows 2,888 kilometres from its source in the German Black Forest to its delta in the Black Sea and passes through ten countries, irrespective of political borders. Of all European rivers it was the Danube that created a basis for conflicts in Central Europe in the 1980s. The two hydro-electric dam projects that led to the Gabčíkovo-Nagymaros and Hainburg cases raised economic, political and environmental questions. This was not only in Austria, Czechoslovakia and Hungary, but also in other countries of the Western and Eastern Blocs. GNV and Hainburg became historical milestones of civil engagement, political opposition and transnational protest. When Czechoslovakia and Hungary agreed on the project in 1977, damming this section of the Danube had already been discussed for decades.[38] Hydro-electric dams were considered an industrial investment by

36 Christian Lahusen, 'Transnationale Kampagnen Sozialer Bewegungen. Grundzüge einer Typologie', *Forschungsjournal Neue Soziale Bewegung* **15** (1) (2002): 40.

37 Della Porta and Caiani, *Social Movements and Europeanization*, p. 162.

38 Fitzmaurice, *Damming the Danube*, p. 3.

the communist system. 'Big mega-projects ... were an advertisement of the building of the communist industrialisation', János Vargha says. Before becoming an activist and founding member of *Duna Kör* in 1984 the biologist worked as an environmental journalist and collected information about damming projects of the communist regime.[39] Not only did its vast size generate huge environmental concerns, but the fact that 140 kilometres of the Danube form a natural border between Czechoslovakia and Hungary also increased the complexity of the GNV controversy. Opposing this project the so-called *Danube Movement* was formed. It consisted of the three main groups: *Duna Kör*, the *Dunáért Alapítvány* (Foundation for the Danube), and *Kékek* (Blues). The *Duna Kör* with Janos Vargha as its most prominent face led the single-issue movement. The critics of GNV reach back much further, though, than the founding of the *Danube Movement*. 'I think that all the environmental movement activities have a history of criticism from scientists and engineers', Vargha says. Among the critical voices were the engineers Mihael Erdelyi, György Hábel and István Molnár, as well as the architect and urbanist Károly Perczel, who, among others, had published critical articles as early as the 1970s.[40] Even top-ranked scientists criticised the project plans. According to Vargha, Sándor Szalai, a renowned professor of sociology and member of the presidium of the Hungarian Academy of Sciences, criticised it behind the scenes.[41] Nevertheless, no group other than *Duna Kör* could mobilise more people to openly disagree. The committee of *Duna Kör* consisted of two biologists, Béla Borsos and Vargha (who was then working as a journalist), András Szekfű, sociologist, and László Vit, who was working as an engineer. Recruitment worked via letters of recommendation from an existing member.[42] The difficulty of reconstructing the preparation and planning activities of civil actors in the 1980s is due to the nature of underground activism. The German historian, Hubertus Knabe was one of the few scientists who tried to reconstruct the rapidly changing landscape of environmentalists during Hungary's transformation period. He set one main starting point of the *Danube Movement* in January 1984, when opponents of the GNV met in *Rakpárt Klub* in Budapest. Back then it was considered the

39 Interview with János Vargha by the author, Budapest 2017. Archive of the author.

40 Vargha Interview, 2017. Also Perczel Károly (ed.), *A bős-nagymarosi vízlépcső regionális terve (háttértanulmányokkal)* (Budapest: VÁTI, 1978).

41 Vargha Interview, 2017.

42 Article by János Vargha and Béla Borsos, 'Duna Kör', 30. Nov 1988. In: OSA 205–4–140 Box 37.

headquarters of the HNF.[43] About fifty people remained there after the official event ended. Vargha remembers the foundation of the 'Danube committee' on that evening with the simple words: 'Somebody said, we should do something'.[44] Some of them went to the private apartment of Ferenc Langmár and discussed what could be done next against GNV. Further public debates were organised, at the BME and at the Karl Marx University (now known as Corvinus University). The young journalist Ádám Csillag, who later produced the documentary '*Dunaszaurus*' dealing with the GNV, attended these very first debates by invitation of Anna Perczel, daughter of the architect and GNV critic Károly Perczel. He also recorded them. Besides the university groups, the Association of Hungarian Writers was also hosting a debate.[45] First as '*Független Duna Kör*' and later as '*Duna Kör*' the group set several protest activities against the GNV. One of the first was a petition that was handed over to the Hungarian National Assembly and the government, demanding that construction stop. The petition was printed and distributed together with some background information on the risks of the project. In this edition of *Vízjel*, published on 2 October 1986 the activists warn against severe damage to the drinking-water supply and to agriculture, and other risks for the local population. Calling the halting of the construction a 'matter of life' they consider their demands legitimate through being 'common interests' of all the people who live along the Danube.[46] According to Knabe, the committee increased to 300 members within a few months. Subsequently they tried to become an association, which would have had to be officially connected with the National Office for Environmental Protection and Nature Conservation (OKTH). Although they negotiated for a long time, they could not succeed. Besides Knabe's study, the highly-active agent network of the Hungarian State Security provides insights; the content of the reports needs to be interpreted with critical distance though. Starting in the early 1980s, State Security regularly observed members of *Duna Kör*. Starting with reports on the activities of 'a group, which is against the Gabčíkovo Vízlépcső, the agents reported on a core group of about 160 members. 'Thirty are very active' and 'consistently well-trained', a report said. János Vargha was called a leading 'propagandist'.

43 See also Vargha Interview, 2017.

44 Ibid.

45 Cf. Vargha Interview, 2017. Remark: the BME discussion was restricted to being video recorded by the rector of the university. According to Vargha, Csillag recorded the sound in secret, though.

46 See Knabe, *Umweltkonflikte im Sozialismus*, pp. 187–88.

Together with the engineer Langmár and the lawyer László Sóylom he would spend most of his income on keeping in touch with foreign contacts, traveling abroad and smuggling out propaganda material.[47] According to an agent called 'László János', mistrust towards new members and inner conflicts about the goals, the means and the self-image of the group dominated the discussions of *Duna Kör*. Another report says that some of the core members didn't want to cross legal boundaries. Vargha was criticised for approaching the members of *Charta* 77 and for not being a 'good leader'.[48] The 'radical wing' wanted to use the Hainburg case 'for taking further actions'. As far as we know from these reports, the group organised a personal meeting and went to Vienna and Hainburg. To get in touch with Hainburg activists they would ask an Austrian journalist to accompany them. According to 'Läszlo', they wanted to study the movement in order to find out how environmentalists could act through parliament and institutions. [49] 'We wanted to call the attention [of the Austrian environmentalists] to the problem, that after they stop Hainburg, the Donaukraftwerke [Österreichische Donaukraftwerke AG] will do the same thing just 200 km downstream', Vargha remembers. As main contacts the *Duna Kör* leader mentions, among others, Peter Weish, who made the famous speech 'The spirit of Hainburg', and Günter Schobesberger, as well as the two famous leading activists Freda Meissner-Blau and Günther Nenning.[50] Born in Upper Austria, the biologist Günter Schobesberger had a house in Hainburg, close to the area that was occupied by activists in December 1984. Schobesberger's name was mentioned in an article in the Hungarian newspaper *Magyar Hírlap*, where he was called a 'sympathiser of Duna Kör'.[51] The author and journalist Günther Nenning, who was expelled from the Austrian Social Democratic Party (SPÖ) in 1985, was a famous voice of the protest and wrote a text for the infamous protest group *Konrad-Lorenz-Volksbegehren*. In 1984 the group criticised the Austrian

47 See report of Budapest Police Headquarters (BRFK), in Állambiztonsági Szolgálatok Történeti Levéltára (ÁBTL), Budapest, BRFK III/III–A do. TMB, 'László János' H–70490 I. p. 8.

48 See daily operational information reports (Napi operativ információs jelentés) in ABTL 2.7.1. NOIJ–III/3–2/1/1985.I.14.

49 Ibid.

50 Vargha Interview, 2017.

51 N.N., MUK – mondja a Dunakör. In: Magyar Hirlap, vom 11.2.1991, n. p., In: OSA 205–4–140, Box 37, 'Egyesületek ‚D' Duna Kör 1987–1991', In OSA 205–4–140, Box 37. Original quote: 'Ezt követöen rendöri bisztosítás mellet a Dunakör tagjai és szimpatizánsai elsétáltak a Margitszigetre, ahol Günter Schobesberger osztrák köryezetvédö buzdította a tüntetöket, hogy továbbra is harcoljanak környezetük védelméert.'

water plant and ran a petition with 353,000 signatures.[52] The Austrian Student Union (ÖH) extensively mobilised students against Hainburg. Nevertheless, the campaign *'Rettet die Au'* (Save the wetland) was started in 1983 by the Austrian section of WWF, which later changed its name into World Wide Fund for Nature. Founded in 1967 the Austrian section of WWF was also one of the main NGOs that supported Hungarian environmentalists. In 1987 they financed a bilingual brochure that was presented simultaneously at two press conferences in Vienna and Budapest.[53] *Duna Kör*'s tendencies towards transnationalisation increased when it was revealed that Hungary had started to negotiate with Austria about financing the dam project in 1983. Austria agreed to finance the project in 1986.[54] In 1987 *Duna Kör* launched a petition, which they handed over to the Austrian '*Nationalrat*'. Günter Schobesberger said to Radio Free Europe:

> Under Austrian law, any petition by private persons, organizations, or pressure groups, if endorsed by a member of parliament, will be submitted to the National Assembly, which then has to consider it. This applies to Hungarian citizens as well. So if private persons or environmentalist groups in Hungary object to Austria's financing of the Nagymaros project, they should lodge a protest with the Austrian parliament ... I guarantee you that we shall forward their petition through a friendly member of parliament to the parliament.[55]

Duna Kör members used several occasions to network with environmentalists abroad. Although Hungarians enjoyed greater freedom to travel abroad compared with citizens of other communist countries, the authorities did not always allow them to leave the country. In 1985, when the famous conference on Global Warming took place in Villach (AT), the GNV opponent Judit Vásárhelyi was not granted permission to travel to Austria.[56] When *Duna Kör* received the 'Right Livelihood Award', known as the Swedish Alternative Nobel Prize, Vargha was allowed to go to Stockholm in order to receive the award in the name of the group. Afterwards they underwent several difficulties in getting paid the prize of 25,000 US Dollars. *Duna Kör* was awarded the Right Livelihood Award in 1985, but it took until 1987

52 Veichtlblauer, ETA: Environmental History Timeline Austria.

53 Vargha Interview, 2017.

54 Austrian Federal Audit Office, 'Wahrnehmungsbericht des Rechnungshofes über die Österreichische Donaukraftwerke AG', series Burgenland 1995/1 (Z1 01000/371-Pr/6/95), (Vienna: Österreichische Staatsdruckerei, 1995).

55 Herbert Reed, 'Hungarian "Greens" petition Austrian parliament'.

56 Vargha Interview, 2017.

to find a way of receiving the prize. In the end twelve members altogether received the money in separate portions, paid in Hungarian currency. With this money they established a foundation, the *DunáértAlapítvány* (Foundation for the Danube), printed further information material (especially the anthology *Duna. Egy antológia*, published in 1988) and financed an international conference, which took place from 2–6 September 1990.[57] 'The foundation will assist private citizens or movements who wish to perform acts towards the preservation of ecology, environment or nature with particular regard to the Danube', the activists wrote in a statement.[58] As we can see from the reports of the State Security in 1984 and 1985, the activists of *Duna Kör* were highly interested in getting in touch with the activists of Hainburg. Besides that, they also tried to get to know the Czech position on the water plant. One of the goals was to win over members of the Hungarian parliament, but also to approach the Czech government. In addition, the activists wanted to build up cooperation with environmentalists in Bratislava and Austria, an agent's report says.[59] Not being forced into underground work, the Austrian environmentalists could much more easily establish their movement against the Danube dam. Whereas Hungarian environmentalists were trying to obtain the status of an NGO, the Austrians could build upon an existing network of groups, established NGOs as well as politicians and scientists. In 1971 the international environmentalist association Friends of the Earth (FoE), which was represented in seventy countries, was founded. In 1982 Global 2000 joined FoE as the Austrian member of the association.[60] Next to ÖH and WWF, Global 2000 played an essential role in the Hainburg movement. In the same year the *Vereinten Grünen Österreich* (VGÖ), who later changed their name to *Die Grünen* registered as Austria's first Green party.[61] With Greenpeace International having been founded in 1970, *Greenpeace Österreich* followed in 1983.[62] As we can see from this development, Austrian environmentalists were already acting in an institutionalised and transnationally operating framework. Taking a comparative look at the means of campaigning, it seems that the Hungarian environmentalists were

57 Ibid.

58 György Krassó, Hungarian October Information Centre, Vol 74. London 1987. In OSA 205–4–140 Box 37.

59 Ibid.

60 Veichtlbauer, Environmental History Timeline Austria.

61 Ibid.

62 Ibid.

imitating the protest against Hainburg. Both movements considered media as their most important communication channels to mobilise and inform the public. Press conferences, advertisements and close contact with journalists were essential to their campaign work. The campaign was addressing not only the Hungarian public, but also that of Austria. In January 1986 *Duna Kör* members organised a common press conference together with Greens from Austria and Western Germany. As the location they chose the *Zöldfa étterem* (Green Tree) restaurant in Budapest. 'We tried to inform the public worldwide', Vargha says.[63] In February 1986 *Duna Kör* organised, together with Austrian environmentalists, what they called an educational tour in Budapest. At the last minute, police prevented them from meeting. Radio Free Europe reported that, before this 'Danube Walk' took place, 22 'prominent Austrian and Hungarian figures' had met on 18 January 1986. They issued a statement saying that they 'want to use all the democratic, peaceful, and constitutional possibilities to dissuade the governments from their intent and make them respect the true interests of all Danubian peoples'.[64] A few months later, on 16 April, they published a big advertisement in the Austrian newspaper *Die Presse* costing, according to Vargha, 'a lot of money'. The activists called on people to protest in Austria and express solidarity with Hungarian environmentalists. Among the signatories of the call were thirty prominent Hungarians including dissidents, artists and sportsmen. 'This advertisement was a striking illustration of the extraordinarily close cooperation that has developed between the Hungarian and Austrian opponents of the Nagymaros Dam', a Radio Free Europe reporter stated. According to the article, the costs of this advertisement were covered by donations collected by the Austrian 'Friends of the Danube'.[65] Speaking of transnational relations with Austrian environmentalists, Vargha today identifies several similarities. 'We had the same approach', he says. 'The critics were similar, also scientifically. We and they used all the options we had.' Although, the protest movement was criticising the socialist system, comparing the cases in Hungary and Austria convinced Vargha, that 'a change of system will not solve environmental issues'. 'We will only gain a democratic environment for debate.' He continues, 'Democracy does not mean that the interest of

63 Vargha Interview, 2017.

64 Reed, 'Hungarian "Greens" petition Austrian parliament', 1–4.

65 Ibid.

weak entities [rivers] will be represented.'[66] The development of a green transnational movement made international observers such as Radio Free Europe more enthusiastic. 'Come what may, transborder cooperation between environmentalists and the personal contacts that have developed during the Nagymaros controversy remain; and there will almost certainly also be other issues, particularly in this post-Chernobyl age, to attract their attention',[67] a background report said. Vargha explains the successful mobilisation against the dam in terms of its original non-political nature. Whereas the Green movements in western countries protested excessively against nuclear power, both Hungarian scientists and civil actors stayed calm. 'Because this was the hottest issue (one of the hottest)', Vargha says. 'But the Danube issue was less political, much less political. It had no any direct connection with the military issues.' [68] Apparently, the collective memory of Soviet tanks in Budapest was playing an influential part in protest culture even three decades after 1956.

Seeking trajectories of transnationalism among Austrian and Hungarian environmentalists means not only a methodical challenge for historical research, but also a permanent perspective change. This study has tried to analyse cross-border activism between opponents of the GNV in Hungary and the activists of the Hainburg movement in Austria. By following the means of campaigning, personal networks and organisational boundaries, a short insight has been given into transnational activities, such as demonstrations, press conferences, scientific conferences and informal meetings. Based on State Security reports, previous research and interviews, the study could reconstruct an active network of environmentalists in the 1980s which had its foundation in previous scientific but not openly vocal critics. Knowledge transfer and common campaigning were the main aspects of the cross-border interaction, but financial aid from abroad also played an important role. At this point of research, it seems that the Hungarian side was mostly the 'receiving' part of this relation. Nevertheless, the activists in Hungary and Austria shared the idea of being Danubiens. As inhabitants of this common landscape along

66 Vargha Interview, 2017.

67 Reed, 'Hungarian "Greens" petition Austrian parliament'.

68 Vargha Interview, 2017.

the Central European River they wanted to protect nature from the damages inflicted by mega dam projects. When Austria showed interest in financing the GNV the transnationalisation was further legitimised. In following the same goals and fostering a transnational understanding of nature protection and human responsibility, activists such as the members of *Duna Kör* tried to build up a common public space beyond the Iron Curtain. Contrary to intellectual and dissident dialogue via *Samizdat* and *Tamizdat*, the activists of *Duna Kör* opened the discourse for a broader public and addressed both local and foreign media. Their activities consisted of common demonstrations, bilingual publications and conferences. As Donatella Della Porta described it years later in the context of 'Europeanization from below', the transnational campaign against GNV enabled 'the construction of transnational identities through the recognition of similarities across countries'.[69] Furthermore the activists shaped the dissidents' intellectual ideas of cooperation and solidarity into a dynamic and active social movement beyond the Iron Curtain – a Central Europe from below. Whereas we could identify some similarities between the activists in the West and East, *Duna Kör* needed to overcome quite different barriers in terms of freedom of opinion, professionalisation and institutionalisation than Austrian environmentalists did. Comparing the different political systems – hence spaces of opportunity – as well as the experiences of repression that shaped the collective memory of dissidents and civil actors, the establishment of a transnational network seems even more surprising and worth investigating further. This paper gave insights into a complex but fruitful transnational interaction of environmentalists in Central Europe. Further research related to the abovementioned questions concerning actors, movements and their interactions is needed, though, in order to create a more comprehensive typology.

69 Della Porta and Caiani, *Social Movements and Europeanization*, p. 162.

CHAPTER 7.

TOWARDS A 'EUROPE OF STRUGGLES'? THREE VISIONS OF EUROPE IN THE EARLY ANTI-NUCLEAR ENERGY MOVEMENT 1975–79

Andrew Tompkins

The opposition to nuclear energy in the 1970s was a transnational phenomenon that connected activists from the United States and Japan to counterparts in the United Kingdom, Belgium, the Netherlands, Luxembourg, Italy, Spain, Austria, Switzerland, Sweden, Denmark and, perhaps most visibly, West Germany and France. The preponderance of European countries in these protest networks was no accident: while some have identified important 'American' roots in environmentalism globally,[1] it was within Western Europe that opposition to nuclear energy proved most widespread, contentious and durable.

At first glance, one might therefore plausibly assume that the movement had a consciously 'European' character. After all, Brussels-based institutions were early allies for some environmentalists, supporting the creation of the European Environmental Bureau and even a series of 'Open Discussions on Nuclear Energy' in 1977–78.[2] In West Germany – often regarded as an environmental leader[3] – well-known campaigners like Petra Kelly openly professed hopes that anti-nuclear protest would foster European internationalism.[4] And beginning with the first direct elections in 1979, the European Parliament would welcome a succession of anti-nuclear campaigners associated with local struggles in places like Flamanville (Didier Anger), Fessenheim (Solange Fernex) and Gorleben (Rebecca Harms). Yet, as this chapter will show on the basis of examples from France and West Germany, the grassroots of the broad and vigorous anti-nuclear movement of the years 1975–79 had a much more

1 J. Radkau, *Die Ära der Ökologie. Eine Weltgeschichte* (München: C.H. Beck, 2011).

2 J.-H. Meyer, 'Challenging the atomic community: The European Environmental Bureau and the Europeanization of anti-nuclear protest', in W. Kaiser and J.-H.Meyer (eds), *Societal Actors in European Integration* (Basingstoke: Palgrave Macmillan, 2013), pp. 197–220.

3 F. Uekötter, *The Greenest Nation? A New History of German Environmentalism* (Cambridge, MA: MIT Press, 2014).

4 S. Milder, 'Thinking globally, acting (trans-)locally: Petra Kelly and the transnational roots of West German Green politics', *Central European History* **43** (2) (2010): 301–326.

ambivalent or even antagonistic relationship with Europe and its institutions, with activists rarely even describing their movement as 'European'.

While this might seem surprising from today's perspective, it is less so when one considers the nature of the anti-nuclear movement – and of Europe – in the 1970s. The early anti-nuclear movement was anchored in place-based opposition to specific nuclear facilities, the latter usually planned with state backing. Protest crystallised within 'directly affected' local communities, which forged informal networks with nearby sympathisers and with distant initiatives at other nuclear sites – including abroad. These local struggles tapped into a large reservoir of protest potential that had spread throughout rural areas, university towns and major cities since the late 1960s, and which increasingly tended toward direct action forms of protest. They were also strengthened by nascent environmentalist organisations such as Les Amis de la Terre (ADLT) and umbrella groups like the Bundesverband Bürgerinitiativen Umweltschutz (BBU). Non-governmental organisations (NGOs) such as these, though, were much weaker in the 1970s than today and served primarily to facilitate communication through informal networks rather than to centrally coordinate action via hierarchical structures.[5] Though European and other international institutions can sometimes help transnational social movements to circumvent domestic political blockages, activists have often preferred to challenge Europe-wide policies within familiar, nation-state channels.[6] Even for more recent movements that have 'Europeanised', Brussels has been far more receptive to the lobbying of professionalised NGOs than to contentious forms of protest by decentralised actors.[7]

The loosely structured nature of the early anti-nuclear movement's transnational networks is apparent in an appeal launched for an 'International day of action against nuclear energy' to take place over Pentecost 1979.[8] Drafters of the appeal included not only ADLT, BBU and other national committees from Austria, Belgium, the Netherlands, Luxembourg and Sweden, but also regional groups from the Basque country, Cataluña, Flanders and Schleswig-Holstein, as well as local groups from Belfort, Cattenom, Limoges,

5 D. Nelkin and M. Pollak, *The Atom Besieged: Extraparliamentary Dissent in France and Germany* (Cambridge, MA: MIT Press, 1981), pp. 126–29.

6 D. Imig and S. Tarrow, *Contentious Europeans* (Lanham, MD: Rowman & Littlefield, 2001).

7 P. Monforte, *Europeanizing Contention: The Protest against 'Fortress Europe' in France and Germany* (New York: Berghahn, 2014).

8 'Appel pour la journée international [sic] d'action contre l'énergie nucléaire' (Flyer, supplement to *alerte* no. 3), Feb./Mar. 1979, Lyon, La Gryffe, dossier 'Super-Pholix'.

Malville, Saarbrücken and Würzburg – some of them actually local chapters of ADLT or members of the BBU. Not only did representation thus overlap, but it could be difficult to distinguish between local, regional, national and international organisations, as in the case of the 'Coordination Régionale-internationale contre la centrale nucléaire de Cattenom'. The joint appeal called for simultaneous demonstrations at a number of specific sites around Europe, but noted that these were 'propositions to be confirmed, expanded, or restricted by the regional and national movements'. Tellingly, the appeal was framed firmly in 'international' rather than 'European' terms. Though twenty of the 28 signatories came from within the European Community and all but one from geographic Europe (a delegation from the Melbourne-based Movement against Uranium Mining), the text makes no mention whatsoever of 'Europe'. When anti-nuclear protest boomed in the late 1970s, activists engaged in local struggles were happy to network across borders, but it is remarkable how seldom they referred to 'Europe' or its institutions in doing so. Their 'international' aspirations were often vague and rarely confined to the European subcontinent, much less to the European Community.

The rest of this article will examine three different understandings of 'Europe' articulated by nuclear energy opponents in the late 1970s. Europe was most frequently invoked within border regions, especially at protests in Alsace and Baden that attracted international attention around 1975. Activists there conceived of joint protest first as a counter-narrative of French-German reconciliation 'from below' (rather than among elites), and second as part of a trilateral 'Dreyeckland' that explicitly included non-EC member Switzerland. Another major site of cross-border anti-nuclear protest was in Creys-Malville (France), where a consortium of state-backed European energy companies known as NERSA collaborated to build a Fast Breeder Reactor (FBR). For many activists who participated in the 'international' demonstration held there in 1977, the Europe of NERSA was one of cross-border repression and technocratic collusion with big business. With the first direct elections to the European Parliament in 1979, different relationships to institutional Europe began to emerge, especially for those who went on to join Green parties. However, many grassroots activists continued to eschew institutions and to prefer informal politics: the 'Europe of struggles' to which activists from Gorleben (Germany) and the Larzac plateau (France) referred in a 1979 report was one manifestation of a persistent, extra-parliamentary form of environmentalism. In the long term, these movements would also serve as incubators for national and transnational solidarities that would re-assert themselves more visibly in the 1990s and 2000s within the Global Justice Movement.[9]

Three visions of Europe

Dreyeckland: Europe as post-war reconciliation

Within the anti-nuclear movement, 'European' ideas and narratives were usually most prominent near the continent's internal borders.[10] Indeed, the earliest protests against nuclear energy took place in the borderlands of the Upper Rhine Valley, where French Alsace and German Baden meet at the Rhine River, which flows northward from Swiss Basel. As a border river, the Rhine could potentially provide cooling water for competing nuclear projects in all three neighbouring countries. France acted first, beginning construction on a nuclear power station in Fessenheim in 1971. This sparked the first significant protests against nuclear energy in Western Europe, drawing 1,000–1,500 participants for a non-violent, silent march to the gates of the future power plant on 12 April. The same year, the West German federal state of Baden-Württemberg and its energy company Badenwerk announced plans to build a reactor of their own on the Rhine at Breisach, which was later moved to Wyhl after 60,000 signed a petition opposing it. Another power station in Kaiseraugst, Switzerland, was also already planned.[11] Further power stations were expected to follow, with the French government alone naming potential sites up and down the river in Marckolsheim, Sundhouse, Gerstheim and Lauterbourg.[12] The nuclear projects were central to development plans intended to turn the region into

9 E. Rivat, 'The continuity of transnational protest: The anti-nuclear movement as a precursor to the global justice movement', in C.F. Fominaya and L. Cox (eds), *Understanding European Movements* (London: Routledge, 2013), pp. 61–75; R. Gildea and A.S. Tompkins, 'The transnational in the local: The Larzac Plateau as a site of transnational activism since 1970', *Journal of Contemporary History* **50** (3) (2015): 581–605.

10 e.g. 'Internationale Zusammenarbeit gegen grenznahe Atomanlagen' (Flyer, 6 pp.), 1981, Berlin, PapierTiger, 'AKW - Westeuropa'.

11 P. Kupper, *Atomenergie und gespaltene Gesellschaft: Die Geschichte des gescheiterten Projektes Kernkraftwerk Kaiseraugst* (Zürich: Chronos, 2003).

12 Ministère de l'Industrie et de la Recherche, 'Localisation des centrales nucléaires' (Dossier, 40 pp.), November 1974, Paris, Bibliothèque de documentation internationale contemporaine, F Δ 1190. The inclusion of these sites did not necessarily mean that each was planned, only that they were among the options considered geographically suitable. Activists nevertheless were concerned that the entire region might be given away to nuclear energy infrastructure.

a 'showcase' industrial economy at the heart of Western Europe.[13] However, protest throughout the region was to prove so intense that only the power station in Fessenheim was ever built. Opposition to environmental threats on different sides of this border proved mutually reinforcing and gave regional protests a transnational dimension.

The earliest protests took place in Alsace and were grounded in discourses of pacifism that resonated with certain conceptions of Europe as reconciliation between former enemies. However, in this classic region of 'national indifference',[14] Alsatian activists emphasised locally specific forms of internationalism rather than European institutions. Esther Peter-Davis, one of the leaders of the 1971 Fessenheim march, was an Alsatian woman with international connections to the United States through her husband, Garry Davis, an American soldier who renounced his citizenship in 1948 and declared himself 'first citizen of the world'. Through her in-laws in New York,[15] Esther met John Gofman, a biologist who supplied her and a circle of friends with reports about the dangers of nuclear energy, which they supplemented with further materials from French- and German-language publications to create a brochure about the issue that they distributed throughout Alsace.[16] Jean-Jacques Rettig, who joined Esther at these early protests, was an Alsatian whose principled opposition to nuclear energy stemmed largely from a historically rooted opposition to militarism in a region swapped four times in 75 years between France and Germany. His family members had fought for different sides in different wars and his wife Inge was born German but later naturalised as a French citizen. All this led him to reject war and nationalism while simultaneously embracing regional identity.[17] For Esther and Jean-Jacques as Alsatians, opposition to nuclear energy was tied to a post-war peace project anchored in a regional vision of French-German reconciliation. Critically though, as Jean-Jacques puts it, this reconciliation was 'bottom-up' as well as 'top-down': 'It wasn't just

13 B. Nössler and M. de Witt (eds), *Kein Kernkraftwerk in Wyhl und auch sonst nirgends: Betroffene Bürger berichten* (Freiburg: Inform-Verlag, 1976), p. 257.

14 Tara Zahra, 'Imagined noncommunities: National indifference as a category of analysis', *Slavic Review* **69** (1) (2010): 93–119.

15 S. Milder, *Greening Democracy: The Anti-Nuclear Movement and Political Environmentalism in West Germany and beyond, 1968–1983* (Cambridge: Cambridge University Press, 2017), pp. 32–33.

16 Comité pour la Sauvegarde de Fessenheim et la plaine du Rhin, 'Fessenheim : vie ou mort de l'Alsace' (Brochure, 69 pp.), Feb. 1971, Saales, A. Tompkins private archives.

17 J.-J. Rettig, 'Eine persönliche Umweltgeschichte, Familiengeschichte und Regionalgeschichte im Elsass', 2007, [http://www.bund-rvso.de/rettig-umweltgeschichte.html (accessed 10 Jan. 2020)].

de Gaulle and Adenauer who shook hands, but the grassroots as well!'.[18]

By 1974, local environmentalist mobilisations began to bear fruit – not in Fessenheim itself, but in Marckolsheim, where a German company proposed building a chemical plant on the French side of the border (after having been refused the required permits at home). German activists were preparing in parallel for a fight over the nuclear power station in Wyhl, only ten kilometres from Marckolsheim but on the German side. Activists from Alsace and Baden thus decided to link the two struggles, with 21 groups issuing a joint declaration opposing both projects and threatening concerted action to oppose them. When construction equipment was brought to Marckolsheim in September 1974, activists quickly responded by occupying the construction site. They further anchored their protest in their opponents' space by building a 'friendship house' at which they held concerts and informational events, giving their illegal civil disobedience a festive character. After more than five months of site occupation, French authorities gave in to protesters' demands and withdrew authorisation for the chemical plant in late February 1975.

However, construction of the nuclear power station in Wyhl began almost simultaneously. Experiences in France provided a useful 'dress rehearsal' for the protests in West Germany that followed.[19] Together, French and German activists occupied the Wyhl site on 18 February 1975. However, West German police were keen to prevent a repeat of the Marckolsheim protests and evicted the demonstrators only two days later. On 23 February, though, activists linked a mass rally attended by 28,000 people to a second, successful occupation attempt. Rotating in village-based teams, they kept the site occupied for almost nine months. Following the Marckolsheim model, they transformed the site by building another, even larger 'friendship house', in which they regularly hosted events to draw supportive crowds to the site. Authorities were forced to suspend construction temporarily over and over again until the project was ultimately abandoned.

This local transnationalism formed the basis for a compelling narrative of French-German reconciliation 'from below', cast in specifically regional terms. Activists played up their already significant cross-border cooperation at every

18 Jean-Jacques Rettig, Interview with the author, Fréconrupt (19 Apr. 2010).

19 Marie-Reine Haug and Raymond Schirmer, Joint interview with the author, Rammersmatt (17 Apr. 2010).

turn, consciously constructing a 'legend'[20] that invoked past national conflict in order to emphasise present-day grassroots reconciliation. They also borrowed liberally from the distant past in search of transnational symbols, seizing notably on the Peasant's War (*Deutscher Bauernkrieg*) of 1525 as a symbol of regional resistance to outside intervention.[21] The sixteenth-century hero Jos Fritz became the pseudonym of choice for anti-nuclear activists as well as the namesake for a left-wing bookshop in nearby Freiburg. Activists also pointed to Baden's role in the revolutions of 1848 as a supposed precedent for the transnational protests of the 1970s. They thus deployed 'invented traditions' that appealed both to left-leaning students and to more conservative locals who chose to understand resistance as part of their local heritage.[22] Such actions framed cross-border protest in regional rather than national terms.

However, this transnational framing rarely made reference to Europe. Indeed, the word 'Europe' and variations upon it appear rarely in two full-length books published in 1976 and 1982 by the local anti-nuclear initiatives to promote and explain their protests to a broader audience: there are only three passing references to European institutions (alongside 21 further references to geographical Europe). The books refer far more frequently to the regions of Baden and Alsace (178 and 154 occurrences, respectively).[23] Interestingly, the second book also makes frequent reference to 'Dreyeckland' (67 occurrences), a term invented in 1977 to describe the region that was home to the protests against Fessenheim (France), Wyhl (Germany) and Kaiseraugst (Switzerland).

An inversion of the usual term *Dreiländereck* (meaning the 'corner' or meeting point of three countries), the literal meaning of *Dreyeckland* ('the country of three corners') reified the region's supposed state of liminality and asserted authenticity through the use of Old German spelling (*drey* instead

20 P. Kenney, 'Opposition networks and transnational diffusion in the revolutions of 1989', in G.-R. Horn and P. Kenney (eds), *Transnational Moments of Change: Europe 1945, 1968, 1989* (Lanham, MD: Rowman & Littlefield, 2004), pp. 207–223, here pp. 210–11.

21 R. Kießling, 'Der Bauernkrieg', in E. François and H. Schulze (eds), *Deutsche Erinnerungsorte*, vol. 2 (München: C.H. Beck, 2001), pp. 137–153.

22 J. I. Engels, *Naturpolitik in der Bundesrepublik: Ideenwelt und politische Verhaltensstile in Naturschutz und Umweltbewegung 1950–1980* (Paderborn: Schöningh, 2006), p. 358.

23 Ch. Büchele, et al., *Wyhl - Der Widerstand geht weiter. Der Bürgerprotest gegen das Kernkraftwerk von 1976 bis zum Mannheimer Prozess* (Freiburg: Dreisam-Verlag, 1982); Nössler and de Witt, *Kein KKW in Wyhl.* These counts exclude 92 references to Badenwerk (the name of a regional electricity supplier) as well as 94 references to the Badisch-Elsässische Bürgerinitiativen (the name of the local protest coalition of 21 groups).

of *drei*). As a label, 'Dreyeckland' was applied liberally to cultural production as well as to political activism: it was a song by the Alsatian singer François Brumbt, the title of an album of local protest music and the namesake of a radio station (successor to the ecological pirate station Radio Verte Fessenheim). Dreyeckland was no conventional nation-state, and its defenders embraced the purely imagined, aspirational character of their community. As an article in the Wyhl squatters' newspaper *Was Wir Wollen* explained it, '"Dreyeckland" doesn't exist; it is only an illusion. One cannot regard these three neighbouring corners of three European nation-states as something united, as one country (Baden, Alsace, Northwest Switzerland). Dreyeckland is the idea of a political and cultural unit, perhaps also a social unit.' The deeper meaning of this regional project revolved primarily around a desire for grassroots reconciliation between French and German citizens, gesturing to a broader (but still local) internationalism through the inclusion of Switzerland. As *Was Wir Wollen* went on to explain, Dreyeckland's unfulfilled potential was largely the product of persistent German mistreatment of Alsatians, as evidenced by the casual arrogance, militarism and even Nazi sympathies of contemporary Badeners out at the pub in Alsace.[24] Dreyeckland thus sought to overcome legacies of war by drawing on shared local experiences of cross-border protest.

While the utopian space of 'Dreyeckland' had greater resonance for activists than 'Europe' in the 1970s, this regional story of post-war reconciliation under environmentalist auspices is in some ways compatible with popular narratives of European integration as a peace-building process. It is thus unsurprising that these protests have taken on more European meaning in collective memory as EU institutions have taken on greater importance in citizens' lives. Axel Mayer was a young anti-nuclear activist in the 1970s who subsequently became the regional manager of an environmentalist organisation in Freiburg. Reflecting in 2010 on the protests in and around the region, he argued that protesters had articulated 'one of the first European visions'. Well before the Schengen Agreement and the Maastricht Treaty formally abolished certain border controls within Europe, activists themselves had broken down borders and challenged old divisions:

> 35 years before was the end of the war and there was always this story of the French-German hereditary hatred ... There was this phrase [in regional dialect]: '*Mir kejje mol d'Granze üewer e Hüffe un tanze drum erum*', that is, 'We throw the border onto

24 R. Burkhart, 'Dreyeckland - Wo liegt das? Was soll das? Wer will das?', *Was Wir Wollen* **15–16** (1 Dec. 1977): 2–4.

> a pile and dance around it'. ... In principle, it was the overpowering [or] overcoming of the border and a bit of living [*gelebtes*] 'Europe'.[25]

In this sense, environmental protesters now see themselves as having advanced a cause that Europe's institutions subsequently embraced.

NERSA: Institutional Europe as the enemy

The positive transnational dynamic of anti-nuclear protest in Dreyeckland also had more ambiguous consequences for subsequent protests elsewhere. The successful site occupation in Wyhl led the West German state to go out of its way to frustrate other anti-nuclear occupations, leading to violent escalations in Brokdorf (November 1976) and Grohnde (March 1977). French authorities followed suit at protests in their own country, leading to fatal violence at a 1977 protest against the 'Superphénix' Fast Breeder Reactor in Creys-Malville.[26] This nuclear facility, to be constructed by the transnational consortium NERSA (*Centrale Nucléaire Européenne à neutrons Rapides, Société Anonyme*),[27] came to epitomise the Europe that anti-nuclear activists opposed. Protesters expressed alarm that FBR technology would produce plutonium, 'the most toxic substance man has ever made', which could then be mixed with depleted uranium to produce MOX fuel for conventional nuclear reactors – or simply used to build atomic weapons. They thus opposed Superphénix as the 'cornerstone of European nuclear programmes'.[28]

NERSA itself upheld the Malville project as a paragon of European cooperation. Superphénix would distribute power to neighbouring countries and serve as prototype for another power station ('SNR-2') along the Dutch-West German border, to be built by a parallel consortium called ESK.[29] Both consortia brought together the French power company EDF, its Italian counterpart ENEL and the West German regional operator RWE; additional partners from Belgium, the Netherlands and later Britain also participated

25 Axel Mayer, Interview with the author, Freiburg (12 Apr. 2010).

26 A. S. Tompkins, 'Transnationality as a liability? The anti-nuclear movement at Malville', *Revue belge de philologie et d'histoire / Belgisch Tijdschrift voor Filologie en Geschiedenis* **89** (3/4) (2011): 1365–1380.

27 A *centrale* is a power station. The clunky acronym is a reflection of the purpose-built nature of the consortium.

28 'Pourquoi refuser Malville ?' (Flyer, 2 pp.), 1976, Lyon, La Gryffe, dossier 'Chooz - Super Pholix - Malville - Nucléaire'.

29 B. Saitcevsky, 'Creys-Malville: les accords de coopération européenne entre producteurs d'électricité', *Revue générale nucléaire* **6** (Nov.-Dec. 1979): 597–98.

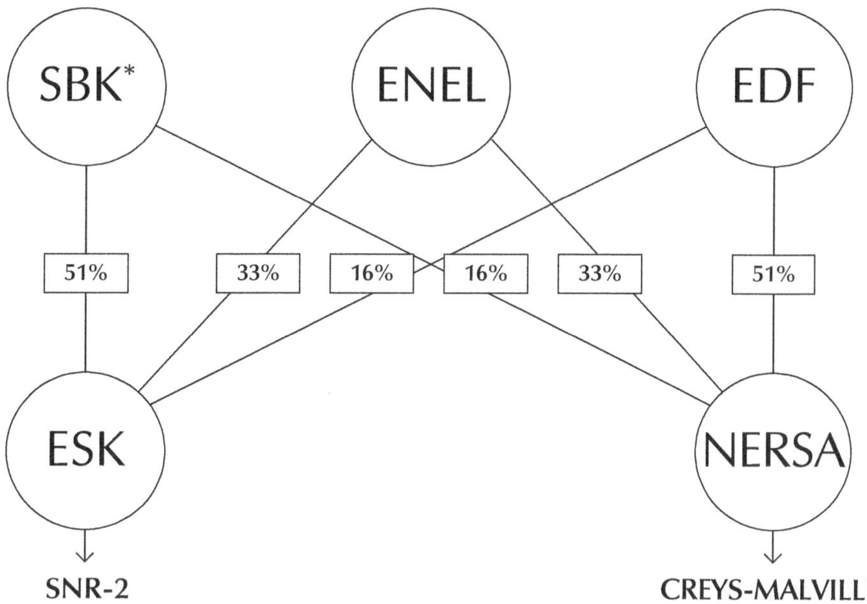

Figure 1.
The cooperation agreements among electricity producers for the development of Fast Breeder Reactors.[30]

* The Schnell-Brüter-Kernkraftwerksgesellschaft mbH (SBK) consortium responsible for the Fast Breeder Reactor in Kalkar (near the German-Dutch border) was jointly owned by companies from West Germany (68.85%), the Netherlands and Belgium (14.75% each) and the United Kingdom (1.65%).

through a joint entity, SBK (see Figure 1.). NERSA received loans from the European Investment Bank and EURATOM – the latter controlled by the same Commission that since 1967 had been responsible for all of the 'European Communities', including the European Economic Community. Activists thus referred to Superphénix as a 'European Community project' and described NERSA as part of the 'international nuclear mafia'.[31]

In response to this corporate transnationalisation, activists proposed to Europeanise protest: as one German flyer put it, 'in order to put a stop to

30 Based on Saitcevsky, 'Accords de coopération', 598.

31 'Sommer 1977 nach Malville' (Flyer, 2 pp.), 13 July 1977, Freiburg, Archiv der Sozialen Bewegungen in Baden (ASB), 00024359 (12.1.9.II Malville); 'Malville. Erfahrungsbericht von 7 Hannoveranern' (Brochure, 40 pp.), 1977, Stuttgart, Dokumentationsstelle für unkonventionelle Literatur, D 1450.

the activities of these firms who have long worked together at the European level, our French friends call upon all environmentally conscious people in Europe' to protest on 31 July 1977 in Malville.[32] Starting in 1975, demonstrations in Malville had attracted some outside support, but primarily from French activists based in Lyon or Swiss ones from nearby Geneva. After an attention-grabbing, peaceful protest in 1976, organisers launched a broader appeal for the following year. Regional organiser Georges David thus explains that 'the enlargement to the European level only happened very late, actually. Only after 1976. At the European level, we only reached the Swiss. It was only afterward that the Italians and Germans joined us'.[33] For Malville protesters, as for their friends in Dreyeckland, Europe began with their near neighbours – even if they were outside the European Community – rather than in Brussels.

The 1977 mobilisation relied on pre-existing site-to-site links as well as networks associated with particular protest factions. For example, advocates of nonviolent direct action organised a 'serpent of struggles' winding down from the Franco-German border (where they visited friends in Wyhl) through a series of sites related to nuclear weapons (Belfort) and workers' struggles (the Lip watch factory in Besançon) before stopping in Malville on the way to another environmentalist demonstration (against a barrage in Naussac) and an anti-militarist rally (on the Larzac plateau).[34] The entire trip was preceded by an international march for non-violence, led by a coordinating committee that included Dutch, Italian, and German participants as well as 'two Alsatians' and 'one Lotharingian'.[35] In this way, non-violent protesters connected local and regional struggles from across France with activism in neighbouring countries.

The radical left also mobilised for Malville. Among the Trotskyist, Maoist and 'non-dogmatic' Marxist groups that proliferated during the 1970s, the Organisation Communiste des Travailleurs (OCT) and its West German sister organisation, the Kommunistischer Bund (KB), worked to bring as many people as possible to the anti-nuclear protest. In the run-up to 31 July, OCT published a dossier on nuclear energy in its weekly newspaper,

32 'Auf nach Malville' (Flyer, 4 pp.), 1977, Freiburg, ASB, 00024361 (12.1.9.II Malville).

33 Georges David, Interview with the author, Lhuis (27 Jan. 2010).

34 I. Cabut, 'Serpent des luttes, premier anneau: haguenau-la frontière allemande', *La Gueule Ouverte/ Combat Non-violent* 167 (21 July 1977): 3–4.

35 'Internationaler gewaltloser Marsch für Entmilitarisierung' (Brochure, 9 pp.), 14–21 July 1977, Hamburg, Archiv Aktiv, 'Intler Gewaltloser Marsch 1976–1980'.

l'étincelle.[36] KB organised buses from Hamburg and Frankfurt to take anti-nuclear protesters to Malville. For these groups, European institutions were an extension of national governments, not a potential ally against them. Indeed, a joint, bilingual Mayday issue of both party newspapers opened with a critical commentary on European integration entitled 'Down with the Europe of Schmidt and Giscard!'.[37] An accompanying article described preparations for a 'Europe of Cops' and noted that TREVI discussions were being prioritised over plans for direct elections to the European Parliament.[38]

The Malville demonstration thus brought together a range of activists with different approaches not only to Europe, but also to protest strategy. Non-violent activists pushed for direct action that would 'go all the way' but 'without hitting cops', while radical groups pushed for either militant action by 'the masses' or individual 'self-defence' against police. The local Malville activists formally leading the demonstration sent mixed signals and were unable to give direction to the growing mass movement against nuclear energy. The result was a confused call for direct action that would be 'non-violent' but 'offensive' (as opposed to merely 'defensive'). René Jannin, the Prefect of Isère in charge of policing the demonstration, seized upon the phrase and declared, 'I am not offensive, I am defensive.'[39] Jannin claimed he would 'take the measures necessary' to protect the 'national [public] good' (*bien national*) that the reactor site represented.[40]

As the demonstration approached, authorities and the media stirred up fears that West German demonstrators would cause trouble in Malville. The right-wing press referred to 'columns' of Germans marching from Munich, Frankfurt and Düsseldorf to Lyon, and the state-run television broadcaster insinuated links between ordinary anti-nuclear demonstrators and Red

36 OCT, 'Non au nucléaire', *l'étincelle*, 23 June 1977.

37 'Nieder mit dem Europa von Schmidt und Giscard! À bas l'Europe des Schmidt et des Giscard!', *Arbeiterkampf/l'étincelle* (29 Apr. 1977), p. 1.

38 KB/OCT, 'Das Europa der Bullen', *Arbeiterkampf/l'étincelle* (29 Apr. 1977), p. 3. 'Terrorisme, Radicalisme, Extrémisme et Violence Internationale' (TREVI) was the name and focus of a group that brought together Interior and Justice ministries from across Western Europe. It served as a forerunner to the Justice and Home Affairs pillar of the EU. See E. Oberloskamp, *Codename TREVI: Terrorismusbekämpfung und die Anfänge einer europäischen Innenpolitik in den 1970er Jahren* (Berlin: De Gruyter Oldenbourg, 2017).

39. *Super-Pholix* **12** (1977): 6–11.

40 Antenne 2, 'Interdiction manifestation Creys-Malville' (News broadcast), 28 July 1977, Paris, Institut national de l'audiovisuel (INA).

Army Faction (RAF) 'terrorists'.[41] The night before the demonstration, the mayor of a local village paid a visit to a campsite where foreign demonstrators had gathered, commenting that he had 'already been occupied by the Germans once' and did not want to put up with it 'a second time' from their descendants; the following day, Jannin made a similar declaration to a press conference: 'Morestel has been occupied by the Germans for a second time'.[42] Such statements stopped short of open xenophobia, but only just: Jean Rabatel, deputy mayor of La Tour-du-Pin, assured the Minister of the Interior in a letter that Jannin had not once used the derogatory term *boche* to describe the Germans. By way of explanation, he added that Jannin did not 'mistake these rowdies with the Federal Republic of West Germany [sic], with whom we maintain good relations for the construction of Europe'.[43]

Indeed, French authorities were adamant that 'Europe' was on their side and not with the demonstrators, who had travelled from West Germany, Switzerland, Italy, Belgium, the Netherlands and beyond to protest together against a reactor that they regarded as an international threat. The poorly coordinated demonstration ended in disaster, with one death and three serious injuries – all inflicted by the stun grenades and exploding tear gas grenades employed by police.[44] To deflect suspicion away from authorities, Interior Minister Christian Bonnet himself went on the nightly news to answer questions about the demonstration. Closely echoing Jannin's previous statements, Bonnet began by saying that the Fast Breeder was *un capital national* designed to guarantee France's energy independence, adding that 'the European Communities have just declared themselves in favour of Fast Breeders'. News anchor Jean-Claude Bourret then asked Bonnet about the protesters, framing the question in a way that linked West Germans with violence.[45] Bonnet responded by stating that the violent demonstrators were 'undeniably groups of anarchist persuasion who disregard borders and who have… "tried their hand" at this elsewhere, notably in West Germany,

41 'Malville', *L'Aurore*, 31 July 1977; 'Creys-Malville' (News broadcast), 31 July 1977, Paris, INA.

42 'Une rumeur: "les allemands"', *Libération*, 1 Aug. 1977.

43 Jean Rabatel, letter to Ministre de l'Intérieur, 1977, Grenoble, Archives départementales (AD) de l'Isère, 6857 W 36.

44 The 31-year-old physics teacher Vital Michalon was killed by the blast of a stun grenade near his chest. One German demonstrator, one French demonstrator and one French police officer had to have limbs amputated after stun grenades exploded near them.

45 'On a beaucoup parlé précisément de ces étrangers, notamment d'une forte participation allemande. Ceux qui ont attaqué les forces de l'ordre n'étaient qu'une infime minorité … Est-ce que vous avez une idée précise de ce qu'ils représentent?'

and who we are identifying through cooperation among members of the [European] Community'.[46] The French state clearly regarded Europe as an instrument to serve its interests.

If France was using European cooperation against demonstrators, it took its cue directly from NERSA. Months before the July 1977 demonstration, a consortium representative had reported back to the Prefect of Isère about his visit to West Germany, where he had studied protests against the nuclear power station construction sites in Brokdorf and Grohnde.[47] The report advised on everything from the proper placement of barbed wire to the undesirability of water cannon, but above all it encouraged police to use screening (*filtrage*) and barricading (*barrage*) procedures well beyond the site's perimeter as a means of controlling access to the demonstration route. French police did precisely that, blocking vehicles six kilometres out and stopping the march a kilometre from the site. The police report also credits NERSA with supplying a film about German protests, which was shown to all unit commanders prior to the Malville demonstration.[48] Perhaps even more than activists realised at the time, NERSA rather accurately embodied the technocratic collusion and repressive potential that the radical left criticised with regard to 'Europe'.

Activists' efforts to counter a perceived international threat with an international demonstration thus largely failed, with authorities digging up old French-German hostilities to pit populations against one another. Yet former regional organiser Georges David argues that this was only possible because activists themselves had failed to sufficiently emphasise the European nature of the issue. Following the examples of Marckolsheim and Wyhl, regional organisers pressed for local leadership, arguing that those most directly affected should ultimately decide on questions of strategy. However, locals in Malville had little experience of protest and were unprepared to lead supporters who greatly outnumbered them. David now argues that this strategic error might have been avoided if protesters had framed Superphénix as 'not only a local issue but a national and even international one'. Indeed, he argues that, while the demonstration itself may have been a failure, it did have some success in building international cooperation: 'We were

46 TF1, 'Journal de 20 h' (News broadcast), 31 July 1977, Paris, INA.

47 NERSA, 'Compte rendu' (Report sent to the Prefect of Isère), 26 May 1977, Grenoble, AD Isère, 6857 W 35.

48 Gilbert Roy, 'Rapport' (Summary police report, with appendices), 5 Aug. 1977, Grenoble, AD Isère, 6857 W 36.

not in a position to capitalise on all that energy, which was… important, if only at the level of Europe! When one thinks about it, in some way that "prepared" Europe, the solidarity that was unleashed with a process like Malville. The problem was that that solidarity was broken by a problem of organisation'.[49] Protests against Malville continued for two decades after the 1977 demonstration, though the organisers never again called for a mass demonstration. Instead, leadership was ultimately ceded to a coordinating committee, 'Europeans against Superphénix', which consisted of professionalised environmentalist NGOs of the kind that emerged in the 1980s.[50] This long-term cross-border collaboration ultimately managed to project a more 'European' frame of opposition than the one-off 1977 demonstration that had been so easily divided along national lines.

A 'Europe of struggles': Building alternatives to institutional Europe?

In September 1979, a delegation from the Larzac plateau embarked on a 1,500-kilometre journey to Gorleben in northern West Germany. The farmers of the Larzac had by that time been fighting against the expansion of a military base onto their land for nearly eight years, using creative, non-violent protests that attracted attention both nationally and internationally. In Gorleben, the struggle against a nuclear facility was much more recent: on 22 February 1977, the Minister President of Niedersachsen, Ernst Albrecht, announced plans to build an integrated nuclear waste disposal site in Gorleben, almost directly on the border with East Germany. Visiting from the Larzac, farmer Pierre-Yves de Boissieu and activist Joseph Pineau discovered many similarities between the two struggles: the methods of the West German government mirrored those of its French counterpart ('disdain for the opinion of the populations concerned', misinformation, recourse to force). So too did those of local activists (who used tractor processions, rallies, and resistance 'on the ground'). Just as the existing military base meant the Larzac farmers had to contend with an invasive army presence, so too did Gorleben residents live in the midst of a heavy border police presence. A certain synergy between the two struggles seemed apparent. Reporting back to the Larzac after their visit, these delegates concluded with an appeal that alluded to the recent first elections to

49 Georges David, Interview.

50 Les européens contre Superphénix, 'Superphénix : le dossier' (Brochure), 1994, Lyon, CEDRATS, dossier 'Super Pholix – Super phénix – Stop Malville'.

the European Parliament: 'After the Europe of parliamentarians, it is time to make the Europe of struggles and the Europe of peoples'.[51]

This 'Europe of struggles' was a further call for extra-parliamentary protests to cultivate transnational connections around key struggles. The Larzac was the perfect centrepiece for a 'Europe' so conceived, having already established itself as a major hub of protest within France. In 1973 and 1974, the farmers hosted rallies on the plateau that drew crowds of more than 100,000. Locally, the Larzac networked with nearby non-violent activists (most famously the Gandhian disciple Lanza del Vasto) and with the Occitan regionalist movement. At the 1973 rally, the Larzac farmers also symbolically 'married' their struggle to that of the striking workers of the Lip watch factory in Besançon.[52] Building on this success, the farmers launched a programme of 'Larzacs everywhere' in 1975, allying themselves with other local groups opposing the Fontevraud military base as well as nuclear power stations in Blayais, Malville and Plogoff. The farmers also attracted national attention with a dramatic, 700-kilometre tractor procession to Paris in 1973, which they repeated on foot in 1978. In the capital itself, they engaged in provocative acts of civil disobedience, bringing sheep to graze under the Eiffel Tower in 1972 and camping along the Seine in 1980. This kind of networking was a promising start for a 'Europe of struggles'.

The actions of the Larzac farmers were an inspiration abroad as well. Indeed, Gorleben activists bounded onto the national stage in West Germany with a Larzac-style tractor procession, which travelled from the affected district of Lüchow-Dannenberg to the regional capital of Hannover, arriving on 31 March 1979. The ultimate success of the march – with 100,000 people converging on Hannover from ten different directions (Figure 2.) – owed much to the coincidence of the Three Mile Island (TMI) nuclear accident in Harrisburg, Pennsylvania on 28 March 1979, just days before the march reached its conclusion. Yet the procession had also been choreographed precisely to build momentum over a period of more than two weeks. Under the circumstances, protesters newly activated by TMI thus had an immediate and visible outlet to express their concerns.

51 Pierre-Yves de Boissieu and Joseph Pineau, 'Durant l'été…', Oct. 1979, J. Pineau private archives, dossier 'Larzac en RFA'; Martin Wetter, 'Larzac-Rundbrief Nr. 1', 13 Dec. 1979, J. Pineau private archives, dossier 'Larzac en RFA'.

52 Xavier Vigna, 'Lip et Larzac: conflits locaux et mobilisations nationales', in Artières and Zancarini-Fournel (ed.), *68: Une histoire collective* (Paris: La Découverte, 2008) pp. 487–494; Donald Reid, *Opening the Gates: The Lip Affair, 1968–1981* (London: Verso, 2018).

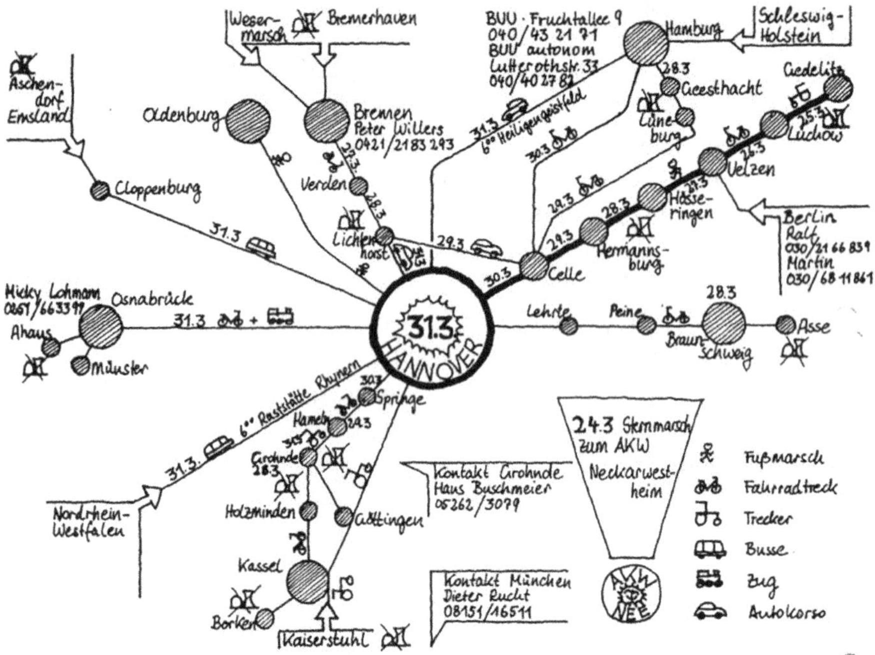

Figure 2.

Routes of Gorleben marches converging on Hannover on 31 March 1979. Bürgerinitiative Umweltschutz Unterelbe (Hamburg), 'Aufruf zur Demonstration zum Abschluß des Trecks der Gorlebener Bauern am 31.3. in Hannover' (Flyer, 2 pp.), 1979, Berlin, APO-Archiv, p. 38.

Key Gorleben activists were very familiar with the Larzac struggle. The latter had been an explicit inspiration to Walter Mossmann, a protest singer who had been active in Wyhl before joining protests in Gorleben.[53] Wolfgang Hertle, a non-violent activist and editor of the monthly newspaper *Graswurzelrevolution* who later moved to Gorleben, had likewise been attuned to developments on the Larzac from an early stage, even writing his doctoral dissertation about the French farmers.[54] Other activists encountered

53 Freia Hoffmann and Walter Mossmann, 'Bürger werden initiativ 1 [Nordhorn/Larzac] und 2 [Wyhl/Wasserburg]' (Manuscripts for radio broadcast), 30 Sept. and 7 Oct. 1973, Amsterdam, Internationaal Instituut voor Sociale Geschiedenis (IISG), Bro 1132-19; W. Mossmann, *Realistisch sein: Das unmögliche Verlangen. Wahrheitsgetreu gefälschte Erinnerungen* (Berlin: Freitag, 2009).

54 W. Hertle, *Larzac, 1971–1981. Der gewaltfreie Widerstand gegen die Erweiterung eines Truppenübungsplatzes in Süd-Frankreich* (Kassel: Weber Zucht & Co., 1982).

the Larzac through ordinary tourism, as residents held 'open farm' events during vacation periods or organised summertime political events.[55] Ethical consumers might purchase Occitan wine from a politically conscious sales network that included more than eighty groups across West Germany.[56] One such group from Hamburg also sold Lip watches and published information about Lip, Larzac and the La Hague nuclear fuel reprocessing site in France as well as Solidarność in Poland.[57] The Hamburg group overlapped with a 'Larzac Circle of Friends', some of whom later published a book of interviews from the Larzac.[58]

All these different groups promoted the Larzac in West Germany to audiences from their own factions of the anti-nuclear movement. In the summer of 1979, non-violent activists, politically engaged wine merchants and left-wing activists working in parallel helped organise around 100–150 Germans to join a group of over 1,000 volunteers who helped with renovation and construction projects on the Larzac under the direction of the farmers.[59] On the plateau, these Germans hosted informational events about Gorleben, which was then fast emerging as a hub of protest in their own country.[60] The Larzac newspaper *Gardarem lo Larzac* published several articles on Gorleben, and German groups reciprocated with articles on the Larzac, usually paired with those on Gorleben.[61] The September 1979 Gorleben visit by Pierre-Yves de Boissieu and Joseph Pineau was a response to this and an attempt to consolidate the link between these two key struggles. The following year, a joint delegation from the Larzac and from Plogoff toured West Germany from 29 April to 7 May, visiting Gorleben just as activists there launched an occupation that turned the construction site into an 'anti-nuclear village'.[62]

55 Hervé Ott, Interview with the author, St-Martin du Larzac (18 Sept. 2010).

56 *Das Fass ist voll: eine Region wehrt sich* (Neu-Isenburg: Verein zur Förderung der Deutsch-Okzitanischen Freundschaft, 1978).

57 'Freundeskreis-Lip-Info/P(r)OVO-Info' (Newsletter), Oct. 1980, Amsterdam, IISG, ID ZK 47369. The name P(r)OVO (*Politische Ökonomie/Offensive Verkaufs-Organisation*) played on memories of the Dutch Provos, a group active in the late 1960s.

58 H. Burmeister and V. Tonnätt, *Zu kämpfen allein schon ist richtig: Larzac* (Frankfurt: Jugend & Politik, 1981).

59 'Plus de cent Allemands sur les chantiers', *Gardarem lo Larzac* **46** (Sept. 1979).

60 'Gorleben-Larzac même combat' (Flyer for event at l'Hôpital du Larzac, 1 p.), 10 Sept. 1979, Millau, Bibliothèque municipale, IZ 62.

61 'Larzac veut leben, Gorleben soll vivre', *Gardarem lo Larzac* **48** (Nov. 1979); 'Gorleben-Larzac Der gleiche Kampf!', *tageszeitung*, 26 Sept. 1979, 8.

62 C. Frey, 'Wachsam in Holzpalästen', *Die Zeit*, 30 May 1980, 54.

After their tour of West Germany, the Larzac farmers made an additional out-of-town trip on 20 May 1980 to Strasbourg. Invited by the Occitan *député du vin* Emmanuel Maffre-Baugé, they attempted (apparently without success) to get the European Parliament (EP) to discuss their case.[63] This represented a rare, direct encounter between these protest movements and European institutions, but it was not one that encouraged the former to abandon their extra-parliamentary approach. The Larzac farmers had brought with them the military service papers of 1,030 Frenchmen, which had been collected as part of a civil disobedience campaign against the military base. After unsuccessfully attempting to present the papers to EP president Simone Veil, the farmers reportedly deposited the entire collection between flowerpots in the corridor.[64] A year later, Veil's office was forced to defend itself from accusations of handing the papers to the French Ministry of Defence after several individuals were reportedly prosecuted for abandoning their military papers.[65] The attempt to appeal to European institutions for support against state opponents had thus largely failed, implying a continued need to build a 'Europe of struggles'.

When it was invoked in 1979, the idea of a 'Europe of struggles' was deployed primarily rhetorically: it was a convenient moniker for informal networks of protest that already existed and which might be strengthened. After Mitterrand's government cancelled the Larzac military base expansion (together with the Plogoff nuclear power station) in 1981, the farmers continued to cultivate ties of solidarity, but mostly outside of Europe.[66] Protest in Gorleben developed along its own trajectory thereafter and, while the Larzac continued to inspire individuals and organisations in West Germany, it lost its centrality to protest movements there. Yet the idea of a 'Europe of struggles' as an alternative to the formal institutional structures of the EU

63 The archival service of the European Parliament was unable to find any pertinent item on the EP's agenda proposed by Maffre-Baugé or anyone else. On Maffre-Baugé as a *député du vin*, see A. W. M. Smith, *Terror and Terroir: The Winegrowers of the Languedoc and Modern France* (Manchester: Manchester University Press, 2016).

64 J.-C. Hano, 'Les paysans du Larzac veulent porter leur affaire devant le Parlement Européen', *Le Monde*, 22 May 1980.

65 Veil's office claimed to have returned the papers to the individuals concerned and to have 'intervened to prevent any legal action against these persons'. Jaak Vandemeulebroucke, Written question 52/81 to President of the European Parliament, 17 Sept. 1981, Luxembourg, EP Archives, PE1.QP//QB-052/81/0020; Simone Veil, Draft response to written question 52/81 to Vandemeulebroucke, 18 Dec. 1981, Luxembourg, EP Archives, PE1.QP//QB-052/81/0030.

66 Gildea and Tompkins, 'Transnational in the local', 599–602.

resonates clearly with later claims that 'Another Europe is possible'. The latter slogan was principally associated with the European Social Forum (ESF), the continent's counterpart to the World Social Form (WSF) and a key venue for activism within the Global Justice Movement (GJM) of the late 1990s and early 2000s.[67] Another key GJM slogan, 'the world is not a commodity', served as the rallying call for a major gathering of French and European activists on the Larzac in 2003 (the thirtieth anniversary of the first rally held there in 1973).

Indeed, the Larzac rose to prominence within the GJM movement in 1999 following the attention-getting 'dismantlement' of a McDonald's construction site near the plateau by veteran Larzac activist José Bové. The McDonald's protest was a key moment in battles over the World Trade Organisation's role as an enforcer of neoliberal norms.[68] Bové later travelled to the protests against the 1999 WTO meeting in Seattle and to the WSF in Porto Alegre, Brazil. After several years of continued extra-parliamentary protest, he ran successfully for a seat in the European Parliament in 2009. There, he sits in the same political grouping as Rebecca Harms, a leading activist from Gorleben who became an MEP in 2004. Not all activists associated with the Larzac and Gorleben today (much less those who participated in these struggles in the 1970s) agree with this embrace of formal politics at the European level. However, thirty years after a 'Europe of struggles' was first proclaimed, it seems to have become a more solid reality, even as it has made certain accommodations with the 'Europe of parliamentarians' to which it was initially opposed.

The shift of much environmental protest from the streets into the parliaments has been a slow process, and one that would have been largely inconceivable to many anti-nuclear activists in the late 1970s. As this chapter has shown, their own understandings of Europe were for the most part non-institutional or even openly hostile to European bodies. Just as significantly, many of their appeals to European institutions fell on deaf

67 D. Della Porta (ed.), *Another Europe: Conceptions and Practices of Democracy in the European Social Forums* (London: Routledge, 2009).

68 Specifically at issue were WTO rules permitting the US to slap high tariffs on Roquefort cheese (produced on the Larzac) in retaliation for the EU's refusal to import hormone-treated beef from the US.

ears, as an incident surrounding the aforementioned 'international day of action' in 1979 demonstrates. On the weekend of Pentecost, simultaneous demonstrations took place as planned around Europe, including one in which thousands of activists marched together across borders to protest the French power station in Cattenom (Lorraine). As at previous anti-nuclear protests in nearby Dreyeckland,[69] French border police singled out 'recognisable demonstration participants' for harassment, turning away those coming from Germany at checkpoints in Perl, Ittersdorf, Goldener Bremm, and along the Saarbrücken highway.[70] Following press coverage of these incidents, a concerned member of the Bundestag and of the (then-appointed) European Parliament, Hajo Hoffmann, attempted to hold French authorities to account. In both parliamentary bodies, he formally asked the German government, the European Commission, and the European Council whether they found it 'compatible with the spirit and letter of the European treaties' that activists had been blocked from entering France 'because they wanted to peacefully demonstrate against the planned nuclear power station in Cattenom and to draw attention to the transnational dangers'.[71]

The answers Hoffmann received illustrate why many anti-nuclear activists felt no great affinity to the institutions of Europe and, indeed, regarded them as extensions of the state. The European Commission responded that those blocked at the border were 'manifestly' not intending 'to exercise an economic activity … in the sense of the treaty' and thus could not benefit from its provisions.[72] The Commission's answer echoed almost verbatim the one previously given by the German government, suggesting coordination.[73]

69 A.S. Tompkins, *Better Active Than Radioactive! Anti-Nuclear Protest in 1970s France and West Germany* (Oxford: Oxford University Press, 2016), pp. 93–94.

70 At Goldener Bremm checkpoint (where the French and German governments later signed an agreement to reduce border controls), activists blocked the border with a sit-in for approximately 30 minutes as the local TV news looked on. Some demonstrators then returned to the Perl checkpoint, reinforced by another 400 people who all marched on foot to Apach – the French town neighbouring Schengen – and from there to Luxembourg city, where 2,000-3,000 anti-nuclear activists protested in front of the French embassy. 'demonstration in thionville - abschlussbericht 21.30 [sic]' (police telex), 4 June 1979, Koblenz, Bundesarchiv (BArch), B 106/107375.

71 Hajo Hoffmann, Written question 232/79 to European Council, 5 June, Luxembourg, EP Archives, PE0.AP.QP.QE.1979//E-0232/79/0020; Written question 233/79 to European Commission, 5 June, Luxembourg, EP Archives, PE0.AP.QP.QE.1979//E-0233/79/0010; Deutscher Bundestag, Drucksache 8/2948, Bonn (8 June 1979).

72 Response to written question 233/79 to Hoffmann, 30 July 1979, Luxembourg, EP Archives, PE0.AP.QP.QE.1979//E-0233/79/0030.

73 The Commission's draft responses can also be found in West German Interior Ministry files in Koblenz, BArch, B 106/107375.

In the Bundestag, Hoffmann told Staatsminister Klaus von Dohnanyi of citizens' frustration with such arguments, particularly given the 'more generous attitude' shown to 'football contacts' than to concerned citizens protesting nuclear risks that directly affected them.[74] For its part, the European Council simply did not deign to answer, stating that 'it does not fall within the competency of the Council to respond to [*se prononcer sur*] the question evoked by the Honourable Parliamentarian'.[75]

Most environmentalists' perceptions of European institutions have become much more positive since the period discussed here. Indeed, the first direct elections to the European Parliament in 1979 – held less than a week after the aforementioned 'international day of action' – helped put the idea of Europe and the institutions that act in its name more firmly onto the radar of political activists and European citizens generally. The EP and other European institutions have also consistently demonstrated a willingness to listen to certain kinds of political activists, though these tend to be those with the resources to professionalise and the will to engage in lobbying activities. However, anti-nuclear activists in the late 1970s largely perceived European institutions for what they were at the time: at best, a well-meaning but remote entity with little power; at worst, an extension of national governments.

In contrast to other environmentalist movements, the opposition to nuclear energy was always more vociferous (and, in France and West Germany, much more violent). Anti-nuclear activists pursued many different strategies, but in the first decade of the movement's existence, extra-parliamentary activism predominated. Successive French and West German governments were unrelenting in their support of nuclear energy, and for much of the 1970s confrontational forms of protest (from non-violent civil disobedience to militant 'self-defence') were common. Anti-nuclear protest remained largely place-based, and its transnational character involved site-to-site links among local struggles rather than the centrally coordinated activities of professional non-governmental organisations. If environmental and anti-nuclear activists today find a more receptive audience in Brussels than they did in the 1970s, it is probably because both the institutions *and* the movements have changed.

French and West German anti-nuclear activists in the 1970s do not seem to have consciously worked toward building 'Europe'. However, they

74 Deutscher Bundestag, Plenarprotokoll 8/159, Bonn (13 June 1979).

75 Response to written question 232/79 to Hoffmann, 30 July 1979, Luxembourg, EP Archives, PE0.AP.QP.QE.1979//E-0232/79/0030.

remained locally rooted and (even prior to Schengen) mobile within the subcontinent's spaces, linking struggles far and wide in an attempt to build opposition to nuclear energy everywhere and to exert power over intransigent governments from within and from without. In their far-flung protest networks, in their joint marches and demonstrations and in their day-to-day interactions in border regions, they unconsciously built transnational relationships that have in the end contributed to a 'European consciousness'. Identification with other Europeans, if perhaps not with the institutions of the European Union, has thus been a positive but largely unintended consequence of anti-nuclear activism.

CHAPTER 8.

ENTERING THE EUROPEAN POLITICAL ARENA, ADAPTING TO EUROPE: GREENPEACE INTERNATIONAL 1987-1993.

Liesbeth van de Grift, Hans Rodenburg and Guus Wieman

The European Union is facing a crisis of legitimacy. In recent years, Brussels has found it increasingly difficult to provide adequate solutions to major disasters, such as the financial, Eurozone and refugee crises, and, as a consequence, support for European integration is dwindling. The Brexit vote and the Dutch referendum rejecting an association treaty with Ukraine prove a widespread scepticism towards the Union. But there is one policy field in which Europeans accept Brussels as the primary actor: the area of environmental policy. Here, the legitimacy of the EU is all but undisputed.[1]

Starting in the 1980s, environmental regulation has been one of the most important responsibilities within the European Union's purview.[2] Originally designed as part of the Common Market, environmental policy sought to diminish distortions of competition between member states. It grew quickly in significance, however, and evolved into a separate policy field with its own Directorate-General.[3] This shift from the national to the supranational level has come to be known as the 'Europeanisation' of environmental policy. It is generally considered a reciprocal process, involving national organisations shifting their attentions towards Brussels in order to influence European policymaking, as well as European institutions opening to suggestions from national bodies, a move that in turn increases their legitimacy. Supranational institutions, in particular the European Commission, started encouraging environmental organisations' activity on the European stage in the late 1980s. In order to be effective on this level, most green groups associated

1 A. Lenschow and C. Sprungk, 'The myth of a green Europe', *Journal of Common Market Studies* **48** (1) (2010): 133–154, 133–35.

2 Ph. M. Hildebrand, 'The European Community's environmental policy, 1957 to 1992: From incidental measures to an international regime?', in A. Jordan (ed.), *Environmental Policy in the European Union: Actors, Institutions and Processes* (London: Routledge, 2005), pp. 19–41, 31–32.

3 Hildebrand, 'The European Community's environmental policy', 31–32.

themselves with European umbrella organisations, such as the European Environmental Bureau, or founded their own European policy units.[4]

From 1987 to 1993, between the implementation of the Single European Act and the scheduled completion of the Internal Market, most major environmental organisations began exploring the possibilities of lobbying in the EU and opened offices in Brussels to handle European institutions specifically.[5] This contribution looks at one of these organisations, Greenpeace International, and examines how and why it stepped up its activities in the European political arena by setting up a special European Communities Unit (EC-Unit) at the end of the 1980s. What were the main considerations leading to the establishment of the EC-Unit? Which factors facilitated or hampered its performance? As this chapter will show, the move of Greenpeace to the supranational stage was anything but likely. For many within the organisation, the EC represented nothing more than a capitalist, technocratic organisation of states, the free trade agenda of which had resulted in extreme pollution and environmental degradation. This stance significantly constrained and complicated Greenpeace's transition to the European level between 1987 and 1993, as did the lack of a clear strategy to lobby European institutions as well as the group's organisational structure itself.

Our research is based on extensive work in the archives of Greenpeace International, housed at the International Institute of Social History (IISH) in Amsterdam. In addition, we conducted in-depth interviews with figures key to the establishment of Greenpeace's European strategy. This involved the establishment of a European office, the European Communities Unit (EC-Unit); one person with whom we spoke was directly responsible for the founding and coordination of this office.[6] Our other interviewee was Steve Sawyer, executive director of Greenpeace International between 1988 and 1993. The first interview focused primarily on the internal dynamics of the Greenpeace EC-Unit and Greenpeace's general attitude towards both the EC-Unit and the European project more broadly; the second deepened our insight into Greenpeace International's organisational structure. Though

4 P. Bursens, 'Environmental interest representation in Belgium and the EU: Professionalisation and division of labour within a multi-level governance setting', *Environmental Politics* **6** (4) (1996): 51–75, 67.

5 T. Long and L. Lörinczi, 'NGOs as gatekeepers: A green vision', in R. Pedler (ed.), *The European Union* (Oxford: Oxford University Press, 2009), pp. 170–76.

6 For privacy reasons and in agreement with the person in question, we have left out the name of the founder and first coordinator of the EC-Unit.

these interviews were valuable, the archival research proved decisive in drawing our conclusions.

We begin with a brief outline of Greenpeace International and its history leading up to the late 1980s. We then demonstrate how Greenpeace attempted to influence European policy with the founding of a European Communities unit in particular, and explore the challenges that Greenpeace faced in the elaboration of a unified European strategy. For analytical purposes, we divide these challenges in two categories, though they are naturally interrelated. First, we address the difficulties that arose from the group's unique organisational structure. The second category deals with the tensions surrounding Greenpeace's ideology and identity, which were deemed incompatible with the EC's technocratic free-market orientation. Finally, we discuss how, in spite of these internal struggles, Greenpeace managed to find a place for itself within the common strategy of the environmental movement in Europe.

The profile of Greenpeace

A single, unified environmental movement does not exist: it is extremely diverse and comprises a wide variety of organisations. Within this diversity of NGOs, Greenpeace currently maintains a reputation as a professional protest organisation, combining professional resources and disruptive actions. However, it started out as a radical social movement and the transition has not been an easy one.[7] Its history and ideological profile are key to understanding how the organisation has sought to position itself in relation to international political institutions like the European Community. The history of Greenpeace dates back to 1969, when a group of Canadian and American environmentalists launched a protest against nuclear testing in Alaska by the United States government.[8] In the following years, this small band of activists succeeded in attracting supporters by way of their high-profile demonstrations. The Greenpeace Foundation was officially established in 1971, and the organisation grew rapidly, first with a number of offices in North America, and by the end of the 1970s in Europe as well.[9] In 1979,

7 N. Carter, *The Politics of the Environment: Ideas, Activism, Policy* (Cambridge: Cambridge University Press, 2007), pp. 147–154.

8 This history has been covered extensively in F. Zelko, *Make it a Green Peace: The Rise of Countercultural Environmentalism* (New York: Oxford University Press, 2013).

9 S. Erwood (ed.), *The Greenpeace Chronicles: 40 Years of Protecting the Planet* (Amsterdam: Greenpeace International, 2011), p. 21.

these offices were officially merged under the umbrella of the international 'Stichting Greenpeace Council'; from that moment on, all national offices' cross-border projects would fall under the name of Greenpeace International.

Though formally a single organisation, Greenpeace International was characterised by division. In the words of Executive Director Steve Sawyer, it was a 'mass of contradictions,' where rebellious youth met older, more jaded activists; the organisation was highly centralised but attached great value to autonomy. By all outside appearances Greenpeace was a united front battling against large political institutions. But internally it was often divided over the appropriate course of action.[10] In spite of this, one central principle united everyone involved:

> Greenpeace is committed to creating a green and peaceful Planet Earth, where a diversity of people and cultures live in harmony, sharing the following basic belief: That the primary value and organising principle that must lay at the base of every human endeavour must be the long term viability of the planetary ecosystem, along with the maintenance of the biological and genetic integrity of that eco-system. That the prime imperative for our long-term survival as a species on this planet must be the preservation of the global environment.[11]

To achieve such ambitious goals, a fundamentally critical attitude was – and still is today – considered essential. Ideologically, Greenpeace opposes injustice wrought by the global economic order, such as the growing gap between rich and poor countries, and it views global economic institutions like the International Monetary Fund (IMF) and the General Agreement on Tariffs and Trade (GATT) with much suspicion. The group's ultimate ambition is to establish an alternative system, structured by binding agreements on the use and conservation of scarce natural resources.[12]

Greenpeace's radical stance is reflected in its preferred activist repertoire. In general, it is willing to employ all necessary means, short of violence, to achieve its goals. Its most familiar trademark is the use of spectacular protests to generate public awareness. Activists in inflatable boats manoeuvring themselves between whalers and their prey, chaining themselves to gas pumps to prevent the extraction of fossil fuels and barring the doors behind which

10 S. Sawyer, *Saving the World the Greenpeace Way* (Amsterdam: Unpublished Source, 1991), p. 3.

11 Ibid., p. 4.

12 International Institute for Social History (IISH), Greenpeace International Archives (GPIA), Folder 454, General correspondence of the International (Campaign) Directors. With other related documents 1990, Steve Sawyer, Overview of the Global Environmental Situation and Greenpeace's Role 1990.

international deliberations on potentially harmful treaties are held; with such tactics Greenpeace aims to force others, within the environmental movement as well as within national and global institutions, to focus on the issues it deems important. Within the environmental movement, Greenpeace aims to lead by strengthening its connections with smaller grassroots groups, thus being prepared for battle when the much-anticipated 'environmental revolution takes off'.[13] Political independence is also a central tenet within the organisation, meaning that no permanent commitments or alliances are to be made with political parties or institutions. This was motivated primarily by the fear of being co-opted by the establishment, like most 'green' parties which, according to Greenpeace, have become embedded in existing power structures, preventing them from ever realising fundamental change.[14] As a precaution against 'selling out,' Greenpeace activists are prohibited from participating in political activities that could give the impression of partisan preference.[15] Furthermore, Greenpeace refuses any form of financial support from governments and relies entirely on private donations.

In the early 1970s, Greenpeace's lack of formal organisational structure meant that anyone was able to establish a new regional or national office without central oversight. Though the Vancouver office, considered the birthplace of Greenpeace, is the group's oldest branch, it never operated as its headquarters. Moreover, in its nascent years no formal agreements dictated the use of the Greenpeace name and brand. According to Frank Zelko, 'so long as all new groups remained faithful to Greenpeace's brand or non-violent direct action, they could do their own thing'.[16] This situation proved unmanageable, however, and in 1979, following a series of lawsuits, Greenpeace formalised its organisational structure and was renamed Stichting Greenpeace Council (SGC), which exists to this day. SGC comprises representatives of all national and regional offices and determines the budget, organisational strategy and appointment of members to the Board of Directors. Given the diverse composition of the Council, it should come as no surprise that finding consensus has often proved a difficult and laborious task.

13 IISH, GPIA, Folder 454, General correspondence of the International (Campaign) Directors. With other related documents 1990, Steve Sawyer, Overview of the Global Environmental Situation and Greenpeace's Role 1990.

14 Ibid.

15 IISH, GPIA, Folder 139, Minutes, agenda and working papers of the meetings of the Executive Committee September 1989, Minutes of the Executive committee, 29 Sept. 1989.

16 Zelko, *Make it a Green Peace*, p. 302.

Greenpeace has always defined itself as a 'campaign-led organisation'.[17] This has meant that specific projects structure the organisation's hierarchy and institutional architecture. All other components of the organisation – from national offices to management bodies – have primarily served the interests of (international) campaigns, which are more or less autonomous, with little central governance from the international office. Despite a series of organisational reshuffles, campaigns and campaign directors have remained the backbone of the organisation, structuring all other activity. Consequently, international campaign directors have been among the most influential people within Greenpeace International.[18] Other branches, such as the Treaties and Conventions project and the scientific and communications sections, have played a supporting role.

In the 1980s, three major developments forced Greenpeace to revamp its mode of operations, which would eventually lead to a more comprehensive reorganisation in the early 1990s. First, Greenpeace had grown exponentially over the course of the 1980s, both in terms of its manpower and its financial resources. The budget, which had hovered around a million dollars in the 1980s, would increase to roughly $140 million by the mid-1990s; former Executive Director Sawyer noted an annual growth of about 35 per cent between 1980 and 1991.[19] Second, Greenpeace had to reconsider its campaigns. The organisation had initially focused on whaling, nuclear testing, and the protection of vulnerable ecosystems, for example in Antarctica. Over the course of the 1980s, however, it became apparent that there were fewer and fewer 'small' victories to win and that 'bigger' and more systemic challenges would become the new priority for environmental activism; global climate change is the most obvious example. These broader crises would be approached as wars of attrition, for which Greenpeace would need to develop a highly coordinated long-term strategy. Finally, these two developments unfolded against the backdrop of a changing world system, which saw the disintegration of the Soviet Union, the rise of a new world order under the Washington Consensus and a deepening and expansion of European integration. These simultaneous processes forced Greenpeace International to reconsider the position it had staked out for itself in the international political arena.

17 IISH, GPIA, Folder 455, Correspondence of the international campaign directors, Memo from Steve Sawyer to all staff, 15 Sept. 1991.

18 IISH, GPIA, Folder 6383, Documents from the archive of Annelieke Zonne, documents concerning the Structure working group (1992), questionnaire.

19 Interview by the authors with Steve Sawyer, Executive Director of Greenpeace International between 1988–1993 (Amsterdam 27 June 2016).

The road to Brussels

The deepening of European integration has had a serious impact on the green movement. This began as early as the 1960s, as public and political awareness of environmental issues increased. The European Commission published its first environmental action plan in 1973, prompting a sharp rise in the number of environmental measures introduced on both the national and international levels. The signing of the Single European Act in 1986 signified another milestone, as this was the first time that environmental policy had been anchored in a European treaty. Meanwhile, environmental groups became increasingly interested in the possibilities offered to them by the European project. In 1974 they established the European Environmental Bureau (EEB), an umbrella organisation of environmental NGOs joining forces to influence policymaking at the European level. A number of Greenpeace's national offices became affiliated, while numerous other environmental organisations, such as the World Wide Fund for Nature (WWF) and Friends of the Earth (FoE), set up European offices in Brussels and established their own presence there. Greenpeace also considered such a move: its leadership saw this opportunity as part of a much-needed push to establish a more coherent international political identity. In 1988, David McTaggart, one of the founders of Greenpeace International, 'got his education on the EEC', as he himself stated, and began exploring ways in which Greenpeace could influence European policymakers most effectively.[20] McTaggart was ambivalent, however, towards 'traditional' forms of political involvement: 'I ... do not want to participate in the funding of a programme that is "lobbying" in the usual environmentalist's mind. It's normally a complete waste of time.'[21]

A few months later, Greenpeace founded the European Communities Unit, or simply the 'EC-Unit', to develop a distinct style of applying political pressure, although at that time it was not yet clear what this would entail. Once the EC-Unit was in place, the national offices withdrew from the EEB at the turn of the 1990s.[22]

20 ISH, GPIA, Folder 956 Documents concerning the European Communities Unit (EC Unit) 1988–1998, Letter from David McTaggart to the International Board of Directors, 2 Apr. 1988.

21 Ibid.

22 IISH, GPIA, Folder, 5331, Correspondence on EC-Unit 1990, Notes on the EC Trustees Meeting Amsterdam 11 Jan. 1990.

A campaign-led organisation: Gods in their own kingdom

The EC-Unit was something of an oddity in an organisation geared towards radical activism, and scepticism was strong amongst Greenpeace activists towards the EC itself; it took considerable time before the EC-Unit became a valued and respected office within Greenpeace International. Ultimately, a major reorganisation would be required to fully integrate the Unit – and with it, the organisation's European lobbying strategy – into the group's larger organisational structure.

The EC-Unit found a home initially in Greenpeace's Treaties and Conventions project, a section linked to the organisation's executive office founded the year prior and tasked with furnishing campaigns with information about international policy. Greenpeace considered such a branch necessary amid the growing complexity of international institutions and transnational environmental problems.[23] Being part of Treaties and Conventions meant that the EC-Unit did not function as an independent lobbying unit but rather as an outpost of the international campaigns within the EC. The proposal for the establishment of Treaties and Conventions reveals that campaign independence and autonomy remained paramount. This is also reflected in the personnel policies of the EC-Unit: five out of its seven staff members were in fact employed by individual campaigns, while only the coordinator and the office manager were hierarchically subordinate to Treaties and Conventions.[24]

This structure became problematic for the coordinator, who was tasked with formulating a clear and unified lobbying strategy from the whims and demands of campaigners and campaign directors who saw themselves, in his words, as 'Gods in their own kingdom'.[25] This drastically undermined the coordinator's authority, perhaps most strikingly demonstrated when one campaigner appropriated the coordinator's office, claiming that her activities were more important than his to begin with.[26] Co-operation with the national branches was often difficult as well. To some extent this can be attributed to lack of interest and scepticism on the part of the activists themselves. In practice, national offices often embarked on activities directed at European

23 IISH, GPIA, Folder 2588, Correspondence on treaties and conventions, Roger Wilson. 1987–1988, Proposal for treaties and conventions, 1986.

24 IISH, GPIA, Folder 130, Minutes, agenda and working papers of the meetings of the Executive Committee December 1988, Minutes of the SCIPOL meeting, Nov. 1988, pp. 17–18.

25 Interview with the first coordinator of the EC-Unit.

26 Ibid.

institutions without first consulting the EC-Unit.[27] When EC-Unit staff attempted to organise a protest in Brussels together with the national offices in 1991, for instance, they were surprised to find that the Belgian national office had already planned a similar action two days prior.[28]

Not only were the relationships between the EC-Unit and the campaigns and national offices problematic, the workplace environment within the Unit itself was also fraught. There was little collaboration among and consultation between staff members, resulting in sparse knowledge of what colleagues were doing. Rivalries formed. A letter by a former volunteer working at the office to the executive director of Greenpeace complained that the office environment had become unbearably hostile and dysfunctional.

> Since the employees resist a sound structure of hierarchy, responsibility, objectivity and 'control' they condescend to scapegoat hunting and blackmailing, block improvement for the best of all and GP [*sic*] and become intolerant, inflexible and handicapped instead of adjusting to the growth of the business and its need ...[29]

Thus, attempts by Greenpeace to forge a durable lobbying network in Brussels largely failed. It was not only the hostile work environment that was to blame: the lack of knowledge about European institutions on the part of Greenpeace campaigners and those staffed by the EC-Unit was also a serious weakness. Greenpeace International tried to preclude this pitfall by hiring experienced people with knowledge of European institutions from outside the organisation to lead the Unit. But Greenpeace campaigners did not always take these outsider professionals seriously. A unified lobbying effort was further hampered by the unwillingness of campaign directors to contribute to the creation of a streamlined Greenpeace programme, as this was seen as having the potential to compromise campaign autonomy. The first EC-Unit coordinator was, in hindsight, highly critical of these directors who, according to him, were unable to look beyond their own individual goals; any attempt to adjust Greenpeace activities to more closely align with the European agenda, even in order to enhance their impact, met with fierce

27 IISH, GPIA, Folder 940, Political Unit, Weekly updates and correspondence of the Political Division. Jan.–May 1992, Letter from Remi Parmentier to Rebecca Johnson, Ulrich Jurgens and Damon Moglen, 23 May 1992.

28 IISH, GPIA, Folder 958, Documents concerning the European Communities Unit (EC Unit) 1990–1992, Minutes of the internal meeting of the EC-Unit, 28 Mar. 1991.

29 IISH, GPIA, Folder 958, Documents concerning the European Communities Unit (EC Unit) 1990–1992, Letter by Inge Nalbach to the executive director's office, 22 Jan.1990.

resistance.[30] Change threatened to fundamentally alter Greenpeace's brand of campaigning.[31]

Amid these challenges the organisation's leadership began to realise that Brussels was becoming the most important arena for European environmental policy, and that internal bickering was hampering the effective lobbying of European institutions. In 1990 the head of Treaties and Conventions, together with the executive director's office, decided to put the EC-Unit under strict supervision, with the aim of improving its working environment and increasing efficiency.[32] Tensions within the Unit continued to rise, however, and the committee tasked with supervision and oversight eventually considered abolishing the Unit entirely.[33] At the end of 1990, the first coordinator resigned, stating: 'I have increasingly become aware that the conditions under which I could make a valuable contribution to Greenpeace's work are no longer there and are not likely to improve in the near future …'[34]

Identity and ideology: Activism, nationalism and euroscepticism

As has been made clear, the founding of a centralised European lobbying office did not sit well within the existing organisational structure of Greenpeace, which was primarily geared towards protecting the autonomy of campaigns and national offices. The new unit in Brussels was seen as an encroachment. But this was more than merely a problem of discretion and competencies. Identity issues lay at the core of this struggle: the Europeanisation of environmental policy forced Greenpeace to reconsider what it stood for.

Confrontational activism tactics were central to the Greenpeace identity, woven into the organisation's cultural fabric that was shaped by passionate activists with antagonistic attitudes. Rather than a formal hierarchy, merit and accomplishment determined the pecking order within the organisation;

30 Interview with the first coordinator.

31 Interview with Steve Sawyer.

32 IISH, GPIA, Folder 958, Documents concerning the European Communities Unit (EC Unit) 1990–1992, Memo from Roger Wilson to all European Trustees.

33 IISH, GPIA, Folder 957, Documents concerning the European Communities Unit (EC Unit) 1990–1992, Report on EC-Unit Visit 3 Aug. 1990.

34 IISH, GPIA, Folder 958, Documents concerning the European Communities Unit (EC Unit) 1990–1992, Letter of Resignation first coordinator, 11 Aug. 1990.

Executive Director Sawyer, reflecting on the group's early days, recalled that founding members often found themselves in competition with one another. Sawyer's description of his own attitude is illustrative: when someone criticised his strategy, he would retort: 'You think it's a waste of time? Okay, just crawl back into whatever hole you came from, until you can put something on the table that you have achieved.'[35] In short, visible results, primarily in terms of successful campaigns, were key to gaining legitimacy. This meritocratic culture proved difficult for the EC-Unit, as it could not easily demonstrate tangible results. The primary activities of the Unit, after all, were distributing information and establishing a lobbying network. Concrete outcomes were achieved, such as its successful campaign against the use of drift nets (1989–1991),[36] the ban on the import of seal fur and new restrictions on the export of hazardous waste (1991).[37] But because these successes were parts of broader campaigns and because the role of the EC-Unit was not always clearly defined, the Unit often did not receive credit where it was due. This, combined with the relatively high overhead costs of maintaining the Unit, raised additional doubts about its viability.[38]

The position of the EC-Unit on the supranational level equally hampered its status within the organisation, given that Greenpeace is essentially made up of self-governing national offices. This too was more than a simple struggle over discretion: the debate within the organisation over the degree of centralisation needed for successful operations has raged since day one. Among Greenpeace International's leadership, advocates of a more decentralised organisation were viewed with suspicion. In the 1980s and 1990s, key figures often warned about growing nationalism within the environmental movement, with some even arguing that divisions between national offices was one of the biggest threats the organisation faced.[39] If national interests were to get the upper hand, international goals and campaigns

35 Interview with Steve Sawyer.

36 Interview with the first coordinator.

37 Ch. Hey and U. Brendle, *Environmental Organizations and the EC: Action Options of Environmental Organizations for Improving Environmental Consciousness and Environmental Policy in the European Community* (Freiburg: EURES, 1992), p. 10; interview with the first coordinator of the EC-Unit.

38 IISH, GPIA, Folder 5332, EC Unit Correspondence 1992, Memo by Roger Wilson to the EC-Unit concerning a skills sharing workshop; Interview with Steve Sawyer.

39 IISH, GPIA, Folder 454, General correspondence of the International (Campaign) Directors. With other related documents 1990, Steve Sawyer, Overview of the Global Environmental situation and Greenpeace's Role 1990; Folder 6383, Structure Working Group, Documents concerning the Structure Working Group (SWG) 1992, Comments on the questionnaire on structure by Roger Wilson, 1992.

would suffer. Leaders were convinced that an international approach and a strong, centralised Greenpeace International were needed to promote the environmental cause.[40] In 1988, the founder of Greenpeace International, David McTaggart, and Executive Director Steve Sawyer noted in a joint interview that Greenpeace had always had to navigate carefully: 'The eternal struggle is to overcome the inbred provincialism and nationalism and keep Greenpeace focused internationally.'[41]

Against this backdrop the EC-Unit worked to involve the national branches as much as possible. This proved difficult, however, as these offices were used to founding and coordinating their own national campaigns. Moreover, national offices had differing views as to which campaigns should be prioritised. Suggestions by the EC-Unit to better coordinate the efforts of the national offices towards Brussels, for example through regular meetings with those offices whose countries were in line for the EC presidency, found little support.[42] This was illustrative of a fundamental clash within Greenpeace, between the desire to remain a grassroots organisation concerned with local issues and the necessity to centralise in order to have a real impact.

A final ideological predicament for the EC-Unit was the broad resentment within Greenpeace towards the very nature of the European project. While it appears that Greenpeace leadership recognised the importance of European integration for environmental policymaking, many campaigners and supporters were suspicious of what they saw as an attempt to forge a capitalist super-state.[43] Recall these activists' dislike of institutions such as the IMF and GATT; after all, Greenpeace seeks to fundamentally alter the capitalist world order. Moreover, the prospect of the common market failed to enthuse environmental activists, many of whom feared that it was simply a ploy to create bigger markets for industry, which would not bode well for the environment.[44]

The opacity of the decision-making process in Europe further aggravated

40 Ibid.

41 IISH, GPIA, Folder 296, Correspondence of the Board 1988, Article 'Daredevils of the Environment'.

42 IISH, GPIA, Folder 957, Documents concerning the European Communities Unit (EC Unit), Minutes, 7 May 1990; Folder 5351, Documents concerning the EC project. 1991, Memo by Roger Wilson to numerous people at toxics and the EC-Unit.

43 IISH, GPIA, Folder 958, Documents concerning the European Communities Unit (EC Unit), 1990–1992, Letter to Kenneth Collins, 22 May 1991.

44 Ibid.

suspicions towards the EC. The first coordinator's successor, who had himself been a member of the European Parliament, suspected that European law-making was deliberately designed to be complicated and vague so that people would stay blind to its undemocratic nature.[45] Furthermore, EC-Unit staff lacked a detailed understanding of the institutional architecture of the EC and struggled to identify which institutions were most important in influencing policy. Between 1988 and 1990, for instance, Greenpeace campaigners spent the majority of their time trying to influence agenda-setting through MEPs, failing to recognise that the European Parliament had far fewer competencies than its national counterparts in this regard.[46] This ignorance must be understood, of course, in the context of a European Community that was still very much in full flux. Campaigners often complained: 'We simply have no idea how important developments in the community are going to impact upon our campaigns.'[47]

The outsider within Europe

Contrary to the scepticism of many activists within Greenpeace, environmental policy did develop in the 1990s into an important European policy domain. Since the late 1980s, the European Commission and its Directorate-General for Environment (DG XI) had stimulated cooperation with environmental organisations. DG XI actively encouraged environmental NGOs, for example, to establish themselves at the European level and granted financial support to these groups in return for their expertise; as a small organisation when it was established, the Directorate-General was in some ways forced to rely on the input provided by NGOs.[48] Its support for non-governmental groups had a reciprocal effect: the involvement of NGOs in the activities of DG XI amplified support among activists for European environmental policy-making, thereby legitimising its existence.[49] In the other direction, this arrangement provided environmental organisations with direct access to the Directorate-General,

45 Ibid.

46 IISH, GPIA, Folder 5351, Toxics: Documents concerning the EC project. 1991, Comments to Memo on the Development of the EC-Unit, 14 July 1991; Folder 5328, Minutes of the EC-Unit internal meeting 1990–1992, EC Toxics Strategy meeting Draft minutes, 17 Jan. 1992.

47 IISH, GPIA, Folder 5339, Contacts with the EC-Unit in Brussels, Correspondence concerning policy. 1988–1989, Letter by Andy Stirling to Ernst Klatte, 9 June 1989.

48 S. Mazey and J. Richardson, 'Environmental groups and the European Community: Challenges and opportunities', in Jordan (ed.), *Environmental Policy*, pp. 106–121, 114–15.

49 Ibid.

which enabled them to exert influence on environmental policy from within.

European-level developments led Greenpeace to reconsider its strategies. The organisation traditionally employed what is known as 'outside-lobbying': through confrontational protest tactics, it sought to pressure political organisations into changing their policies. In other words, Greenpeace tried to impose its own agenda on national policymakers instead of integrating itself into an already existing agenda, as is the case with 'inside lobbying'. But within the European Community, in part due to the lack of an integrated public sphere, the exercise of public pressure proved much less effective, as European politicians and policymakers were considerably less accountable than their national counterparts. This is primarily due to the fact that European Parliament elections are often dominated by national rather than European issues; it is often suggested that 'outside' strategies are in general less effective in Europe than they are at home in individual Member States.[50]

Greenpeace's stance on governments of any kind, be they national or supranational, has always been highly critical. In the course of its history, the organisation has invested much time and energy in guarding its independence from political institutions, with activists' use of partisan political channels as detestable as the use of violence.[51] Greenpeace maintained this position vis-à-vis the European Commission. It refused financial support from DG XI, even though this meant the end of direct access to the Directorate-General, unlike the WWF and FoE. The ideological independence so typical of Greenpeace was complimented by its so-called 'issue-based approach', in which specific projects determined the agenda of the organisation rather than a long-term strategy. As could be expected, this did not work well in Brussels, and the first EC-Unit coordinator pushed a connection to the European policy agenda as an absolute necessity. Additionally, it proved difficult to establish a productive dialogue with politicians and policymakers while at the same time planning and executing confrontational campaigns.

50 S. Princen, 'Agenda-setting in the European Union: A theoretical exploration and agenda for research', *Journal of European Public Policy* **14** (1) (2007): 21–38, 31–32; J.-H. Meyer, 'Getting started: Agenda-setting in European Environmental policy in the 1970s', in J. Laursen (ed.), *The Institutions and Dynamics of the European Community, 1973–83* (Baden-Baden: Nomos, 2014), pp. 221–42. It is important to observe that none of the environmental NGOs active in Europe resorts exclusively to inside or outside lobbying: see J. P. Richards and S. Heard, 'European Environmental NGOs: Issues, resources and strategies in marine campaigns', *Environmental Politics* **14** (1) (2005): 23–41, 32.

51 IISH, GPIA, Folder 454, General correspondence of the International (Campaign) Directors. With other related documents 1990, Steve Sawyer, Overview of the Global Environmental situation and Greenpeace's Role 1989.

Starting in 1991, Greenpeace International underwent significant organisational changes that had major implications for the EC-Unit. Following its explosive growth in the 1980s, Greenpeace saw a levelling-out and then decline in its revenue at the end of the decade, as expenses kept rising. This necessitated a re-evaluation of the organisation's budget. There were also operational considerations behind the restructure, which should be contextualised within a broader professionalisation of the environmental movement at this time. Greenpeace's executive director felt that, in order to run the organisation more smoothly, decision-making had to be streamlined in such a way as to relieve him of his heavy workload. For the EC-Unit, this seems to have had a positive effect, as the reorganisation led to a more clear-cut position within the organisation as a whole. From 1992, the coordinator supervised all staff of the Unit directly where before they had officially been part of the campaigns.[52]

These organisational changes likely enhanced the effectiveness of the EC-Unit and Greenpeace within the EC. While Greenpeace has remained true to its identity as an activist organisation, it has managed to integrate its focus on specific environmental issues within a European strategy. Today, Greenpeace holds a unique position as a more activist organisation within the wider European environmental movement.[53] The group's financial and political independence allows it to take extreme positions and organise disruptive actions against European policies. By occupying the extreme flank of the green movement, it creates additional space for other environmental organisations to focus on cooperation and negotiation with European institutions. Other environmental groups recognise Greenpeace's role in creating public awareness. They even consider it necessary for strengthening their own bargaining position.[54] For the environmental movement as a whole, the diversity of organisations and strategies has turned out to be a major strength; Greenpeace has managed to find a position that enables it to contribute to shared environmentalist goals without having to sacrifice its identity.[55]

52 IISH, GPIA, Folder 958, Documents concerning the European Communities Unit (EC Unit) 1990–1992, Mail from Roger Wilson to EC Unit staff, 8 Nov. 1991.

53 Richards and Heard, 'European Environmental NGOs', pp. 33–34.

54 Ibid.

55 Ibid., pp. 34–35.

Entering the European arena and becoming an effective player proved a difficult process for Greenpeace. As for most environmental organisations, the shift of environmental policymaking from the national to the European level forced Greenpeace to retool. It tried to adapt to the European policy process without jeopardising its own values and identity. For an NGO like Greenpeace, which combines fierce political independence with a confrontational activist approach, situating itself within the broader European environmental movement proved quite a challenge. Although the decision to become active in Brussels was motivated by the desire within the leadership of the organisation to exert real influence on the international level, there were no clear-cut, pre-established strategies for the creation of a European office. The transition was a gradual and difficult learning process without precedent, of which the outcome was unclear in advance.

This case study suggests that the manner in which and extent to which environmental organisations adapt to the European political arena upon entering it are highly dependent on their organisational culture, identity and internal dynamics. With Greenpeace being organised along thematic (campaigns) and national lines and geared towards visible results, the EC-Unit experienced major difficulties fitting in. Only five years after the establishment of the EC-Unit would it obtain a more autonomous position within the organisation as the result of a general reorganisation of Greenpeace. Moreover, critical attitudes within the organisation towards the European Community, its capitalist outlook and opaque decision-making structures hampered the effective coordination of strategies and actions within Greenpeace. The result was a process of trial and error, which reflected the uneasy shift of Greenpeace from an outsider to an insider role in the European political arena.

PART III.

FROM A PUBLIC TO A POLITICAL SPHERE: THE ROLE OF GREEN PARTIES AND PARLIAMENTARY ACTIVITY IN SETTING AN ENVIRONMENTAL AGENDA

CHAPTER 9.

THE DEVELOPMENT OF GREEN PARTIES IN EUROPE: OBSTACLES AND OPPORTUNITIES 1970-2015[1]

Emilie van Haute

The comparative literature on Green party politics has primarily focused on one specific dimension: their emergence,[2] electoral fortune,[3] organisational developments,[4] ideological positioning[5] or relation to power.[6] Only recently have these dimensions been analysed simultaneously in a comparative study that covers 25 European democracies as well as the European level via case studies,[7] and an additional five countries in comparative chapters,[8] bringing the geographical scope to a total of thirty countries and 71 parties, out of which 37 are analysed in depth (van Haute 2016).

This chapter builds on that study in order to analyse the development of Green parties in Europe. It analyses whether one can speak of a Green party

1 This chapter is a revised version of E. van Haute, 'Conclusions: Green Parties in Europe: Which Family Ties?', in E. van Haute (ed.), *Green Parties in Europe* (London: Routledge, 2016), pp. 315–24.

2 F. Müller-Rommel (ed.), *New Politics in Western Europe. The Rise and Success of Green Parties and Alternative Lists* (London/Boulder, CO: Westview Press, 1989); D. Richardson and C. Rootes (eds), *The Green Challenge. The Development of Green Parties in Europe* (London/New York: Routledge, 1995); M. O'Neill, *Green Parties and Political Change in Contemporary Europe* (Aldershot: Ashgate, 1997).

3 H. Kitschelt, *The Logics of Party Formation: Ecological Politics in Belgium and West Germany* (New York: Cornell University Press, 1989); F. Müller-Rommel, 'Green parties under comparative perspective', *ICPS Working Papers* **99** (1994).

4 Th. Poguntke, 'The "new politics dimension" in European Green parties', in F. Müller-Rommel (ed.), *New Politics in Western Europe. The Rise and Success of Green Parties and Alternative Lists* (London/Boulder, CO: Westview Press, 1989), pp. 175–94; Kitschelt, *The Logics of Party Formation*; B. Rihoux, *Les partis politiques: organisations en changement. Le test des écologistes* (Paris: L'Harmattan, 2001).

5 J. Burchell, 'Evoloving or conforming? Assessing organisational reform within European green parties', *West European Politics* **24** (3) (2001): 113–34.

6 F. Müller-Rommel and Th. Poguntke (eds), *Green Parties in National Governments* (London/Portland, OR: Frank Cass, 2002).

7 Austria, Belgium, Bulgaria, Croatia, Czech Republic, Estonia, Finland, France, Germany, Greece, Hungary, Ireland, Italy, Latvia, Lithuania, the Netherlands, Poland, Portugal, Romania, Slovakia, Slovenia, Spain, Switzerland, Sweden, the United Kingdom.

8 Cyprus, Denmark, Luxemburg, Malta and Norway.

family in Europe[9] and investigates how the more or less integrated nature of the Green party family on various dimensions constitutes obstacles and opportunities for Green parties to contribute to a European public sphere.

The concept of a (European) public sphere is highly debated in the literature.[10] Koopmans and Erbe[11] identify three possible forms of European public sphere, understood as the emergence of transnational communication and mobilisation:

1. The emergence of a supranational European public sphere, i.e. 'interaction among European-level institutions and collective actors', made difficult to achieve due to linguistic and cultural barriers;
2. Vertical European public sphere, or Europeanisation: linkages between the national and the European levels;
3. Horizontal European public sphere, or linkages between different Member States.

Eriksen[12] distinguishes between a general public sphere ('communicative spaces of civil society in which all may participate on a free and equal basis'), transnational segmented publics ('policy networks constituted by a selection of actors with a common interest in certain issues, problems and solutions') and strong publics ('legally institutionalised and regulated discourses specialised in collective will-formation at the polity centre').

Our main argument in this contribution is that national Green parties, because they can be regarded as a party family, constitute transnational segmented publics that contribute to a horizontal Europeanisation of the public sphere. It also argues that their collaboration in the European Green Party and the Group in the European Parliament contributes to a supranational European public sphere. These contributions are facilitated by some opportunities but made more difficult by barriers or obstacles linked to the more or less integrated nature of the Green party family on various dimensions.

The chapter highlights some opportunities that point towards a European

9 P. Mair and C. Mudde, 'The party family and its study', *Annual Review of Political Science* **1** (1998): 211–29.

10 E. Dacheux, *L'Europe qui se construit. Réflexions sur l'espace public européen* (Saint-Etienne: Publications de l'université de Saint-Etienne, 2003).

11 R. Koopmans and J. Erbe, 'Towards a European public sphere? Vertical and horizontal dimensions of Europeanized political communication', *Innovation: The European Journal of Social Science Research*, **17** (2) (2007): 97–118, 101.

12 E. O. Eriksen, 'An emerging European public sphere', *European Journal of Social Theory* **8** (3) (2005): 341–363, 349.

Green party family and that can facilitate or contribute to the development of a European public sphere: an electorate that shares similar characteristics, core common ideological positions, a specific organisational model and a cohesive group in the European Parliament. However, it also points to obstacles to that development: divergent electoral performances and relation to power, and the weakness of the European Green Party. Before examining these opportunities and obstacles, we provide a brief overview of the origins and developments of Green parties in Europe.

Origins and development

The taxonomy of parties based on their origins refers to the idea of a common cleavage or conflict on which they emerged. These common roots are clearly discernible in the case of Green parties, at least for some of the parties considered. In Western Europe, there was a clear trend towards the emergence of new political issues revolving around environmental concerns and opposition to nuclear energy, but also around pacifism, human rights and radical democracy. These issues initially pushed forward by environmental movements were gradually politicised, as they were not effectively addressed by existing parties. It opened up the political space for the emergence of Green parties (Table 1). In that sense, most Green parties have roots outside parliament, with notable exceptions such as *GroenLinks* in the Netherlands.

With their origins outside parliament, the transformation of these groups or movements into political parties was in most cases a matter of dispute and generated some tensions. The threshold of declaration was not easily overcome, as some parts of the movements were reluctant to transform into parties and to enter the political sphere. In most countries, the founding of a political party was preceded by one or several efforts to organise politically. As reviewed in Table 1, Green parties first appeared in the 1970s in the UK (1973), France (1973), Germany (1979), and Belgium (1974; 1982). The movement then extended to Sweden, Ireland, Portugal and Spain in the early 1980s, and then to Austria (1982), Switzerland (1983) and Finland (1988). The Netherlands, Italy and Greece saw their Green parties develop in the late 1980s (respectively, 1990, 1986 and 1988). In most cases, the threshold of authorisation (participation in general elections) was passed less than three years after the establishment of the party, with the exception of *GroenLinks* that first passed the threshold and took part in elections before being formally founded as a political party, and of the Greens in Greece and Portugal.

Table 1. Origins and Development of Green Parties at the National Level

	Party	Origins (Declaration)[1]	Authorisation[2]	Representation[3]	Relevance[4]
AT	Die Grünen	1982 (pre-), 1986	1983 (pre-), 1986	1986 (28)	- (only Land level)
BE	Groen	1982	1981	1981 (22), 2007 (7)	1999 (4)
BE	Ecolo	1974 (pre-), 1980	1977 (pre-), 1981	1981 (33)	1999 (4)
CH	GPS	1983 (pre-), 1987	1979 (pre-), 1987	1979 (pre-), 1987 (27)	- (only at Cantonal level)
CH	GLP	2004 (pre-), 2007	2007	2007 (7)	-
DE	All/The Greens	1979 (pre-), 1980	1980	1983 (31)	1998 (8)
EL	OP	1988 (pre-), 2002	1990 (pre-), 2007	NO	-
ES	Conf. de los Verdes	1984	1986	2004 (4)	-
FI	GL	1988	1983 (pre-), 1991	1983 (pre-), 1991 (23)	1995 (8), 2007 (8)
FR	EELV	1973 (pre-), 1984	1973 (pre-), 1986	1997 (17)	1997 (5), 2012 (2)
IE	Green Party	1981	1982	1989 (22)	1997 (4)
IT	FV	1986 (pre-), 1990	1987 (pre-), 1992	1987 (21)	1993 (8 - via alliances)
NL	GroenLinks	1990	1989	1989 (25)	- (only at local and provincial levels)
PT	PEV	1982	1987	1987 (27)	-
SE	MP	1981	1982	1988 (3), 1994 (20)	- (but deals from outside government)
UK	GPEW	1973 (pre-), 1985	1974 (pre-), 1987	2010 (1999 in Scottish Parliament)	-

	Party	Origins (Declaration)[1]	Authorisation[2]	Representation[3]	Relevance[4]
BG	Zelenite	2008	2009	-	-
BG	ZPB	1989	1990 (interruption 2009–13)	1990 (1), 1997 (2)	-
BG	*PC Ecoglasnost*	1990	1990	1990 (1), 1995 (6)	-
CZ	SZ	1989, 1993	1990 (interruption 1996–98)	1992 (4), 2006 (4)	-
EE	ER-EER	1989 (pre-), 1991	1992 (interruption 1999–2007)	1992 (4), 2007 (4)	
HR	ZL-ORaH	2004, 2013	2007	-	-
HR	*HSZ*	1996	1996	-	-
HR	*ZS*	1996	1996	-	-
HR	*ZH*	2001	2003	-	-
HU	LMP	2009	2010	2010 (4)	-
HU	*MZP*	1989	1990	-	-
HU	*ZA-ZDS-ZB*	1993	1994	-	-
LT	LZP	1989	1990 (interruption 1996–2011)	1990 (2), 2012	1990 (2)
LT	*LVZS*	2001 (pre-), 2012	2004 (pre-), 2012	2012 (2)	-
LV	LZP	1990	1993	1995 (3), 2002 (12)	1993 (6), 2002 (9)
PL	Zieloni	1988 (pre-), 2003	1991 (pre-), 2005	-	-
RO	PER		1992	1992 (8)	-
RO	PV-MVDA	2006, 2009, 2011	2008	-	-
RO	*MER-FER*	1990	1990	1990 (2), 1996 (4)	1991 (2)
SI	ZS	1989	1990	1990 (3)	1990 (3)
SI	SMS-Zeleni	2000	2000 (interrupted in 2014)	2000 (4)	-
SI	*ZA*	1995	1996	-	-
SI	*TRS*	2011	2011	-	-

Notes: (pre-) denotes that the threshold was passed that year with a pre-existing political organisation; 1: Year of foundation of the party at the national level; 2: Year of first participation in national elections; 3: Year of first seats in national parliament (lower Chamber) – number of years of uninterrupted presence in parliament between brackets (end 2014 as reference point); 4: Year of first governmental participation at the national level – number of consecutive years in government between brackets.ce: van Haute, 'Conclusions', in *Green Parties in Europe*, pp. 317–18, courtesy of Routledge.

In Central and Eastern Europe, environmental groups and parties emerged quite rapidly after 1989; yet they were confronted with rapid waning. This can be explained by a combination of factors: the politicisation of their core themes faced more difficulties due to the prevalence of economic issues, the absence of post-materialist attitudes and the incorporation of environmental issues by other challengers.[13] Furthermore, establishing a party and participating in elections are much more regulated in these countries.[14] Consequently, the thresholds of declaration and authorisation were difficult to pass on a structural basis, and some parties have simply not been able to consistently participate in elections (see for example the intermissions in Bulgaria, Czech Republic, Estonia, Latvia or Slovenia emphasised in Table 1).

Once founded, Green parties have met very distinct electoral fates across Europe. In Northern and Western Europe, they have established themselves as relevant electoral actors, even if they still rarely cross the ten per cent mark. Elsewhere, Green parties struggle to perform electorally and remain weak, with some exceptions. Especially in Central and Eastern Europe, Green parties have been marginalised (with the notable exception of Latvia), despite the second wave of party foundation in the 2000s.

Electoral fortunes are linked to the sociological composition of their voter basis (see below) but are also related to the capacity of parties to enter parliament, as voters may be discouraged to cast a vote for parties that have low probabilities of getting a seat in parliament. In most cases in Northern and Western Europe, the delay between the first participation in national elections and the first seats in parliament was null or below five years (see Table 1). In other cases, it took longer for the Greens to gain parliamentary representation, as in Ireland, Sweden and especially France, the UK and Spain (not to mention Greece where the Greens have not passed this threshold yet).

The interval between the first participation in elections and the first seats in parliament (threshold of representation) very much depends on electoral rules in the various countries (type of electoral system, existence of a threshold, etc.) (see Table 1). In countries using proportional representation or two-round runoff voting (where Green parties can benefit from alliances), Green parties were much more successful in entering national parliaments or re-entering

13 E.G. Frankland, 'Central and Eastern European Green parties: Rise, fall and revival?', in E. van Haute (ed.), *Green Parties in Europe* (London: Routledge, 2016), pp. 59–91.

14 J.-B. Pilet and E. van Haute, 'Criteria, conditions, and procedures for establishing a political party in the Member States of the European Union' *Report to the European Parliament*, Policy Department C (PE 431.512) (2012).

them after important electoral setbacks, than in majoritarian systems such as in the UK. In these cases, Green parties often passed the threshold of representation at infra- (e.g. Land level in Germany in 1979 vs. federal level in 1983) or supra-national (European) levels first, since some countries adopt more proportional rules at these levels. In Northern and Western Europe, if most parties have secured a continuous presence in parliament since their access, important electoral defeats can also mean a step backwards. In Italy, Spain or Ireland, Green parties lost their parliamentary representation in the 2000s and have not (yet?) managed to regain it, contrary to Belgium (for *Groen*) and Sweden. In most Central and Eastern European countries, gaining parliamentary representation is still the exception rather than the norm, with the notable outlier being Latvia, where the Greens experienced twelve steady consecutive years of parliamentary representation.

Lastly, the threshold of relevance is related to the capacity of parties to have an input on policies from within or outside government (Table 1). The question of participation in power has been and still is a matter of debate for most Green parties, as was the debate of movement vs. party in the early days. It is especially the case at the national level, as several Green parties have successfully passed the threshold of relevance at the sub-national or local levels (for instance in Austria, Switzerland and the Netherlands). In the cases of governmental participation, given their modest electoral size, Green parties in Northern and Western Europe have often been in a situation of junior coalition partner. It has put them in a difficult bargaining position, hard to reconcile with their policy-oriented grassroots base. Policy gains related to governmental participation have been variable and difficult to quantify.[15] Yet the costs of governmental participation have been high for most Green parties and have led to electoral setbacks and parliamentary exit. However, Green parties in Western Europe have, with few exceptions, recovered electorally from their post-incumbency major setbacks. Participation in government is becoming a standard feature, just as their representation in parliament did in earlier years. In Central and Eastern Europe, short-lived governmental participation has been achieved in the early days but not since (again, with the exception of Latvia).

15 C. Little, 'Green parties in government', in van Haute (ed.), *Green Parties in Europe*, pp. 265–79.

Sociological composition

Green parties in Europe have faced diverse electoral performances but their electorate shares common characteristics that distinguish them from voters of other parties. Sociologically, the ideal-typical green voter is young, non-religious, female, urban and educated. Besides, the Green vote can be seen as an issue-based vote that transcends old politics: values that are related to new politics are determinant. Lastly, Green voters also share an 'activist' profile: they tend to be more involved in new forms of political participation, which reflects the original connection of Green parties with new social movements.[16]

Green voters today share the same characteristics as the Green electorate in the early years of Green parties. This points toward a relative stability of Green parties' core electoral supporters, despite fluctuating electoral results. What remains less clear is the capacity of Green parties to maintain the protest component in the Green vote. With a normalisation of Green parties' governmental participation and the emergence of new challengers trying to capitalise on protest sentiments, will Green parties lose their protest element or will they manage to remain the promoters of a societal and political revolution?

Ideological and programmatic positions

Green parties have never been ideologically homogeneous.[17] Yet, they share some distinctive features.[18] As expected, the environment is clearly the most salient issue for Green parties. Yet, some parties emerged as single-issue parties (UK Greens), while other developed from the very start a comprehensive programme and project (e.g. *Ecolo* or *Groen* in Belgium). The environmental issue is still the most salient issue for Green parties today, and they tend to own the issue over the other parties. However, the proportion of their manifesto dedicated to environmental issues has decreased over time to leave more space for other issues. On the environment itself, the Greens have tended to adopt more pragmatic positions (e.g. EELV in France or *GroenLinks* in the Netherlands). In Central and Eastern Europe too, sustainability, ecology and environmental issues are at the core of the

16 C. Close and P. Delwit, 'Green parties and elections', in van Haute (ed.), *Green Parties in Europe*, pp. 241–64.

17 Müller-Rommel and Poguntke, *Green Parties in National Governments*.

18 G. Price-Thomas, 'Green party ideology today: Divergences and continuities in Germany, France and Britain', in van Haute (ed.), *Green Parties in Europe*, pp. 280–97.

Green parties' manifestos. It is especially the case among the newly founded parties of the 'second wave' that receive clear support from the European Green Party in designing manifestos.

Green parties were overall initially reluctant to position themselves on the left/right scale, which was considered as 'old' politics. Yet in most cases they adopted a clear left anchorage that was progressively more assumed, to the point that electoral or governmental alliance with (centre-)right parties is sometimes more taboo for the Greens than for some social-democratic parties. Over time, Green parties have shifted closer to the centre in some cases (e.g. Germany), or further to the left in other cases (e.g. UK Greens), mainly guided by the party's place in their national context. There are a few exceptions to the left-wing positioning, and they are mainly located in Central and Eastern Europe. For instance, the Estonian or Latvian Greens tend to position to the right of the centre, as does the PER in Romania. It translates in their manifesto in a support of entrepreneurship, private property, national identity or more conservative positions. These positions are very much linked to the space occupied by the party in the national party system.

On socio-economic issues, Green parties overall favour issues related to social justice, welfare and solidarity over issues related to the economy *stricto sensu* (again, with some of the above-mentioned CEE exceptions). Finally, the position of Green parties on Europe illustrates a clear shift in most parties from anti-EU toward more moderate or even supportive positions (with the exception of the Finnish and the Belgian Greens). The pro-EU positions are exacerbated in some cases where the competition for recognition by the European Green Party is fierce.[19]

Thirty years of political competition have had an impact on the ideology and positions of Green parties.[20] The three parties analysed by Price-Thomas, the German and French Greens and the Green Party of England and Wales, are characterised by a lot of similarities in their position on six themes (ecology: the concepts of nature and growth; radical democracy; egalitarianism: women's rights, migration and pacifism), and these similarities make them 'sufficiently distinctive from that of other types of party'[21] to label them as part of one common party family. More specifically, Green parties share core common ideological positions on environmental and ecological issues and

19 Frankland, 'Central and Eastern European Green parties'.

20 Price-Thomas, 'Green party ideology today'.

21 Ibid., 292.

participatory democracy, whereas divergences appear on growth, feminism and pacifism. More generally, divergences can be linked to the specific integration of Green parties in their national political systems. In particular, Green parties that have passed the threshold of representation and relevance have lost part of their radical edge and are less critical of the state, even if they still do promote more direct forms of democratic participation. They do not oppose economic growth but remain sceptical of military activity. Electoral setbacks and challenging participations to power have raised internal debate as to the ideological route in which to engage.

Organisational structure

Over the last three decades, Green parties have undergone deep organisational transformations toward institutionalisation, professionalisation and 'normalisation', mainly in Western and Northern Europe. With the expansion of party goals from policy-oriented to increasing vote- and office-seeking goals, Green parties in these countries have evolved from movements to amateur-activist parties, to fully fledged organisations that look more like conventional parties (Table 2).

The Green case(s) could be taken as an illustration of the 'iron law of party institutionalisation'.[22] However, Green parties have kept some of their initial characteristics. The party-on-the-ground is still committed to the initial emphasis on policy-seeking goals and the amateur-activist model, anti-authority and anti-hierarchy. This internal feature is prone to generate tensions with higher levels in the party (party-in-public-office in particular), especially when Green parties face hard choices such as governmental participation or compromises on policies, or new electoral competitors.

Obstacles to the development of a European-wide political sphere for Green parties

Despite these common features, there are obstacles to the development of a European-wide political sphere for Green parties: the divergent electoral performances and relation to power, and the weakness of the European Green Party.

22 P. Ignazi, 'The iron law of party institutionalization'. Paper presented in the Workshop 'Challenges to Established Party Organization? Theory and Practice of Green and Alternative Left Party Organization', *ECPR Joint Sessions of Workshops*, Warwick (1998); Rihoux, *Les partis politiques: organisations en changement.*

Table 2.
Presence or Absence of Organisational Features in Green Parties, Founding Years–1998

Country	Party (year of statutory text)	Collective leadership		Rotation rules		Separation of office and mandate		Amateur leadership		Gender parity rules	
		1980s	1998	1980s	1998	1980s	1998	1980s	1998	1980s	1998
Germany	Die Grünen (1980–1998)	x	x	x	x	x	x	x			x
Netherlands	GroenLinks (1991–1998)			x	x	x	x				
Netherlands	De Groenen (1993–1998)	x	x		x		x	x	x		x
Belgium	Ecolo (1981–1998)	x	x		x	x	x	x			
Belgium	Agalev (1982–1998)	x	x		x		x	x			
Ireland	Green Party (1983–1998)	x	x		x		x	x			x
United Kingdom	Green Party of England and Wales (1977–1998)		x		x		x	x			x
Finland	Vihreä Liitto (1987–1998)	x		x	x		x				x
Austria	Die Grünen (1987–1998)	x		x	x	x				x	x
France	Les Verts (1985–1998)	x	x	x	x	x	x	x			x
Sweden	Miljöpartiet de Gröna (1982–1998)	x	x	x	x	x	x	x		x	x
Italy	Federazione dei Verdi (1986–1998)	x					x	x		x	
Switzerland	Fédération des partis verts de Suisse/Parti écologiste suisse (1983–1998)		x	x	x			x			x
Luxembourg	Gréng Alternativ Partei/ Dei Gréng (1983–1998)	x	x	x	x		x	x			x

Note: an 'x' denotes the presence of the rule in the party statutes;
Source: B. Rihoux, 'Green party organisations: The difficult path from amateur-activist to professional-electoral logics', in van Haute (ed.), *Green Parties in Europe*, pp. 298–314, here p. 301. Courtesy of Routledge..

Figure 1.

Green Parties in National Governments, 1995–2014.

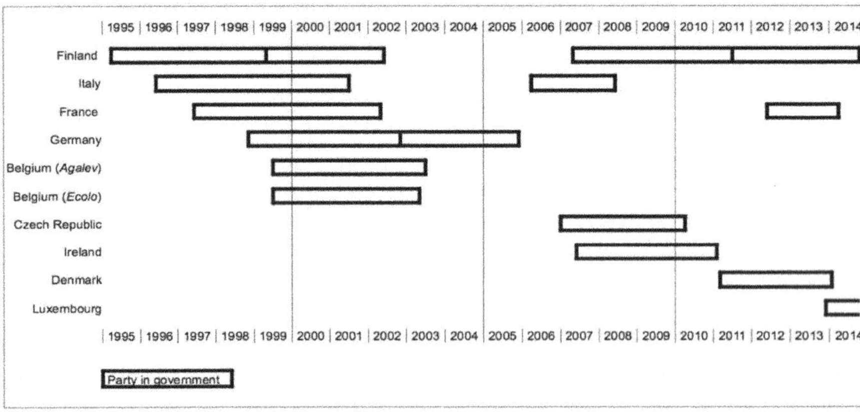

Source: Little, 'Green parties in government', in van Haute (ed.), *Green Parties in Europe*, pp. 265–79, here p. 267. Courtesy of Routledge.

The divergences in electoral performances result in an asymmetrical relation to power. Some Green parties have an extensive experience of government at the national level, such as in Belgium, Finland, France, Germany or Italy, while others have none or limited experience (Denmark, Luxemburg, but also most Green parties in Central and Eastern Europe) (Figure 1). This is crucial, as presence in parliament or government strongly affects what Green parties are, how they organise and how they position themselves ideologically. However, participation in government is increasingly becoming a standard feature and is less discussed or a source of internal tensions. A growing number of national Green parties with government experience might imply a reduced heterogeneity of ideological positions and organisational structures Europe-wide.

At the European level, the diversity of national parties in terms of size and weight in the parliamentary party group, experience in parliamentary representation and participation in power in their national systems, but also in terms of organisational structures, ideological positions and electoral support, generates difficulties in coordination. However, Bardi et al.[23] argue that

23 L. Bardi, E. Bressanelli, E. Calossi, L. Cicchi, W. Gagatek and E. Pizzimenti, *Political Parties and Political Foundations at European Level. Challenges and Opportunities* (Brussels: EP, 2014).

Figure 2.

Cohesion of the Greens/EFA Group in the EP, 1984–2014.

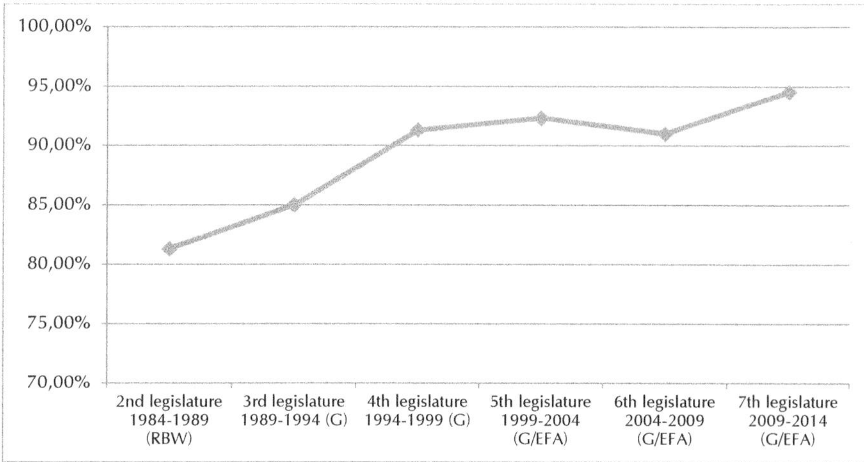

Source: Brack and Kelbel, *The Greens in the European Parliament*, p. 221, based on VoteWatch Europe.

European-wide party organisations have emerged, composed of similar faces as national party organisations: the parties at national level correspond to the 'party on the ground', Europarties to the 'party in central office' and the groups in the EP (together with the Ministers in the Council of Ministers and the Commissioners in the European Commission) to the 'party in public office'. Van Hecke[24] argues that these faces operate at different levels: the national level for national political parties, the supranational level for the groups in the EP and the transnational level for the Europarties.

Among the various faces of the Greens at the European level, the group in the EP shows signs of advanced cooperation.[25] Over time, the group has increased its level of cohesion in terms of votes in the EP, to become the most cohesive group in the EP and to remain united in a wide range of policy areas (Figure 2). The relative weakness of Green parties from Eastern and

24 S. van Hecke, 'Do transnational party federations matter? (… and why should we care?)', *Journal of Contemporary European Research* **6** (3) (2010): 395–411.

25 N. Brack and C. Kelbel, 'The Greens in the European Parliament: Evolution and cohesion', in van Haute (ed.), *Green Parties in Europe*, pp. 217–38.

Southern Europe compared to their Western counterparts is an important challenge but it can also constitute an asset for the Greens in the EP, as it helped maintain cohesion within the group.

Conversely, the European Green Party remains rather weak. First, identifying Green parties at the national level as potential members of the EGP can be a source of tensions. Some members were banned (*Los Verdes* in Spain), while others compete for recognition (Zelenite vs. ZPB in Bulgaria; LMP vs. ZB in Hungary). Tensions can also arise when some get recognition while others do not, as in Denmark, Romania, Slovenia or Croatia, where several parties claiming to belong to the Green family coexist at the national level but only one gets the recognition of the EGP.

Second, the EGP faces difficulties in establishing itself as the dominant actor in the European elections. European elections remain 'second-order national elections' dominated by national parties and national issues. Europarties face difficulties in imposing themselves and making themselves visible in the campaign: national parties campaign under their national logo and with their own manifesto. There is little room for European parties. However, the EGP goes one step further than most other Europarties when it comes to striving to establish a European campaign. It has introduced direct individual membership and is on the path toward granting individual members more direct say in their decision-making processes, following a similar trend among party organisations at the national level.[26] Among these decision-making processes, the EGP seized the opportunity of the Lisbon Treaty regarding the designation of the European Commission President. At the 2014 European elections, it organised an online, European-wide open primary to select their candidate to be the next European Commission President. Although the Greens had a very low probability of holding this position, they organised a process to select their two leading candidates. The open primary was organised between 10 November 2013 and 28 January 2014. Four contenders were running. Despite being arguably the most ambitious system to date among the political parties at European level, this attempt was largely considered as a failure, as fewer than 23,000 citizens participated, EU-wide. This shows that the path towards strong Europarties is still long and paved with difficulties, from resistance from national parties to institutional barriers to a full recognition as parties at the European level,

26 S.E. Scarrow, P.D. Webb and T. Poguntke, *Organizing Political Parties. Representation, Participation, and Power* (Oxford: Oxford University Press, 2017).

to a low visibility for European citizens for whom the reference point is still very much the national public sphere when it comes to electoral processes.

After the emergence of Green parties in the 1970s and 1980s, scholars have tried to explain the birth of these new parties in what appeared at the time as 'frozen' party systems in Western Europe. Scholars linked this emergence to the diminished saliency of old cleavages and the emergence of a new conflict dimension.[27] Using Lipset and Rokkan's[28] cleavage theory, they pointed toward the emergence of 'new politics' as opposed to 'old politics', and the development of new issues and values.[29] Social movements pushing these issues progressively turned into parties, especially when 'old' parties failed to integrate these issues.[30] Therefore, the emergence of Green parties has been interpreted as the first sign of the 'defreezing' of party systems.[31] In line with this interpretation, Green parties have been classified as a new party family. Poguntke[32] stresses that Green parties are by no means alike, but he argues that they share a distinct, 'new politics' feature that translates in their organisation, programme and electoral base. He identifies two sub-groups in the 'new politics' family: the moderates and the fundamentalists. O'Neill identified four types of Green parties based on ideology (Eco-socialists vs. pure Green) and behaviour in the system (anti-party vs. pragmatic).[33]

Using a large number of dimensions and cases on which to assess the existence of a distinct party family, van Haute points at common origins, as well as strong similarities in the sociological composition of Green parties.[34] These are two crucial elements that directly refer to Lipset and Rokkan's

27 Müller-Rommel, *New Politics in Western Europe*; Poguntke, 'The "new politics dimension" in European Green parties'; for a counter argument, see D.-L. Seiler, 'Comment classer les partis verts en Europe?', in P. Delwit and J.-M. De Waele (eds), *Les partis verts en Europe* (Brussels: Complexe, 1999), pp. 43–70.

28 S.M. Lipset and S. Rokkan, *Structures de clivages, systèmes de partis et alignement des électeurs: une introduction* (Brussels: Editions de l'université de Bruxelles, 2008).

29 Müller-Rommel, *New Politics in Western Europe*, p. 5.

30 Müller-Rommel, 'Green parties under comparative perspective'.

31 R.J. Dalton, S.E. Flanagan and P.A. Beck (eds) *Electoral Change in Advanced Democracies* (Princeton: Princeton University Press, 1984).

32 Poguntke, 'The "new politics dimension" in European Green parties'.

33 O'Neill, *Green Parties and Political Change in Contemporary Europe.*

34 Van Haute, 'Conclusions: Green parties in Europe'.

classic conception of party family.[35] Yet, any comparative or classification exercise cannot ignore the fact that Green parties have grown out of their respective national context and have been exposed to these contexts from their foundation. That context affects their fate, in particular their capacity to overcome the threshold of representation and governmental participation. This is crucial, as presence in parliament or government in return strongly affects what Green parties are, how they organise and how they position themselves ideologically. Ideologically and organisationally, Green parties are thus more diverse. Nevertheless, this chapter has emphasised core basic organisational features and positions on issues that could be considered as part of the ideal-type or the genes of Green parties. Furthermore, these divergences might decrease as more and more Green parties experience the threshold of representation and relevance.

The path to a European public sphere for Green parties is still long. The European Green Party is probably a step further in that direction compared to other Europarties, with its experience of primaries and the high degree of cohesion within the group in the EP. However, the strength of national parties remains a strong barrier, as does the absence of unified public sphere at the voter level due to the language diversity. These factors are obstacles to the EGP running a European-wide electoral campaign and establishing a direct dialogue with citizens in Europe.

35 S. M. Lipset and S. Rokkan (eds), *Party Systems and Voter Alignments: Cross-national Perspectives* (Toronto: The Free Press, 1967).

CHAPTER 10.

WILL EUROPE EVER BECOME 'GREEN'? THE GREEN PARTIES' PRO-EUROPEAN AND FEDERALIST TURNING POINT SINCE THE 1990s[1]

Giorgio Grimaldi

During the early 1980s, some Green movements and parties emerged, mostly in Western Europe. Such new political forces combined a growing concern for ecology and the adoption of environmental protection as a political priority with specific campaigns, for instance against nuclear energy. They also advocated more democratic, transparent and decentralised political institutions, capable of promoting nonviolence, minority rights and a social transition towards an ecological economy based on renewable energy sources.[2]

Europe proved to be an important political arena for Green parties: not only did it enable some of them (for instance, the German Greens – *Die Grünen*) to lay the foundations for political ascent inside their countries, but it also allowed them to slowly develop a distinctive profile. After an initial period mainly characterised by strongly critical positions against the European Community (EC), since the early 1990s they have gradually adopted a more constructive, pragmatic and reformist stance on both Europe and the European integration process.[3]

1 I wish to thank Dr. Lucilla Congiu for her careful revising and editing of this paper.

2 G. Grimaldi, 'Prospects for ecological federalism', *L'Europe en formation* **363** (1) (2012): 301–23; G. Grimaldi, 'Thirty years of challenges. The Green Parties' transnational cooperation and their dilemmas and choices over European integration', in D. Preda and D. Pasquinucci (eds), *Consensus and European Integration. Consensus et intération européenne. An Historical Perspective. Une perspective historique* (Brussels, Berlin, Frankfurt am Main, New York, Oxford, Wien: Peter Lang, 2012), pp. 197–210; G. Grimaldi, 'From rejection to support for a supranational Europe. The evolution of the Greens' positions on European integration up to the early 1990s', in G. Grimaldi (ed.), *Political Ecology and Federalism: Theories, Studies, Institutions* (Turin: Centre for Studies on Federalism, 2012), pp. 265–304, G. Grimaldi, 'The Greens and the European Union from Maastricht to Lisbon: from national diversity to unity for the Democratic and Federal development of Europe', in Grimaldi (ed.), *Political Ecology and Federalism: Theories, Studies, Institutions*, pp. 305–347.

3 For more detailed analyses, see E. Bomberg, 'The Europeanisation of Green Parties: Exploring the EU's impact', *West European Politics* **25** (3) (2002): 29–50; E.H. Hines, 'The European Parliament and the Europeanization of Green Parties', *Cultural Dynamics* **15** (3) (2003): 307–325; E. Bomberg and N. Carter, 'Greens in Brussels: Shapers or shaped?', *European Journal of Political Research* **45** (1) (2006): 99–125; N. Carter, 'Mixed fortunes: The Greens in the 2004 European Parliament election', *Environmental Politics* **14** (1) (2005): 103–111.

It took a long time for the European ecologist parties to come to a common and shared 'green' vision of Europe. For many years their ideological differences, as well as their prevailing lack of focus on European integration, made it possible for the Green movements to converge only on a vague appeal to a Europe of regions and peoples, a demilitarised Europe built from the bottom up, outside NATO and in opposition to the institutional and bureaucratic framework of the EC.[4]

Die Grünen, which in 1983 established themselves as the first new political force since 1949 in the Federal Republic of Germany (FRG) to overcome the threshold of five per cent necessary to enter the *Bundestag* (5.6 per cent of the vote), were for a long time the critical avant-garde of ecologists as concerned European integration. On the other hand, the entry of some Green Eurodeputies into the European Parliament (EP) led all the Greens to a more complex and pragmatic attitude, aimed at adding some ecological issues to the European political agenda.

Yet, this pragmatic tendency still coexisted with an opposite approach in the majority of European Green parties, particularly in the German Green Party, that included two main trends: a reformist one, the so-called *Realos* (Realistics), open to alliances with other parties as well as to environmental and social measures consistent with the preservation of natural resources; and a radical one embodied by the *Fundis* (Fundamentalists) – the dominant faction within the party in the 1980s – which promoted leftist policies and opposed any compromise with European and national institutions. An analogous dichotomy was also present, though in different ways and to a lesser degree, in the Green parties of other countries, and it certainly had a deep influence on the development of European cooperation among these political actors.[5]

Three factors were also important in promoting either a pro-Europeanist and Eurofederalist attitude, or a Eurosceptic and hostile one in the European Greens:

- the national political contexts in which the various Green movements and parties had developed;
- the change occurring in the international political situation since 1989, with the end of the Cold War;

4 The EC/EU was initially neglected by the Green parties; cf. S. Hix and C. Lord, *Political Parties in the European Union* (London: Macmillan, 1997), p. 26. See also G. Grimaldi, *Federalismo, ecologia politica e partiti verdi* (Milan: Giuffré, 2005).

5 B. Doherty, 'The Fundi-Realo controversy. An analysis of four Green Parties', *Environmental Politics* 1 (1) (1992): 95–120.

- the specific and relevant role played by some prominent politicians in addressing collective choices and positions.

The Greens' reflection on European institutions has developed at three levels strictly interlinked with one another, as well as with the European dynamics and the specific situations of the national Green parties:

- at national level, through the internal debate inside the various Green parties;
- at institutional European level, firstly in the GRAEL (Green-Alternative European Link), a subgroup of the Rainbow Group in the EP (1984–1989), then in the Green Groups in the EP (1989–1994; 1994–1999) and later on in the Greens/European Free Alliance (G/EFA) in the EP (1999–2004; 2004–2009; 2009–2014; 2014–present);
- at European level, but outside the EP, through transnational and regional cooperation, at first in the European Coordination of Green Parties (ECGP, 1984–1993), next in the European Federation of Green Parties (EFGP, 1993–2004) and subsequently in the European Greens or European Green Party (EGP, since 2004).[6]

First the EC and then the European Union (EU), both fragile, incomplete, in *fieri* and *sui generis* multi-level political organisations, gradually became a battleground for environmental struggles. Nowadays Green parties are a medium-small political family with a rather cohesive profile, organised at European level and coordinated by a world federal political body called Global Greens Coordination (GGC).[7]

This article analyses the key stages of the development of the European Greens' political vision on European integration from 1979 to today, focusing on the activity of the most relevant Green parties as well as on some Green prominent representatives, in order to highlight the changes that led

6 On the EGF, see G. Grimaldi, 'Il Partito verde europeo', in G. Levi and F. Sozzi (eds), *Unione politica 'in progress'. Partiti e gruppi parlamentari europei 1953-2014* (Padua: CEDAM, 2015), pp. 115–126.

7 On the Global Greens, see their official website: http://www.globalgreens.org On the history and evolution of Green parties, see F. Müller-Rommel (ed.), *New Politics in Western Europe. The Rise and Success of Green Parties and Alternative Lists* (Boulder, San Francisco and London: Westview Press, 1989); D. Richardson and C. Rootes (eds), *The Green Challenge. The Development of Green Parties in Europe* (London: Routledge, 1995); P. Delwit and J.-M. de Waele (eds), *Le partis verts en Europe* (Bruxelles: Editions Complexe 1999); F. Müller-Rommel and T. Poguntke (eds), *Green Parties in National Governments* (London, Portland, OR: Frank Cass, 2002); E.G. Frankland, P. Lucardie and B. Rihoux (eds), *Green Parties in Transition. The End of the Grass-roots Democracy?* (Aldershot: Ashgate, 2008); E. van Haute (ed.), *Green Parties in Europe* (London and New York: Routledge, 2016).

the ecologists, since the end of the 1990s, to become a pro-European and Eurofederalist political force.

1979–1984: The emergence of green politics in the EC

In addition to the first EP elections by universal suffrage, the year 1979 saw the first electoral experience of a variety of ecologist formations in Europe, as well as the emergence of a number of leading groups that would turn into Green parties within the next few years.

Even though they performed quite well, the Green coalitions won no seats in the 1979 elections. In 1980, along with a number of alternative and radical parties, they created the Coordination of Green and Radical Parties in Europe (CEGRP)[8] which adopted as its common, basic paper, the *Declaration* of the international ecological organisation *Action Ecologique Européenne*, later known as ECOROPA, established in Paris in June 1976 by a number of scientists and ecologists from around the continent.[9] In October 1983, following their rift with the radical wing of the CEGRP, the Green parties gave birth to a new transnational group, the ECGP, which on 23 January 1984, in Brussels, presented the *Joint Declaration of the European Green Parties*, a paper detailing the basic principles of a common political programme for the 1984 EP elections. Here the Greens criticised the European 'unecological … and centralised power structure in economy as well as in society', and asked for 'a reconstruction of the relationship between the human race and the rest of nature', as well as 'between the rich and the poor', also by highlighting such political issues as 'peace and defence agriculture, antinuclear action, sustainable economy, women's and human rights, the Third World'. They undertook to work together at international level, by extending cooperation to East Europe and to other continents, and described themselves as an alternative to the traditional parties. As for the EC political set-up, they advocated 'a federal structure based on regions rather than on nation-states', where diversity would be 'taken into account and highly respected'.[10]

The debut of the ECGP was immediately marked by a severe conflict between the German Greens – who supported the entry into the Coordination

8 S. Parkin, *Green Parties. An International Guide* (London: Heretic Books 1989), p. 258.

9 T. Dietz, 'Similar but different? The European Greens compared to other transnational party federations in Europe', *Party Politics* **6** (2) (2000): 199–210, here 200.

10 *Joint Declaration of the European Green Parties*, Brussels, 23 Jan. 1984, in Parkin, *Green Parties. An International Guide*, pp. 327–29.

of an alternative cartel of four Dutch parties, the *Groen Progressief Akkoord* (GPA), which also included radical and leftist formations – and other 'pure'[11] Green parties, which were reluctant to admit such groups into the EGC and preferred instead to let in another Dutch Green Party, *De Groenen*. As a result of this choice, the German Greens decided to leave the ECGP.

During the first Congress of the ECGP (Liège, 31 March–1 April 1984), a 'technical alliance' was formed in order to obtain reimbursement for the expenses of the upcoming EP elections. On that occasion, the Greens also drafted a *Declaration* that was signed in Paris on 28 April 1984 and became their common manifesto for the 1984 EP elections. The *Paris Declaration*, which advocated a 'new Europe, neutral and decentralised, with autonomous regions each [with] their own cultural independence', was based on the following seven points:

1) opposition to the presence of nuclear weapons in Eastern and Western Europe, total disarmament and dissolution of both military and power blocs;
2) promotion of environmental policies respecting the ecological balance and fight against every kind of pollution and degradation of nature and the countryside;
3) advancement of women's equality in all areas of social life;
4) development of measures against unemployment and the reduction of the welfare state;
5) implementation of policies towards the Third World based on equitable relations;
6) defence of the free expression of people's fundamental rights as a basic condition to build an ecological society;
7) promotion of environmentally friendly agriculture and defence of jobs in medium- and small-sized rural production units.[12]

The GRAEL's experience (1984–1989) within the Rainbow Group in the EP

In the 1984 EP elections Green parties improved their performance compared to 1979, especially in such countries of Central and Northern Europe as the FRG and Belgium (eight per cent of the poll), as well as the Netherlands, France and Luxembourg (four to six per cent of the vote).[13]

These good results allowed the European Greens to enter the EP for the first

11 On the distinction between pure reformist (the Belgian, British, Finnish, Swedish, Irish, Swiss and French Greens) and radical/alternative Green parties (the Dutch *GroenLinks* and the Green formations of Germany, Luxembourg, Austria), see F. Müller-Rommel, 'The Greens in Western Europe. Similar but different', *International Political Science Review* **6** (4) (1985): 483–499.

12 *The 'Paris Declaration'*, in Parkin, *Green Parties. An International Guide*, pp. 329–30.

13 W. Rüdig, 'The Greens in Europe: Ecological parties and the European elections of 1984', *Parliamentary Affairs* **38** (4) (1985): 56–72.

time with twelve representatives. The newly-elected MEPs (mostly German, Belgian and Dutch) formed a little group called GRAEL; yet, since it was too small to have access to the EP funds and committees, it had to join the Rainbow Group.[14] The GRAEL was the first 'green experience' within the EP. Actually dominated by the German Greens (seven of twelve seats), it was more active on peace, women's and workers' rights than on ecological issues. The dogmatic and deeply hostile attitude of most German MEPs towards the EC also characterised the GRAEL, which not only lacked a uniform approach to Europe, but was also adversely affected by its own division into three factions:[15] a minority of pro-European MEPs; a majority of Eurosceptic *Fundis* and a few pragmatic, single issue-oriented Greens (for instance the high-born farmer Friedrich-Wilhelm Graefe zu Baringdorf, engaged in the reform of the Common Agricultural Policy, or Undine-Uta Bloch von Blottniz, strongly dedicated to nonviolence, sustainable ecology, and antinuclear action).[16]

The GRAEL, which was completely independent from the ECGP, mainly under pressure from the German Greens, prevented the creation of a Green international network, due to the high number of inner conflicts as well as to the individualism of some MEPs. Euroscepticism prevailed in the GRAEL's ranks, as proved both by its vote against the creation of the Delors Commission in January 1985, and by its splitting at the moment of voting on Spain's and Portugal's accession to the EC,[17] as well as on the increase in the EP's powers or the Single European Act (SEA).

14 The Rainbow Group represented the 'green' updating of the Technical Coordination Group (TCG), an alliance of independent groups and members started in 1979. Inside the Rainbow Group three subgroups were established, on the basis of political affinity: the GRAEL – the largest one; the Eurosceptical Danish party *Folkebevægelsen mod EU* (People's Movement against the EU); the European Free Alliance (EFA), a clustering of regionalist movements.

15 See W. Rüdig, 'Green Parties and the European Union', in J. Gaffney (ed.), *Political Parties in the European Union* (London: Routledge, 1996), pp. 254–272, here p. 262.

16 See 'Der Koloss Europa. Die Grünen und das Europäische Parlament. Analysen und Halbzeitberichte der Europagruppe', *Grüner Basis-Dienst* **10** (1986). On the GRAEL, see K.H. Buck, 'Europe: The "Greens"and the "Rainbow Group" in the European Parliament', in F. Müller-Rommel (ed.), *New Politics in Western Europe. The Rise and Success of Green Parties and Alternative Lists*, pp. 176–194.

17 For an analysis of the positions of the European Greens and the Green parties on the enlargement of the EC and the EU, see G. Grimaldi, 'I Verdi e gli allargamenti delle Comunità europee e dell'Unione europea 1973–2004', in A. Landuyt and D. Pasquinucci (eds), *Gli allargamenti della CEE/UE 1961-2004*, vol. 2 (Bologna: Il Mulino, 2005), pp. 1099–1126.

The *Grünen* and European integration in the 1980s: The stage of radicalism and the contribution of Petra Kelly

Owing to their strong left-alternative and grassroots identity – unsurprisingly they were defined a 'non-party party'[18] – in the early years the *Grünen*, by far the most important European Green party, did not deal with the integration issue.[19] They were especially afraid that the EC could turn into a military and economic superpower: for this reason, in February 1984 their MEPs voted against the *Draft Treaty* drawn up by Altiero Spinelli and approved of by the EP. The party's first official position against the EC was adopted during the Congress of Karlsruhe (3–4 March 1984), where the *Grünen* introduced the idea of a 'Europe of regions', seen as a sort of radical-democratic confederation.[20] The EC, instead, was bluntly defined as a set 'of bureaucracies, bombs and butter mountains'.[21]

The German Greens remained deeply opposed to European integration until the mid-1990s.[22] Indeed, in December 1986 the West German Green Party (WGGP) was the only political force in the *Bundestag* to vote against the ratification law of the SEA, denouncing both a democratic deficit and a bias towards the EC Council, and accusing the EC itself of being a capitalistic superpower which exploited the resources of Southern countries. Its radical foreign policy mirrored an alternative view of 'ecological peace' (*Ökopax*), based on a strong criticism of the modern industrial society, as well as on a clear identification of the linkage between environmental damage and militarism.[23] This vision urged the German Greens to be more critical of

18 A. de Petris and T. Poguntke (eds), *Anti-Party Parties in Germany and Italy. Protest Movements and Parliamentary Democracy* (Rome: LUISS University Press, 2015).

19 In the German Greens' programme for both the 1980 and the 1983 federal elections, the European Community was not even mentioned. On the development of a U-turn change of the German Greens' position on European political integration, see G. Grimaldi, 'The "German Greens"' long march from the opposition to the European Communities to a struggle for a more democratic, federal and ecological European Union 1979–2016', in G. Levi and D. Preda (eds), *Euroscepticisms. Resistance and Opposition to the European Community/European Union* (Bologna: Il Mulino, 2019), pp. 333–349.

20 H.A. Leonhardt, 'Zur Europapolitik der Grüne', *Zeitschrift für Politik* **2** (1984), pp. 192–204, here p. 193; Die Grünen, *Global Denken - vor Ort handeln: Erklärung der Grünen zur Europawahl 1984* (Bonn, 1984), p. 9.

21 Die Grünen, *Global Denken - vor Ort handeln*, p. 38.

22 E. Bomberg, 'The German Greens and the European Community: Dilemmas of a movement-party', *Environmental Politics* **1** (4) (1992): 160–185; C. Roth, 'No European superpower', in *Green Leaves* (Bulletin of the Greens in the European Parliament) **2** (May 1991): 2–3.

23 Die Grünen, *Friedensmanifest* (Bonn, 1980), p. 2.

the USA than of the USSR and the German Democratic Republic (GDR) in their programme for the 1987 national elections;[24] for the same reason, after the fall of the Berlin Wall, they came out against a quick German reunification and proposed a 'third way' between the annexation of the GDR by the FDR and the preservation of the status quo – the creation of a federation of regions, a solution aimed at preventing 'the recrudescence of German expansionism and nationalism'.[25] Therefore, in August 1990, the WGGP voted against the Treaty of Unification.

Yet Petra Kelly (1947–1992),[26] one of the founders and a prominent leader of the party, was not totally averse to the idea of European integration. After graduating from the European Institute at the University of Amsterdam with a thesis on this very subject (1971), Kelly started working in Brussels, at both the European Commission and the European Economic and Social Committee (1972–1983). In the same years, she was also involved in many peace, women's and ecological movements, and was a correspondent for the Young German Federalists's international magazine 'Forum Europe'.[27] As a Green member of the Subcommittee for Europe in the *Bundestag*, she expressed pro-European views, but radically opposed the EC. As for the unification of Germany, she was for a federal rearrangement of the country grounded on regional states.[28]

The first Green group in the EP: From the end of the Cold War to the Maastricht Treaty and the new global challenges

The first *Common Statement of the European Greens* was signed during the 5th ECGP Congress held in Paris in March 1989, in the run-up to the EP elections. This *Statement*, a sort of compromise between the above-mentioned fundamentalist German and Dutch views and the reformist attitudes of the

24 E.G. Frankland and D. Schoonmaker, *Between Protest and Power: the Green Party in Germany* (Boulder, San Francisco, Oxford: Westview Press, 1992), p. 136.

25 T. Shull, *Redefining Red and Green. Ideology and Strategy in European Political Ecology* (Albany: State University of New York Press, 1999), p. 78.

26 On Petra Kelly, see S. Richter, *Die Aktivistin: Das Leben der Petra Kelly* (München: Deutsche Verlags-Anstalt, 2010); V. Cavanna, *Petra Kelly. Ripensare l'ecopacifismo. Vita e attualità della donna che ha fondato i Grünen tedeschi* (Rimini: Interno4, 2017).

27 S. Milder, 'Thinking globally, acting (trans-)locally: Petra Kelly and the transnational roots of West German Green politics', *Central European History* **43** (2) (2010): 301–326.

28 P.K. Kelly, 'A Green view of a German reunification and Europe's future' in P.K. Kelly, *Nonviolence Speaks to Power*, ed. by G.D. Paige and S. Gilliatt (Honolulu: Center for Global Nonviolence, 1992).

Belgian, French and Italian Greens, seemed more interested in criticising the EC than in suggesting an alternative European institutional set-up. Indeed, despite their participation in the EP elections, the Greens considered the EC an intrinsically anti-democratic organisation and pledged to democratise it, first of all by demanding full access for every citizen to information on the decision-making processes of both the European Commission and the European Council of Ministers.

In 1989, the Green movements in Europe were at the peak of their success, so it is no surprise that in the EP elections they managed to perform brilliantly almost everywhere[29] and to elect thirty MEPs from all EC countries except Greece. The most striking national results were those of the British Greens – who got 14.9 per cent of the vote (but won no seat due to the majority system adopted in their country) – and the French Greens, who obtained 10.6 per cent of the poll and also managed to elect nine MEPs.[30] The brilliant achievement of the Greens was the starting point of a new trend rather than a mere protest against the EC as a whole.[31] Moreover, the French and Italian Greens' electoral success altered the balance of power within the EP and led to the birth of the first independent Green Group in the EP (GGEP). This group had a less 'Nordic' and more 'Mediterranean' composition than in the past, especially due to the entry of several French MEPs, and this clearly emerged when the French, Belgian and Italian Greens refused to join the Rainbow Group with the regionalists and the Danish anti-Europeanist movement. According to the Statutes of the GGEP, its two Co-Presidents had to be a man and a woman, in order to guarantee gender equality.

The GGEP proved to be more solid and compact than the GRAEL, in spite of the persistence of significant internal differences. Besides, unlike GRAEL it was closely connected with the ECGP, whose Secretariat was hosted in its Brussels seat. The issues of peace and disarmament became the GGEP's specific field of action inside the EP, thanks also to the important role played within the EP's Intergroup on 'Peace and Disarmament' by Solange Fernex, who chaired it, as well as by Alexander Langer. However, the greatest break with the GRAEL's tradition was the choice to put at the top

29 M.N. Franklin and W. Rüdig, 'On the durability of Green politics. Evidence from the 1989 European election study', *Comparative Political Studies* **28** (3) (1995): 409–439, here 411.

30 J. Curtice, 'The 1989 European election: Protest or Green tide?', *Electoral Studies* **8** (3) (1989): 217–230.

31 D. Pasquinucci and L. Verzichelli, *Elezioni europee e classe politica sopranazionale 1979–2004* (Bologna: Il Mulino, 2004), pp. 132–133.

of the GGEP's political agenda an EC institutional reform oriented towards democratic federalism.

In April 1992 the GGEP voted against the Maastricht Treaty,[32] raising three fundamental issues relating to the democratic deficit inherent in the Treaty itself:

- the need for a unified institutional structure called to decide on all matters of Community competence;
- the effective and democratic functioning of the EP;
- the assignment of a mandate to the EP to draw up a draft constitution, to be submitted to national Parliaments within 1996.

Adelaide Aglietta and Alexander Langer: two Green MEPs deeply engaged in the fight for a federal Europe

Inside the GGEP, the Italian MEPs Alexander Langer and Adelaide Aglietta were the first to strongly promote the need to build a European federation in order to ensure peace, an ecological change of economy and society, respect for human rights and peaceful coexistence among peoples.

Adelaide Aglietta (Turin 1940–Rome 2000) began her political engagement in 1974 within the Italian Radical Party, a small libertarian formation characterised by a European federalist view, and in 1976 became its national secretary as well as the first woman to lead a political party in the history of the Italian Republic. In 1989 Aglietta was a cofounder of *Verdi Arcobaleno* (Rainbow Greens) and was elected MEP for two successive terms (1989–1994; 1994–1999). Inside the EP, Aglietta was directly involved in the EU institutional reforms, fighting for a European democratic constitution as well as for civil and political rights, especially against death penalty and human rights violations in the world. Inspired by Altiero Spinelli, the father of European federalism,[33] who used to say that 'Europe does not fall from the sky',[34] Aglietta warned against the risk that a failure of the draft constitution could involve and urged continuing working in the direction of

32 A. Aglietta, '"No" to ratification of democratic vacuum', *Crocodile* (Nov. 1991), p. 11; H. Breyer, F.-W. Graefe zu Baringdorf, C. Roth and W. Telkämper, *Europa Ja – Maastricht Nein. Dokumentation der Verfassungsbeschwerde gegen die Maastrichter Verträge* (Bonn: Europagruppe Die Grünen, 1993).

33 A. Aglietta, 'Il disegno federalista', *Metafora verde* I (1) (1990): 45–46.

34 'Europe does not fall from the sky' ('L'Europa non cade dal cielo') is also the title of a collection of essays by Altiero Spinelli; see A. Spinelli, *L'Europa non cade dal cielo* (Bologna: il Mulino, 1960).

the *Ventotene Manifesto*,[35] especially in a period when powerful resurfacing of national self-interests and racism threatened to prevent the formation of a Europe of the peoples.[36]

The South Tyrolean journalist, teacher, social activist and politician Alexander Langer (1946–1995) approached European federalism by different routes. In his youth he was first involved in catholic associations, then in the left-wing political organisation *Lotta Continua* as well as in various ecological and pacifist movements. In subsequent years, not only did he play a leading role in the birth of the Green movement in Italy, but he was also a tireless organiser of various ecological, humanitarian and non-violent initiatives and campaigns, especially related to the issue of the North-South divide and the need to overcome the 'ethnic' barriers existing between peoples and individuals. As an MEP (1989–1995), Langer mostly devoted himself to fighting for the democratisation of the European institutions and the creation of a federal Europe.

Langer advocated a federalist reform of Europe based on a concurrent shift in power and competences both downwards (reinforcement of autonomy and local self-government) and upwards (creation of supranational institutions). In his opinion, the EC had to change in order to meet that 'need for Europe' which was so widespread among the people, 'but always with a view to a federalist European integration of the continent',[37] as well as according to specific guidelines: the primacy of the political union over the economic one; openness to Eastern enlargement and to interregional cooperation with the Mediterranean region;[38] the creation of a 'pan-European community'; strong social and legal guarantees; decentralisation; democratisation; disarmament; linguistic and cultural pluralism; an alliance between the North and South of the world; both consumer and productive self-restraint, 'a condition so that the planet can have a future'.[39] Federalism was an effective way to counter the

35 The *Ventotene Manifesto* is a political statement drawn up by Altiero Spinelli and Ernesto Rossi in 1941, while they were prisoners on the Italian island of Ventotene.

36 A. Aglietta, 'L'Europa non cade dal cielo. Bilancio di una legislatura al Parlamento Europeo (1989-1994) di Adelaide Aglietta', suppl. of *Notizie Verdi,* 31 Jan. 1994: 30.

37 A. Langer, 'Pan-european Federalism', *Green Leaves* (Bulletin of the Greens in the European Parliament) 2 (May 1991): 3.

38 A. Langer, 'Ethnicity and co-existence in the East Mediterranean' (speech delivered at the Joint International Conference 'Palestine, the Arab World and the Emerging International System: Values, Culture and Politics', Birzeit-Jerusalem-Nablus, 1993), in *Alexander Langer Foundation-June 2001* (Bolzano: Alexander Langer Foundation, 2001), pp. 18–26.

39 Langer, 'Pan-european federalism'.

effects of the rise of new nation-states, to guarantee democracy, participation and the recognition of ethnic minorities – especially in some areas previously subject to communist control such as the Balkans and the Caucasus – and to create prospects for peace in devastating conflicts such as those in the Middle East.[40] From this point of view, the Maastricht Treaty showed all its limitations and betrayed the reluctance of governments to build a Europe of citizens; in Langer's opinion, on the contrary, it was necessary to invest the EP with the role of a Constituent Assembly which should draw up a constitutional project for a united Europe, to be submitted to referenda in all member countries.[41]

A few months before his tragic death,[42] Langer, as Co-President of the GGEP, drew up a draft plan to outline the Group's position on the occasion of the Intergovernmental Conference (IGC) that was held in 1996 in order to review the Maastricht Treaty. In this draft he exposed the key changes necessary to make the EU progress towards a federal supranational democracy: full co-decision power on institutional reforms of the EP to be submitted to member states for ratification; abolition of veto power; flexible and differentiated integration of the new Eastern former communist countries. A Green Europe was necessary to carry out an ecological and social conversion, setting sustainability and social justice as the main criteria for all EU policies, in order to strengthen a European social model ensuring transparency, democratisation at any level and adequate protection of both citizens' and residents' rights. Such a Europe should necessarily be federal and based on democratic control of the European Monetary Union (EMU) as well as on common ecological, economic, foreign and security policies, to be implemented by majority vote at a European level and aimed at both the defence of international law and preventing conflicts by disarmament and progressive demilitarisation.[43]

40 Langer, 'Ethnicity and co-existence in the East Mediterranean', 20.

41 A. Langer, 'L'Unione Europea bussa alle porte. Davvero a Maastricht si può dire solo sì?', *Azione Nonviolenta* **12** (1992): 4–7, at 6.

42 Langer committed suicide on 3 July 1995.

43 Langer 'Draft proposal for the political position of the Green Group in the EP at the Intergovernmental Conference in 1996', April 1995, in A. Langer, *The Importance of Mediators, Bridge Builders, Wall Vaulters and Frontier Crossers* (Bolzano/Forlì: Alexander Langer Foundation/Una Città, 2005), pp. 203–227; on Langer, see G. Grimaldi, 'Alexander Langer: speranze e proposte per un'Europa federale', *I Temi* **26** (2001): 9–40; G. Grimaldi, 'Alexander Langer "costruttore di ponti" tra i popoli', in C. Malandrino (ed.), *Un popolo per l'Europa unita. Fra dibattito storico e nuove prospettive teoriche e politiche* (Florence: Olschki, 2004), pp. 193–212; Alexander Langer Foundation: http://www.alexanderlanger.org (accessed 15 May 2020).

From the European turning point to the challenges of today: A green European future?

Although in the second GGEP (1994–1999) Euroscepticism increased again, due to the entry of Green MEPs from Austria, Finland and Sweden,[44] it lasted only until the Austrian and Finnish Greens adopted more pro-Europeanist stances. Both the Swedish Greens, led by their Eurosceptical founder Per Gahrton, and the Irish Greens opposed the EU, questioning especially the Common Foreign and Security Policy (CFSP) and the EMU. Ultimately, in the GGEP there was an 'agreement to disagree' on institutional issues, since it was impossible to reconcile opposing views on the European integration.[45]

In May 1998, the EP met in extraordinary session to deliver its final judgement on the EMU as well as the introduction of the single currency in eleven out of fifteen member countries, and the majority of the GGEP voted in favour of the Euro. However, there was one 'no' vote by Carlo Ripa Di Meana, a former European Commissioner for the Environment (1983–1993) and also spokesperson of the Italian Greens (1993–1996), who shortly after left the GGEP.

In 1999, the German economist Michaele Schreyer was the first Green to join the European Commission led by Romano Prodi (1999–2004) as a Commissioner responsible for Budget, Financial Control and the Fight against Fraud.

After the 1999 EP elections,[46] a co-operation between ecologists and progressive regionalists led to the creation of the G/EFA Group,[47] still active – indeed it was reconstituted in 2004,[48], 2009[49] and 2014,[50] – and based upon a common vision of a federal Europe of peoples and regions.

44 J. Burchell, 'No to the European Union: Miliöpartiet's success in the 1995 European Parliament elections in Sweden', *Environmental Politics* **5** (2) (1996): 332–38.

45 E. Bomberg, *Green Parties and Politics in the European Union* (London, New York: Routledge, 1998), pp. 119–120.

46 F. Müller-Rommel, 'Les écologistes: de l'anonymat au succès électoral', in G. Grunberg, Pascal Perrineau and Colette Ysmal (eds), *Le vote des quinze. Les élections européennes du 13 juin 1999* (Paris: Presses de Sciences Po, 2000), pp. 163–180.

47 On the EFA, see the official website: http://e-f-a.org/home/ (accessed 15 May 2020).

48 Carter, 'Mixed fortunes: The Greens in the 2004 European Parliament election'.

49 N. Carter, 'The Greens in the 2009 European Parliament election', *Environmental Politics* **19** (2) (2010): 295–302.

50 On the current G/EFA Group in the EP, see the official website: http://www.greens-efa.org/ (accessed 15 May 2020); on its organisation and cohesion, see N. Brack and C. Kelbel, 'The

The former 1968 student leader and prominent French-German MEP Daniel Cohn-Bendit – he was elected twice in the ranks of the German Greens (1994 and 2004) and twice in those of the French ecologists[51] – and Joschka Fischer (Foreign Minister of Germany between 1998 and 2005 as well as the undisputed leader of the German Greens),[52] were among the staunchest supporters of a federal Europe.[53]

Since the 2000s, after a long and complex path, the EGP has certainly become a stable political party, positively oriented towards a federal development of the EU. Yet, the Greens' impact on both national and European politics is still limited, even though, since the mid-1990s, some Green parties have entered national European governments either inside centre-left alliances or within broader and heterogeneous coalitions. In a nutshell, there is still too great a gap between the need for a European ecological renewal and the Greens' capacity to obtain larger consensus and representation on the political arena.[54]

Greens in the European Parliament. Evolution and cohesion', in van Haute (ed.), *Green Parties in Europe*, pp. 217–237.

51 Cohn-Bendit proved to be a charismatic, just like Monica Frassoni, the 2009–19 co-president of the EGP (https://europeangreens.eu/, accessed 15 May 2020), and former co-chair of the G/EFA Group in the EP (2002–2009), previously engaged in the European Federalist Movement.

52 In his Berlin speech *Quo Vadis Europe?* at Humboldt University (May 2000), Fischer proposed the vision of a European federal state based on a democratic Constitution and on a political 'centre of gravity' (i. e. a core group of nation states) provided with full both executive and legislative powers; see J. Fischer, 'From confederacy to federation. Thoughts on the finality of European integration', Speech by Joschka Fischer at the Humboldt University in Berlin, 12 May 2000, http://ec.europa.eu/dorie/fileDownload.do?docId=192161&cardId=192161 (accessed 15 May 2020).

53 On Green contribution inside the European Convention on the Future of Europe (2002–2003), see G. Grimaldi, 'I Verdi e la Convenzione europea', in A. Landuyt and D. Pasquinucci (eds), *L'Unione europea tra Costituzione e governance* (Bari: Cacucci, 2004), pp. 299–334. In January 2005, 80% of the G/EFA Group voted 'Yes' to the European Constitution Draft approved by the EP by a large majority. Shortly after, an extraordinary EGP Council Meeting in Brussels (February 2005) voted in favour of the Treaty establishing a Constitution for Europe: only four Green parties out of 32 voted 'no' (the Swedish, Danish, Norwegian and Greek Greens); see European Green Party (EGP), 'Adopted Resolution of the European Green Party on the EU Constitutional Treaty' (EGP Extraordinary Council Meeting, Brussels, February 17th, 2005), http://www.heide-ruehle.de/heide/artikel/298/doc/reso_verfassung_european_greens.pdf (accessed 15 May 2020). At the EGP Second Congress in Geneva (Oct. 2006), the Joint Declaration 'A Green Future for Europe' was approved; EGP, 'Joint Declaration "A Green Future for Europe"' (2006).

54 W. Rüdig, 'The Greens in the 2014 European elections', *Environmental Politics* **24** (1) (2015): 56–162; W. Rüdig, 'Green parties and elections to the European Parliament, 1979–2019', in L. Ward (ed.), *Greens for a Better Europe. Twenty Years of UK Green Influence in the European Parliament, 1999–2019* (London: London Publishing Partnership, 2019), pp. 3–48.

CHAPTER 11.

A TOUCH OF GREEN AMID THE GREY. EUROPE DURING THE FORMATIVE PHASE OF THE GERMAN GREENS FROM THE 1970s TO THE 1980s: BETWEEN REJECTION AND REFORMULATION

Silke Mende

'We are fighting for the cohesion of Europe': that was the slogan used by the German Green Party for its platform positions on 'Europe' during the 2017 electoral campaign. We are informed that this entails 'continuing the path of European integration', including during difficult periods, because:

> Only together can we solve the problems that transcend borders, first and foremost the climate crisis and terrorism, but also tax evasion and unemployment. Only together can we make globalisation more equitable, create a modern ecological economy, and ensure peace. That is why we want to make the European Union more social, ecological and democratic.[1]

This comes across as contradictory, given the history of their relation to the European project. On the one hand, from their very beginnings the Greens have consistently called for renewed emphasis on direct democracy as part of a 'Europe from below', and they have also emphasised the transnational dimension of subjects such as the environment and peace. On the other hand, the Green Party's perspective on Europe has changed profoundly over the past four decades: during the formative phase of the 1970s and 1980s, they could hardly expect that one day their supporters would be the most fervent advocates of the European project. On the contrary, there were sceptical voices that were critical of its 'costly, disorderly, and confusing bureaucracy'.[2]

1 'Nur zusammen können wir grenzüberschreitende Probleme lösen, allen voran die Klimakrise und den Terrorismus, aber auch Steuervermeidung und Arbeitslosigkeit. Nur zusammen können wir Globalisierung gerecht machen, eine ökologisch moderne Wirtschaft schaffen und Frieden sichern. Deshalb wollen wir die Europäische Union sozialer, umweltbewusster und demokratischer machen'. Bundestagswahl 2017, Grüne Argumente von A bis Z, p. 39, https://www.gruene.de/programm-2017/a-bis-z/wir-kaempfen-um-europas-zusammenhalt.html (accessed 10 Oct. 2017).

2 Bundesvorstand der Grünen (ed.), *Global denken – vor Ort handeln! Erklärung der Grünen zur Europawahl am 17. Juni 1984* (Cologne: Farbo-Team, 1984), p. 6: 'eine aufgeblähte, kostspielige und unüberschaubare Bürokratie'.

It is not possible in this article to retrace the complete path that transformed the Greens from Eurosceptic to Euroeuphoric; the objective is instead to consider their initial phase, which is to say the formative period of the German Greens during the 1970s and 1980s. The 'European public sphere' – a central concept of this collection – will play a dual role.[3] I will first underscore the influence of international, transnational and European elements in the process of founding the Greens. I will then sketch out their different positions with respect to Europe: what were their viewpoints regarding European institutions in Brussels and Strasbourg? What were their own utopias for another Europe, for an alternative 'European public sphere'?

The creation of the Greens within a 'European public sphere': The impact of international and transnational elements

Today the German Greens, *Die Grünen*, are seen as a model ecological party, a high-performance prototype.[4] Many of their central goals – such as the famous drive to abandon nuclear technology – have become part of Germany's political agenda, and the Greens have participated in government as members of various coalitions: on the federal level with Gerhard Schröder's Social Democrats starting in 1998, and, as is currently the case in Baden-Württemberg, with the Social Democrats and later the Christian Democrats. However, in spite of this undeniable success, they are far from being the historic pioneers of ecological parties, as the first one was created not in Europe but Australia, while the first European Green party was founded in January 1973 in Great Britain. This ecological party, which was named *People* and chiefly included famous defenders of nature, was more on the conservative side of the political spectrum, and had little relation to new social movements. With regard to elections, the first acclaimed success with a major cross-border impact took place not in Germany but France, when René Dumont, the ecologist presidential candidate, garnered 1.3 per

3 For the concept of a 'European public sphere', see especially Jan-Henrik Meyer, *The European Public Sphere. Media and Transnational Communication in European Integration 1969–1991* (Stuttgart: Steiner, 2010); and Robert Frank, Hartmut Kaelble, Marie Françoise Lévy and Luisa Passerini (eds), *Building a European Public Sphere. From the 1950s to the Present. / Un espace public européen en construction. Des années 1950 à nos jours* (Brussels/Bern: Peter Lang, 2010).

4 This is also seemingly the case for the environmental history of Germany in general, at least at first sight. See Frank Uekötter, *The Greenest Nation? A New History of German Environmentalism* (Cambridge: MIT Press, 2014).

cent of the vote in 1974.[5] Three years later, it was once again France that provided impetus for the creation of Green parties across Europe, as Green candidates found renewed success in the regional elections of March 1977, a development that was observed with great interest by future German Greens such as Joschka Fischer.[6] Transnational and mutual perceptions subsequently played a significant role in the formative phase of the first Greens.

Nevertheless, it was the German Greens who would soon be elected to parliaments, and who achieved relatively consistent electoral success from that point forward: the Greens of Bremen were the first to serve in the parliament of a German Land, followed by those from Baden-Württemberg in 1980.[7] This was only a prelude to other electoral successes at various levels of the political system. Finally, in 1983, the Greens joined the Bundestag, which profoundly changed the political spectrum in the FRG, as a fourth actor was able to enduringly establish itself within the parliamentary system. The success of the Greens in the Federal Republic was due in large part to a political system whose proportional voting makes it fairly accessible to new political parties, unlike in France for instance.

Yet aside from the similarities and differences in the formation of the Greens in the FRG, as compared to other countries in Europe and across the globe, the German Greens had deep roots in the 'European public sphere' that was developing at the same time. To begin with, the first Greens, along with the sociocultural world from which they came, were part of a larger series of movements and circles that were transnational and European in nature. These were for the most part social movements that engaged with transnational problems, and pursued a fairly international agenda, as nuclear clouds do not stop at borders, and the arms race could not be checked by a single nation state acting alone. The contributions in this collection show the highly transnational character of many protest movements, in addition to a genuine internationalisation of the protest sites themselves. For instance the *Dreyeckland* – the famous regional triangle between Germany, France, and Switzerland characterised by major antinuclear protests – along with Larzac

5 For the French situation, see Alexis Vrignon, *La naissance de l'écologie politique en France. Une nébuleuse au cœur des années 68* (Rennes: PUR, 2017). For representations of 'Europe' among the ranks of the French Greens, see especially Vrignon, pp. 252–255.

6 See Joschka Fischer, 'Warum eigentlich nicht?', in Fischer, *Von grüner Kraft und Herrlichkeit, Reinbek bei* (Hamburg: Rowohlt, 1984), pp. 88–98, here p. 89.

7 For the history of the first Greens in the FRG and their origins, see Silke Mende, *'Nicht rechts, nicht links, sondern vorn'. Eine Geschichte der Gründungsgrünen* (Munich: Oldenbourg, 2011).

in south-west France, deeply impacted the formation of the future German Greens, and created a genuine transnational and even European spirit, although 'European' often meant 'French-German'.[8] One could thus argue that the alternative 'European public sphere' so dear to the first Greens had its roots in the larger context of new social movements and alternative circles. An undated paper drafted by the 'Young European Federalists' explains:

> The citizen's initiatives are a reaction to the inertness of the state apparatus and major organisations. This development is present in all states within the European Community. In border areas in particular, citizen's initiatives represent a movement that is transnational in its awareness. Given that environmental protection and radioactivity do not adhere to borders, various international committees have been formed ... Ten thousand Dutchmen in the Kalkar marketplace, and as many Alsatians at the Wyhl construction site, have raised European awareness more than the traditional attempts of mainstream European education.[9]

The example of the 'Young European Federalists', a transnational pro-European network of young activists, brings us to a handful of influential protagonists in the German ecological movement, whose personal backgrounds were international and European: among the first Greens, this notably includes Petra Kelly and Roland Vogt. Kelly, who was from Bavaria, grew up and studied in the United States, and later at the University of Amsterdam's Institute of Europe, where she intensely delved into European politics.[10] In 1972 she began to work for the European Commission in Brussels as part of the European Economic and Social Committee, which also addressed environmental questions. Like Roland Vogt, she was an important member of the 'Young European Federalists', a youth organisation advocating for Europe,

8 See Andrew Tompkins, *Better Active than Radioactive. Anti-Nuclear Protest in 1970s France and West Germany* (Oxford: Oxford University Press, 2016), here pp. 80–82; and Stephen Milder, *The Anti-Nuclear Movement and Political Environmentalism in West Germany and Beyond, 1968–1983* (Cambridge/New York: Cambridge University Press, 2017).

9 Loseblattsammlung der JEF, p. VI/9f. (Petra-Kelly-Archiv [PKA], Nr.: 3115): 'Die Bürgerinitiativen sind damit eine Reaktion auf die Unbeweglichkeit des Staatsapparates und der großen Organisationen. Diese Entwicklung ist in allen Staaten der Europäischen Gemeinschaft zu verzeichnen. Die Bürgerinitiativen stellen besonders in Grenzregionen eine Bewegung mit transnationalem Bewusstsein dar. In der Erkenntnis, daß Umweltschutz und vor allem Radioaktivität keine Grenzen kennen, haben sich verschiedene internationale Komitees gebildet ... 10.000 Holländer auf dem Marktplatz von Kalkar und entsprechend viele Elsässer auf dem Bauplatz von Wyhl haben mehr europäisches Bewusstsein hervorgebracht als die traditionellen Versuche europäischer Bildungsarbeit'.

10 For Petra Kelly's biography, see Saskia Richter, *Die Aktivistin. Das Leben der Petra Kelly* (Munich: DVA, 2010). For her engagement with Europe, see the detailed article by Robert Camp, '"Für ein Europa der Regionen. Für eine ökologische europäische Gemeinschaft". Über die Europapolitikerin Petra Kelly', in *Die Grünen in Europa. Ein Handbuch*, published by the Heinrich Böll Foundation (Münster: 2004), pp. 12–29.

with branches in various European countries. As to Vogt, he also coordinated cooperation between ecological and radical-democratic parties in Strasbourg. He emphasised the importance of Europe as a factor in his political biography, as part of his candidacy for the leadership position of the Green party (*Bundesvorstand*) in 1981. Vogt spoke of his own experiences in Larzac in 1974, which led to a research project with the weighty title of '*Möglichkeiten gewaltfreien Systemwandels in Westeuropa unter besonderer Berücksichtigung der zunehmenden Staatsqualität der Europäischen Gemeinschaft*' ('The Possibilities for a Peaceful Transformation of System in Western Europe in Light of the Increasingly State-like Nature of the European Community'). This project, Vogt explained, led him to Wyhl, where he ultimately became an antinuclear activist. He undertook initiatives and actions in Wyhl itself, but also in Kaiseraugst (Switzerland), Brokdorf, Malville (France), Kalkar and Gorleben.[11]

In addition to the influence of these networks, which were based on ecological movements, there were more restricted discussion circles of a more elitist character. One of the most influential was ECOROPA, founded in 1976. This European organisation brought together a large number of representatives from different European ecological movements, often including fairly eminent actors. In addition to protest activities, this group sought to address ecological challenges in a more theoretical fashion. The German participants in its discussions included the famous Bavarian novelist Carl Amery, who had left the SPD a few years earlier.[12] They were joined by other social democrats who were more from the left wing of the party, and who also took an interest in ecological questions, such as Freimut Duve and Johano Strasser. There were also futurologists such as Ossip K. Flechtheim or Robert Jungk. A few representatives from other countries also participated, at least temporarily, such as the conservative British ecologist Edward Goldsmith, Italian Aurelio Peccei from the *Club of Rome*, Paul Blau and Freda Meissner-Blau from Austria, and the French ecologists Brice Lalonde and Solange Fernex. Finally, they were joined by the Swiss writer and philosopher Denis de Rougemont, who coined the famous phrase a 'Europe of the regions'.

The idea to participate in the first direct elections for the European Parliament in 1979 emerged quickly within this group. In June 1976, Carl Amery recounted a meeting that had taken place in France:

11 Kandidatenvorstellung Roland Vogt (PKA, Nr.: 2487).

12 See Silke Mende, 'Auf der Suche nach der verlorenen Orientierung – Carl Amery: Ein grüner Bewegungsintellektueller zwischen konservativer Bewahrung und progressiver Veränderung', *Revue d'Allemagne et des pays de langue allemande* **46** (2014): 365–379.

> The shared opinion was that we should leave behind our respective shells, that it was not a Europe of the fatherlands (*Vaterländer*), but a Europe of small countries (*Heimaträume*). We even wondered whether the elections for the European Parliament were not an ideal opportunity to present such a vision of Europe as part of an independent list.[13]

This idea became a reality three years later, when a German ecological list named *'SVP – Sonstige Politische Vereinigung Die Grünen'* ran alongside ecological or alternative lists from four other European countries during the first direct elections for the European Parliament,[14] garnering nearly 900,000 votes. Particular legal conditions allowed for the candidacy not just of parties but also of 'simple' lists, which were not even organised as parties. The first direct elections for the European Parliament thus opened an extraordinary 'window of opportunity' for all those who were still sceptical about forming actual parties with an entire organisational 'apparatus' – an oft-criticised characteristic of 'classical' parties. The elections were also an attempt to focus the different ideological and organisational forces of the new social movements – although many groups on the left preferred to remain on the sidelines.

Although the German list, with 3.2 per cent of the vote, was ultimately unable to join the European Parliament due to the five per cent clause, this was nevertheless a remarkable success. The list received a significant reimbursement for its electoral campaign, which it used to establish the organisation for the future party in the FRG. The French list, 'Europe Ecologie', also won nearly 900,000 votes, or 4.4 per cent. Even though no Green candidate joined the Parliament in Strasbourg, these commendable results motivated ecological movements to form parties and present candidates during elections. The first direct elections for the European Parliament also provided an opportunity to form the first organisational group on the European level, which included six Green-alternative parties: the German Greens were joined by the Green-Alternatives from Belgium, France, Great Britain, Italy and the Netherlands,

13 'Die allgemeine Ansicht war, daß wir aus unseren jeweiligen regionalen Schneckenhäusern heraus müssen, daß es nicht um ein Europa der Vaterländer, sondern der Heimaträume geht. Ja, man überlegt sich schon, ob nicht möglicherweise bevorstehende europäische Parlamentswahlen der ideale Anlaß sein könnten, ein solches Europa auf einer unabhängigen Liste der Öffentlichkeit anzubieten'. Carl Amery to Bund Naturschutz in Bayern e. V., z. Hd. Herrn [Hubert] Weinzierl, 1 June 1976, p. 1 (Monacensia: NL Carl Amery, Mappe: Verlage an Amery: Ökologie). Weinzierl was the chairman of 'Bund Naturschutz Bayern'.

14 These were the Green-alternatives from Belgium, France, Great Britain and Luxembourg. See Ferdinand Müller-Rommel, 'Green parties and alternative lists under cross-national perspective', in Müller-Rommel (ed.), *New Politics in Western Europe. The Rise and Success of Green Parties and Alternative Lists* (Boulder: Westview Press, 1989), pp. 5–19.

who took part in *P.E.A.C.E.* A 'European bureau' was opened in Strasbourg, and served as a nerve centre in providing the European Parliament with ecological positions through extra-parliamentary work.[15]

In the eyes of many actors, Europe was not just an opportunity and source of motivation, but also a true need. For instance, in 1978 Petra Kelly noted:

> The upcoming European elections in 1979 are vitally important for the entire alternative movement – we must act transnationally and demonstrate that we need a balance based on confidence and not terror, that we need cooperation and not confrontation.[16]

This observation – along with the others cited earlier – has underscored some of the representations of Europe that marked the thought of the first Greens. In the second part of this essay, I will expand this subject by sketching out the Green Party's positions with respect to institutional Europe, and present some of their ideas for an 'alternative Europe'.

The Greens and Europe: Between reality and imagination

In the first part of this essay, I emphasised the influence of transnational experiences and genuinely European biographies, such as those of Roland Vogt and Petra Kelly. The latter is of special interest, for as someone who collaborated with the European Commission, she ended up being in the 'lion's den', so to speak. Since she was also a key figure at the intersection between different groups and networks of new social movements, her personal accounts have had a strong influence on the opinions of other members from alternative circles.[17] For instance, when she spoke about 'this crazy masculine bureaucracy',[18] or about her personal experiences in what for her was the 'soulless routine' of the European Commission's daily life,[19] she also shaped

15 Camp, 'Für ein Europa der Regionen', p. 20.

16 Petra Kelly, 'Die Zukunft gehört der Güte', in *Frauen zum Lebensschutz. Vorträge auf dem in Troisdorf durchgeführten Deutschen Umwelttreffen 1978* (PKA, Nr.: 2940), pp. 13–16, here p. 16: 'Die kommenden Direkt-Wahlen zum Europäischen Parlament im Jahre 1979 sind für uns in der Gesamtalternativenbewegung lebenswichtig – wir müssen transnational handeln und demonstrieren, daß wir ein Gleichgewicht des Vertrauens, nicht des Schreckens, daß wir Zusammenarbeit und nicht Konfrontation brauchen'.

17 For this and the following, see Camp, 'Für ein Europa der Regionen'.

18 Petra Kelly to August Haußleiter (und weitere Vorstandsmitglieder), 13 Nov. 1979 (PKA, Nr.: 950), p. 1: 'Ich habe seit 1972 diese irre Männerbürokratie in Brüssel durchhalten müssen und wollte nicht mit gesenktem Kopf dieses Pflaster hier verlassen'.

19 Petra Kelly to Manfred Quickert, 15 Mar. 1981 (PKA, Nr.: 2310), p. 1: 'und ich sitze nach einem seelenlosen EG Alltag (mit Asbeststudien, Stellungnahmen zu Mikrowellen, zu Behinderten in

the image that other Greens had of Europe, and vice versa. There are also elements of similar discourses in the communications of the Green list for the 1979 elections, which assert: 'Brussels and Luxembourg should not be a retirement or annuity for discarded functionaries'.[20] The Greens began speaking in 1979 of the need for a 'radical transformation of the European Community'.[21] Such a discourse, which was not necessarily anti-European, but instead emphasised a different conception of Europe, was also evoked by new social movements in both the FRG and France.[22]

This representation of an institutionalised Europe – negative in the eyes of most, and henceforth marked by classic stereotypes – prompted fairly different reactions to the question of how to engage with this Brussels-based Europe. As we saw earlier, criticism of Europe as it existed was a central motivation in running for elections, in an effort to 'take part in organising a future Europe in keeping with our own ideas'.[23] However, one's vision of Europe as it existed could also provoke the opposite effect, for many groups from the extra-parliamentary left did not intend to run for office in 1979, as they criticised the European Parliament's lack of competence,[24] or objected to traditionally institutional Europe. To cite Thomas Ebermann, an influential member of the Hamburg-based communist group *Kommunistischer Bund (KB)*:

> We had good reasons, based on the common sense of the left, to not pay much attention to these elections. What did we have to do with Europe? No election was as banal or unimportant as this one, or further reinforced the idea of a European superpower … At the time we ignored the European elections less out of a grandiose political and strategic calculation, than out of a habit of not taking interest in them.[25]

Europa usw.) wieder um 1 Uhr morgens vor meinem Schreibtisch und tippe in die tiefe Nacht hinein … '

20 Mitteilungsblatt der 'Sonstigen Politischen Vereinigung' (Achberger Kreis, AUD, FIU, GAZ, GLU, GLSH und BIs) zu den Europawahlen am 10. Juni 1979 (PKA, Nr.: 2476): 'Brüssel und Luxemburg dürfen nicht zum Altenteil für abgehalfterte Altfunktionäre werden'.

21 Die Grünen – Alternative für Europa, Was wir wollen… Einige Programmaussagen, p. 1: 'DIE GRÜNEN fordern daher einen radikalen Wandel der Europäischen Gemeinschaft …'.

22 See Tompkins, *Better Active*, p. 81.

23 Tätigkeitsbericht des Vorstands des Bundesverbandes Bürgerinitiativen Umweltschutz e. V. (BBU) für die Zeit vom 20.11.1976 bis zum 26.11.1977, p. 8 (Archiv BBU), p. 10: Ausgestaltung eines Europa der Zukunft im Sinne unseres Gedankengutes mitzuwirken'.

24 Manfred Zieran, 'Für die Spaltung', *Pflasterstrand* **94** (1980): 18–20, here 18.

25 Thomas Ebermann, 'Ich und meine Freunde sind bei den GRÜNEN gescheitert', in Michael Schroeren (ed.), *Die Grünen. 10 bewegte Jahre* (Vienna, 1990), pp. 213–221, here p. 215: 'Wir hatten ein gesundes und gutes linkes Motiv, diese Wahl nicht sonderlich zu beachten. Was interessiert uns Europa? Keine Wahl ist banaler, unwichtiger und fördert deutlicher den ideologischen Gedanken an eine Supermacht Europa … Wir haben damals die Europawahl weniger

Yet what visions did these other Greens, who already took a certain interest in Europe, have for an ecological and alternative Europe? First, they were deeply marked by the perception of different crises that was so characteristic of the first Greens. Beyond the ecological crisis, it was the question of pacifism that fuelled different visions of Europe. At the turn of the 1980s in particular, many Greens imagined a Europe united on both sides of the Iron Curtain. The German ecological movement was not alone in speaking regularly about a 'neutral, non-aligned Europe without nuclear weapons',[26] stretching from 'the Atlantic to the Urals',[27] or at a minimum from 'Poland to Portugal', according to the famous phrase.

The same keywords were present, which at the time were fairly hazy and lacking in detail. This was true of a 'pacifist Europe of the regions', which was presented as an 'alternative to the European Community of bureaucrats, bombs, and mountains of butter'.[28] Behind this 'Europe of the regions' were recommendations for historic and self-determined – albeit interconnected – units. Among the more concrete demands, the idea of strengthening the weight of the European Parliament in relation to other institutions regularly appeared.[29] Already in the early 1970s, in the context of her own research for the Institute of Europe, Petra Kelly regretted an 'insidious de-democratisation'[30] of institutional Europe – a point that would be part of Green discourse on Europe in the ensuing years. The Commission and Council of Ministers, one read in 1984, 'have played into the hands of the destructive powers of industrial society'.[31]

In broad terms, one could say that the classic ideas of the ecological movement – such as 'decentralisation', 'self-determination', or 'thinking and acting

aus einem großartigen politischen und strategischen Kalkül heraus ignoriert, sondern weil wir gewohnt waren, uns nicht dafür zu interessieren'.

26 Die Grünen Baden-Württemberg, Gegen den Rüstungswahn – für eine Politik des Friedens [1981] (PKA, Nr.: 2487), p. 2: 'Für ein atomwaffenfreies, neutrales und blockfreies Europa'.

27 Broschüre der AL Berlin, Paktfreiheit für beide deutsche Staaten, Atomwaffenfreies Europa vom Atlantik zum Ural, Einheit für Deutschland [1981] [PKA, Nr.: 74 (1)].

28 Bundesvorstand der Grünen (ed.), *Global denken*, p. 38: 'Die Alternative zur EG der Bürokraten, Bomben und Butterberge ist ein friedliches Europa der Regionen'.

29 For instance, the platform for the 1979 European elections: 'Die Grünen, Alternative für Europa', [1979], e.g. p. 10.

30 Petra Kelly, cited in Camp, 'Für ein Europa der Regionen', p. 13: 'schleichende Entdemokratisierung'.

31 Bundesvorstand der Grünen (ed.), *Global denken*, p. 6: 'Die EG-Kommission und der Ministerrat haben den zerstörerischen Kräften der Industriegesellschaft Vorschub geleistet'.

from below' – were transposed onto the European scale: the same topics appear when we explore the Green's vision of their own party organisation, or when they spoke of the FRG's parliamentary system. It is worth noting that the four pillars of the German Greens' platform were mentioned for the first time in their European manifesto for 1979. It begins thus:

> The new European policy should be an overall policy guided by long-term future considerations, and should include four dimensions: it must be ecological, social, and should embody basic democracy as well as be non-violent, for Europe is threatened today by an ecological and economic crisis, military catastrophe, and permanent degradation of both democracy and fundamental rights.[32]

This final example once again clearly demonstrates that the perception of crises and criticism of the European *status quo* went hand in hand with outlines for an alternative Europe – however hazy it may have been at the time.

I would like to end by drawing three conclusions. First, Europe was not central to the political discussions of the German Greens during their formative phase, although transnational experiences deeply marked the perception of many of them. The impact of this international element proved even more important for a handful of influential protagonists who greatly contributed to the creation of a Green 'European public sphere'. This has had a certain importance in the organisation and cooperation of future Green and alternative parties on the European scale. Second, the first direct elections to the European Parliament in 1979 played a major role. These elections were not just an important 'window of opportunity', but also a demonstration that participating in parliament could create opportunities to formulate demands, some of which have been heard today. Third, with respect to the more concrete visions of Europe held by Greens, fairly classic topics and vague concepts were dominant, at least initially. However, this characteristic is not specific to the history of the first Greens in Europe or elsewhere, as it appears to be a common characteristic of all formative periods for new political parties.

32 'Die neue Europapolitik muß eine Gesamtpolitik sein, die von langfristigen Zukunftsaspekten geleitet wird und vier Dimensionen umfaßt: Sie muß ökologisch, sozial, basisdemokratisch und gewaltfrei orientiert sein; denn Europa ist heute bedroht durch die ökologische und ökonomische Krise, durch eine militärische Katastrophe und durch einen ständigen Abbau der Demokratie und der Grundrechte'. 'Die Grünen, Alternative für Europa', p. 1.

CHAPTER 12.

ENERGY AND THE ENVIRONMENT IN PARLIAMENTARY DEBATES IN THE FEDERAL REPUBLIC OF GERMANY, UNITED KINGDOM AND FRANCE FROM THE 1970s TO THE 1990s

Eva Oberloskamp

This contribution will explore the evolution of German, English and French parliamentary discourses regarding the connection between energy and the environment during the final three decades of the twentieth century. The central objective is to identify the historical roots of the discourse on the *energy transition* in Europe's 'big three', as they were expressed in parliamentary debates. The analysis will use a comparative perspective to retrace similarities and differences in the discursive structures of parliamentary debates. The central argument of my contribution is that the similarities between the three cases studied are largely insufficient to demonstrate the emergence of a European discursive space on the parliamentary level. The influence of political ecology on parliamentary debates varied considerably from one country to the other.

Parliament has a complex position with respect to the public sphere. Generally, within a democratic system, parliaments contribute to the formation of the public sphere, although they are also subject to numerous influences emanating from the extra-parliamentary space. While parliamentary debates do not reflect all facets of public opinion in identical fashion, they are nonetheless a place of crystallisation, where various points of view in society are expressed in condensed form. We should nevertheless keep in mind that the permeability of the parliamentary sphere to the concerns of the extra-parliamentary public varies according to the structural specificities of each country. With regard to the discursive connection between energy and the environment, it is important to examine the extent to which the environmental[1] and anti-nuclear move-

1 In accordance with the dominant terminology used in the Anglophone literature on the subject, the term 'environmental movement' will be used hereinafter, despite the fact that it diverges from the terminology of the source language: in all three of the countries studied, relevant sections of social movements intentionally differentiated themselves by considering themselves to be 'ecological movements'.

ments succeeded in being heard by members of parliament. Furthermore, the intensity of the external influences directed toward parliaments varies according to their respective position in each country's political system.

Two developments arose during the 1970s that called into question the energy policies pursued during the *Trente Glorieuses*. First, in many countries the '*ecological turning point*' that occurred around the year 1970 politicised the link between energy and the environment. Second, growing awareness that fossil fuel resources were limited prompted concerns about energy security. The three countries analysed here had different starting points and forms of energy dependence: the Federal Republic of Germany (FRG) had major coal deposits, but imported oil and gas on a large scale, with commercial use of nuclear technology still in its initial stage; the United Kingdom (UK) had considerable coal deposits, had just launched the extraction of petroleum and gas in the North Sea and, during the 1950s, had already put into service nuclear power plants for electricity production, with the sector expanding substantially in the ensuing decade; France had modest deposits of fossil fuels, was dependent on oil imports, and connected its first atomic power station to the grid in 1959, with its nuclear programme subsequently being developed further. Despite these differences, the reactions to the oil crises of the 1970s were similar: development of nuclear energy, energy conservation, emphasis on domestic resources and research programmes in the field of renewable energy.

The body of sources studied mainly includes the plenary session debates of the Bundestag, the House of Commons and the Assemblée nationale. The parliamentary documents were examined using keywords established for each national context, which can be grouped in the following four semantic fields: *energy and air pollution*, *nuclear energy and the environment*, *energy transition*, and *energy, environment, and market forces*. For each of these fields, a qualitative analysis was conducted of central debates selected by way of example.

The chapter will, for each national case, emphasise the evolution of parliamentary discourse, and a series of contextual factors that influenced them. The conclusion will offer a comparative synthesis.

The Federal Republic of Germany

During the 1970s, the subject of air pollution was highly present in West German parliamentary debates. Its importance can be explained by the social liberal government's efforts to promote legislation protecting against pollution.[2] Some debates focused on nuclear energy, but without making it a fundamental question. Only within a minority of the Social Democratic Party did a certain opposition to nuclear power gradually emerge in the late 1970s.[3] The initial drive for considering radical arguments against this energy source came from a hearing of experts organised in 1974,[4] and particularly from the Bundestag's 'Zukünftige Kernenergie-Politik' [Future Nuclear Energy Policy] Enquete Commission, which sat from 1979 to 1983.[5] Renewable energy was initially grasped mostly as an alternative to fossil energy, with the term 'renewable energy' becoming established only in the 1980s.

During the 1980s, all of the parties represented in the federal parliament developed their own thinking on the environment. Multiple reports from the Sachverständigenrat für Umweltfragen [German Advisory Council on the Environment],[6] along with a series of major debates, raised awareness regarding problems of air pollution.[7] At the same time, the German Greens joining the Bundestag in 1983[8] substantially widened the range of arguments regarding nuclear energy, which was fundamentally challenged by the new party. The social democrats gradually developed their own environmental and anti-nuclear conception, which sought to link environmental questions to economic policies and employment.[9] The debates that followed the

2 See Bundestags-Plenarprotokoll (BT-PlPr. 7/74), 18.1.1974, pp. 4677–4691; BT-PlPr. 8/162, 22.6.1979, pp. 12899–12926.

3 See Gerhard Kiersch and Sabine von Oppeln, *Kernenergiekonflikt in Frankreich und Deutschland* (Berlin: Wissenschaftlicher Autoren-Verlag, 1983), pp. 43–48; Martin Tretbar-Endres, 'Die Kernenergiediskussion der SPD Schleswig-Holstein. Ein Beispiel innerparteilicher Willensbildung'. *Demokratische Geschichte* **8** (1993): 347–372.

4 Presse- und Informationszentrum des Deutschen Bundestages (ed.), *Das Risiko Kernenergie. Aus der öffentlichen Anhörung des Innenausschusses des Deutschen Bundestages am 2. und 3. Dezember 1974* (Bonn: 1975).

5 See Bundestags-Drucksache (BT-Drs.) 8/4341, 27.6.1980; BT-Drs. 9/2001, 27.9.1982; BT-Drs. 9/2438, 24.3.1983; Cornelia Altenburg, *Kernenergie und Politikberatung. Die Vermessung einer Kontroverse* (Wiesbaden: Verlag für Sozialwissenschaften, 2010).

6 BT-Drs. 9/872, 6.10.1981; BT-Drs. 10/113, 8.6.1983; BT-Drs. 11/1568, 21.12.1987.

7 See for example BT-PlPr. 10/22, 15.9.1983, pp. 1429–1535; BT-PlPr. 10/204, 13.3.1986, pp. 15692–15712; BT-PlPr. 11/131, 9.3.1989, pp. 9586–9658.

8 On the history of the party, see Silke Mende, *'Nicht rechts, nicht links, sondern vorn'. Eine Geschichte der Gründungsgrünen* (Munich: Oldenbourg Wissenschaftsverlag, 2011); Sebastian Bukow, 'The Green Party in Germany', in Emilie van Haute (ed.), *Green Parties in Europe* (London/New York: Routledge, 2016), pp. 112–139.

9 I would like to thank Felix Lieb, who is preparing a Ph.D. thesis on 'Sozialdemokratische Umwelt- und Energiepolitik 1975–1998', for his suggestions on this topic: https://www.ifz-muenchen.de/forschung/ea/forschung/sozialdemokratische-umwelt-und-energiepolitik-1975-1998/ (accessed 11 Feb. 2019).

Chernobyl disaster in 1986 marked the height of this evolution.[10]

Beginning with the German Greens joining the Bundestag in 1983 and the Social Democrats becoming an opposition party, two competing conceptions formed within the parliament regarding the integration of environmental concerns in energy policy. On one side was the notion of a moderate technological adjustment to environmental requirements, defended by the Christian Democrats and Liberals. They fully acknowledged the existence of environmental challenges, but did not see a need for major changes. The objective of an energy supply compatible with environmental requirements was part of this conception, although emphasis was placed on goals of an economic nature.[11] On the other side were those who supported an *Energiewende* (energy transition), a concept initially defended by the German Greens, but that proved increasingly compatible with the programme of the Social Democrats. The primary concern of its supporters was to abandon nuclear power, with emphasis being placed on renewable energy and energy conservation. The central goal was environmental viability, which was seen as a prerequisite for profitability and long-term energy security. Furthermore, it added the goal of *social acceptability*, which included the idea of a decentralised energy supply organised in accordance with grassroots democracy.[12] It was defenders of *Energiewende* who demanded a liberalisation of energy markets, as they maintained that the current system had a distortion effect, subsequently preventing both greater energy conservation and the entry of renewable energy on the market.[13]

The basic pattern of two competing visions for energy policy within the Bundestag remained unchanged during the 1990s. The scenario of potential dangers connected to global warming gave renewed momentum to parliamentary debates.[14] The existence and gravity of the problem were hardly in doubt, regardless of any partisan interest.[15] Both international policy for

10 See the first major debate on the Chernobyl disaster: BT-PlPr. 10/215, 14.5.1986, pp. 16522–16574.

11 See for instance BT-PlPr. 10/94, 25.10.1984, pp. 6875–6907; BT-PlPr. 11/46, 3.12.1987, p. 3178; BT-Drs. 19/1773, 20.7.1984.

12 See for instance BT-PlPr. 10/94, 25.10.1984, pp. 6878–6881; BT-PlPr. 10/236, 3.10.1986, pp. 18266–18280; BT-PlPr. 11/16, 4.6.1987, p. 1015–1039.

13 See, for instance, BT-Drs. 8/4341, 27.6.1980, p. 75; BT-PlPr. 10/171, 7.11.1985, p. 12778f.

14 Initial impetus for this came from the report by the Sachverständigenrat für Umweltfragen, BT-Drs. 11/1568, 21.12.1987, p. 33.

15 See, for example, BT-PlPr. 12/152, 22.4.1993, pp. 13006–13026; BT-PlPr. 13/27, 16.12.1995, pp. 1860–1906; BT-PlPr. 13/166, 20.3.1997, pp. 14930–14959; BT-PlPr. 14/67, 5.11.1999, pp. 5985–6026.

preserving the climate and Enquete Commissions provided considerable impetus for the debates.[16]

Supporters of a moderate technological adjustment to environmental requirements placed greater emphasis on the need for nuclear energy,[17] although they henceforth had a more favourable view of renewable energy.[18] The latter's chance of accessing the market had improved with the Stromeinspeisungsgesetz (Electric Supply Act) of 1990, which ensured grid access and a guaranteed feed-in tariff for renewable energy. The rejection of nuclear power by supporters of *Energiewende* remained unquestioned.[19] The use of coal, however, was not seriously attacked. The *ökologische Steuerreform* (ecological tax reform) proposed by the German Greens and also by Social Democrats sought to create incentives for reducing the use of fossil fuels.[20] But at the same time, the Social Democrats continued to be indulgent toward coal, due to its economic and social significance, along with its importance for the party's very identity.[21] The German Greens, Social Democrats and, since 1990, the Party of Democratic Socialism/Die Linke, have continued to combine their efforts to exert strong pressure on the government. They submitted a string of questions and parliamentary motions to put forward their conceptions of *Energiewende*,[22] which finally began to be implemented with the Red-Green federal government in 1998.[23]

The evolution of the discourse in the Bundestag was deeply imbued with the exceptional force of the West German environmental and anti-nuclear movement.[24] The latter drew much of its dynamic from the fact that nuclear technology was criticised as essentially a totalitarian and immoral excess of

16 BT-Drs. 11/3246, 2.11.1988; BT-Drs. 11/7220, 24.5.1990; BT-Drs. 11/8030, 24.5.1990; BT-Drs. 12/8600, 31.10.1994.

17 See, for example, BT-PlPr. 13/101, 25.4.1996, pp. 8907–8908.

18 See, for example, BT-PlPr. 12/67, 12.12.1991, p. 5745; BT-PlPr. 12/226, 29.4.1994, p. 19546.

19 See, for example, BT-Drs. 13/4447, 24.4.1996, p. 1.

20 See, for example, BT-PlPr. 11/13, 21.5.1987, p. 769; BT-Drs. 12/1794, 11.12.1991.

21 See, for example, BT-PlPr. 12/179, 30.9.1993, pp. 15475–15477 and 15489–15491.

22 See Sonja Boehmer-Christiansen and Jim Skea, *Acid Politics. Environmental and Energy Policies in Britain and Germany* (London/New York: Belhaven Press, 1991), p. 198.

23 See Edgar Wolfrum, *Rot-Grün an der Macht. Deutschland 1998–2005* (Munich: C.H. Beck, 2013), pp. 214–269.

24 More recently, see Stephen Milder, *Greening Democracy. The Anti-Nuclear Movement and Political Environmentalism in West Germany and Beyond, 1968–1983* (Cambridge: Cambridge University Press, 2017); Andrew S. Tompkins, *Better Active than Radioactive! Anti-Nuclear Protest in 1970s France and West Germany* (Oxford: Oxford University Press, 2016).

modernity, with the history of the Holocaust being implicit.[25] In addition, the movement expanded significantly in scope because the political parties were initially reluctant to appropriate the criticisms made by activists,[26] who subsequently felt they were not represented in state policies,[27] resulting in the movement adopting a strong identity of fundamental opposition.

The FRG's political system was nevertheless able to eventually integrate this new oppositional force thanks to its proportional voting system.[28] The Green Party introduced into parliamentary discourse the concept of *Energiewende,* which was originally developed by the Öko-Institut in Freiburg, an institute of scientific counter-expertise that grew out of the anti-nuclear movement.[29] The lasting presence of a Green party in the Bundestag since 1983 has made *Energiewende* an enduring feature of parliamentary discourse. Furthermore, there has been persistent public interest in environmental subjects in the FRG.[30] The intense societal debates surrounding the decline of forests,[31] the Chernobyl nuclear disaster and, beginning in the second half of the decade, global warming,[32] were major sources of pressure that affected parliamentary debates.

25 See, for example, Hans-Helmuth Wüstenhagen (head of the Citizens' Initiatives for Environmental Protection BBU), 'Nachbemerkung: Widerstand ist Pflicht', in Werner Biermann, *Plutonium und Polizeistaat* (Bonn: SPAK-Publikationen, 1977), pp. 113–114, here p. 113. 'Atommacht Deutschland. Die Wiege stand im 3. Reich', *Atomexpress* 15 (1979): 29–35; 'Sie haben versagt', *Die Zeit*, 23 May 1986 (poem published anonymously by Inge Aicher-Scholl and others).

26 See BT-PlPr. 7/215, 22.1.1976, pp. 14916–14953.

27 See, for example, Hans-Helmut Wüstenhagen, *Bürger gegen Kernkraftwerke. Wyhl – der Anfang?* (Reinbek bei Hamburg: Reinbek Rowoholt, 1977), p. 101; Hans-Helmut Wüstenhagen, 'Bürgerinitiativen, Atomenergie und Wissenschaft', *Blätter für deutsche und internationale Politik* **21** (1976): 1360–1367, here p. 1365.

28 See Karl-Werner Brand, 'Vergleichendes Resümee', in Karl-Werner Brand (ed.) *Neue soziale Bewegungen in Westeuropa und den USA. Ein internationaler Vergleich* (Frankfurt/New York: Campus Verlag, 1985), pp. 306–334, here pp. 323–324.

29 Florentin Krause, Hartmut Bossel and Karl-Friedrich Müller-Reißmann, *Energie-Wende. Wachstum und Wohlstand ohne Erdöl und Uran. Ein Alternativ-Bericht* (Frankfurt: S. Fischer 1980). This book applied the ideas developed by the American environmental activist Amory B. Lovins to the German case. See Amory B. Lovins, *Soft Energy Paths. Towards a Durable Peace* (Harmondsworth: HarperCollins, 1977).

30 See Frank Uekötter, *Deutschland in Grün. Eine zwiespältige Erfolgsgeschichte* (Göttingen: Vandenhoeck & Ruprecht, 2015), pp. 151–168.

31 See Birgit Metzger, *'Erst stirbt der Wald, dann du!' Das Waldsterben als westdeutsches Politikum (1978–1986)* (Frankfurt: Campus Verlag 2015); Roderich von Detten (ed.), *Das Waldsterben. Rückblick auf einen Ausnahmezustand* (Munich: oekom verlag, 2013).

32 See Peter Weingart, Anita Engels and Petra Pansegrau, *Von der Hypothese zur Katastrophe. Der anthropogene Klimawandel im Diskurs zwischen Wissenschaft, Politik und Massenmedien*, slightly revised 2nd edition (Opladen/Farmington Hills: Verlag Barbara Budrich, 2008).

Another factor that jumps out, especially in comparison with the UK, is the relatively minor significance of social conflicts in debates on energy policy, which might have contributed to excluding the environment from the agenda. In addition, liberal markets were less important in discourse in the FRG.

The United Kingdom

In the UK, the topic of air pollution was often present in parliamentary debates on energy during the 1970s.[33] In addition, the Parliament of the UK served very early on as an important discursive space in which the implications of nuclear energy were intensely discussed. These debates did not so much address the question of whether to approve or reject nuclear power, but rather the transition to new types of reactors.[34] New momentum was provided by the report published in 1976 by the Royal Commission on Environmental Pollution, entitled 'Nuclear Power and the Environment',[35] along with the public investigation of the possible construction of a reprocessing plant in Windscale, which was published in 1977.[36] All anti-nuclear arguments were present in these debates; in the Labour Party, Liberal Party and Scottish National Party, Members of Parliament (MPs) from constituencies particularly affected by the nuclear question defended highly critical positions.[37] From the 1970s onward, renewable energy was evoked in con-

33 See for example the debate on Clean Air in Commons Hansard, 19.7.1973, vol. 860, https://hansard.parliament.uk/Commons/1973-07-19/debates/a9518bea-5fe0-45f7-8533-f72cda678181/CleanAir?highlight=air%20pollution#contribution-87cf2a31-90f8-4ecf-aa28-f9051454b2db; the debate on the Control of Pollution Bill, 17.6.1974, vol. 875, https://hansard.parliament.uk/Commons/1974-06-17/debates/b743717c-51fe-4c5f-92bb-e191bc4c7b74/ControlOfPollutionBillLords?highlight=air%20pollution#contribution-67d6619a-05c1-4700-9788-42c675d4ff92 (accessed 20 July 2018). Considerable impetus for this came from a number of reports by the Royal Commission on Environmental Pollution.

34 For a general overview, see Stuart Butler and Robert Bud, *United Kingdom. Short Country Report* [on the History of Nuclear Energy and Society], http://www.honest2020.eu/d36-short-country-reports pp. 22–25 (accessed 19 July 2018).

35 *Royal Commission on Environmental Pollution, Sixth Report. Nuclear Power and the Environment*, London 1976, https://www.google.de/url?sa=t&rct=j&q=&esrc=s&source=web&cd=1&ved=0ahUKEwj9_oe_n4_aAhVOJ1AKHU3mDPoQFggnMAA&url=http%3A%2F%2Fwebarchive.nationalarchives.gov.uk%2F20110322143804%2Fhttp%3A%2Fwww.rcep.org.uk%2Freports%2F06-nuclear%2F1976-06nuclear.pdf&usg=AOvVaw1s0-PODXUPRxqSEHxQE7D0 (accessed 28 Mar. 2018).

36 *Die Windscale-Untersuchung. Bericht des Hon. Mr. Justice Parker*, German trans. ed. by Deutsches Atomforum, Bonn 1979.

37 See, for example, Commons Hansard, 28.6.1977, vol. 934, https://api.parliament.uk/historic-hansard/commons/1977/jun/28/energy; 2.12.1977, vol. 940, https://api.parliament.uk/histor-

nection with environmental considerations, as demonstrated by the choice of terms: MPs spoke not only of 'renewable energy', but also of 'benevolent' or 'benign energy'.[38]

The two competing conceptions of energy policy, which have been analysed for the German case, were less explicit, although they emerged much earlier in the UK than in the FRG. On the one hand, conservative MPs and some Labour and Liberal MPs supported a *hard* strategy that relied on fossil fuels, nuclear power and energy conservation, giving priority to economic considerations. On the other were MPs who spoke of *soft energy*, emphasising renewables.[39] However, there were no calls for abandoning nuclear energy, or for the decentralisation or grassroots democracy inherent in the concept of *Energiewende* in Germany. Demands to liberalise energy markets were made by conservatives.[40]

After the fairly turbulent debates of the 1970s, the House of Commons grew less concerned with green energy during the ensuing decade. The bitter economic and social controversies over the privatisation projects for the energy sector proposed by the Conservative government that came to power in 1979, helped relegate environmental considerations to the background.[41] The debates over nuclear energy were primarily about a possible privatisation of the industry. The economic logic – in both a neo-liberal sense and a more traditional sense of the left's emphasis on social conflicts – took precedence over environmental logic, notably in discussions regarding the development of renewable energy.[42] For all that, environmental concerns did not disappear

ic-hansard/commons/1977/dec/02/nuclear-energy; 22.3.1978, vol. 946, https://api.parliament.uk/historic-hansard/commons/1978/mar/22/windscale-inquiry-report (accessed 20 July 2018).

38 See, for example, the debate on Energy in Commons Hansard, 28.6.1977, vol. 934, https://api.parliament.uk/historic-hansard/commons/1977/jun/28/energy (accessed 20 July 2018).

39 See, for example, the debate on Energy in Commons Hansard, 28.6.1977, vol. 934, https://api.parliament.uk/historic-hansard/commons/1977/jun/28/energy; the debate on Nuclear Energy, 2.12.1977, vol. 940, https://api.parliament.uk/historic-hansard/commons/1977/dec/02/nuclear-energy; the debate on Energy Policy and Nuclear Energy, 7.12.1977, vol. 940, https://api.parliament.uk/historic-hansard/commons/1977/dec/07/energy-policy-and-nuclear-energy; the debate on the Windscale Inquiry Report, 22.3.1978, vol. 946, https://api.parliament.uk/historic-hansard/commons/1978/mar/22/windscale-inquiry-report (all accessed 20 July 2018). On Lovins, see footnote 29.

40 See, for example, Commons Hansard, 28.6.1977, vol. 934, https://api.parliament.uk/historic-hansard/commons/1977/jun/28/energy (accessed 20 July 2018).

41 On the beginnings of Conservative privatisation policies, see David Parker, *The Official History of Privatisation. Volume I: The Formative Years 1970–1987* (London/New York: Routledge, 2009).

42 See, for instance, the debate on Energy (Alternative Sources) in Commons Hansard, 25.10.1985, vol. 84, https://api.parliament.uk/historic-hansard/commons/1985/oct/25/energy-alternative-sources (accessed 20 July 2018).

from discussions regarding energy,[43] as there was a certain trans-partisan consensus about the importance of environmental questions – all the more so when Margaret Thatcher began to ascribe considerable importance to environmental policy at the end of the decade.[44]

This created a favourable context during the 1990s for receiving new momentum from international climate protection policy. For conservatives, the combination of privatisation (which represented the 'key to unlock' the 'potential' of renewables), along with the Non-Fossil Fuel Obligation established in 1989 (which ensured that selected producers of nuclear and renewable energy would receive both orders and an agreed-upon purchase price), represented a decisive element in reducing greenhouse gas emissions.[45] The contraction of coal production pursued since the 1980s for economic reasons also took on an additional environmental justification in this context.[46] Labour MPs, on the other hand, asserted that 'the government's obsession' for privatisation was preventing the UK from honouring its international commitments in environmental matters.[47] There was ambivalence surrounding nuclear energy. On the one hand, since fundamental criticism had fallen silent, it was seen as being respectful of the environment;[48] on the other, the government's privatisation efforts underscored its problems of profitability, which gave this energy source a bad image from an economic point of view.[49] With regard to debates on coal, they primarily focused on economic and social matters, even though its environmental impact was not denied.

43 See the debate on Acid Rain in Commons Hansard, 11.1.1985, vol. 70, https://api.parliament.uk/historic-hansard/commons/1985/jan/11/acid-rain; the debate on Nuclear Power Stations (Safety) 23.5.1986, vol. 98, https://hansard.parliament.uk/Commons/1986-05-23/debates/548065c6-8c3a-41a3-9d3d-3a046080d7ac/NuclearPowerStations(Safety)?highlight=chernobyl#contribution-2c2cc158-917a-456b-879f-d6728737458a; and the debate on World Climate Change, 10.11.1989, vol. 159, https://api.parliament.uk/historic-hansard/commons/1989/nov/10/world-climate-change-1 (accessed 20 July 2018).

44 See Dieter Helm, *Energy, the State, and the Market. British Energy Policy since 1979* (Oxford: Oxford University Press, 2003), p. 346.

45 Commons Hansard, 20.2.1990, vol. 167, https://api.parliament.uk/historic-hansard/commons/1990/feb/20/electricity-industry (accessed 20 July 2018), quote from Malcolm Moss.

46 See Helm, *Energy*, pp. 346–352.

47 Commons Hansard, 20.2.1990, vol. 167, https://api.parliament.uk/historic-hansard/commons/1990/feb/20/electricity-industry (accessed 20 July 2018), quote from Frank Dobson.

48 See for example Commons Hansard, 25.6.1992, vol. 210, https://hansard.parliament.uk/Commons/1992-06-25/debates/d90b6a6e-8bcd-4681-bd9e-0738ab0d0bb5/CommonsChamber (accessed 20 July 2018).

49 On the partial privatisation of the nuclear sector completed during the 1990s, see Helm, *Energy*, pp. 186–203.

In the UK, the environmental and anti-nuclear movements did not have as widespread an impact as they did in the FRG. There were indeed groups that mobilised against the construction of nuclear installations, contributed to the Windscale investigation and took a close interest in alternative energy policies,[50] although they were less numerous and more moderate than in Germany. British ecologists generally remained in the shadows of established nature protection associations, while the anti-nuclear movement was largely absorbed during the 1980s by the pacifist organisation Campaign for Nuclear Disarmament. For all that, certain MPs were open to the arguments of activists, and transferred them to parliamentary debates.

At the same time, the UK's election by majority vote did not permit the establishing of a Green political force that could become a part of the national parliament and develop its own dynamic as a political actor.[51] Political ecology gradually ran out of steam during the 1980s, while the concept of *energy transition*, which was steeped in the anti-nuclear movement, did not become an enduring part of parliamentary debates in the 1980s and 1990s. Nature conservation was nevertheless firmly rooted in English society, and there was social pressure not to lose sight of these questions in the political sphere. As a result, environmental topics did not completely disappear from the agenda of parliamentary debates on energy.

A distinctive feature of British discourse – one that contributed to weakening political ecology – was the broad hegemony of economic thinking: its logic was present in both discourses steeped in Thatcherist ideals of privatisation and free markets, as well as discourses of the left giving priority to social conflicts.

France

During the 1970s and 1980s, debates on energy policy in the Assemblée nationale gave little consideration to environmental implications: there were no major debates about air pollution or the dangers of radiation from

50 Important organisations were, inter alia, Friends of the Earth, Greenpeace, Conservation Society, Scottish Campaign to resist the Atomic Menace and Anti-Nuclear Campaign. See Heinz Rothgang, *Die Friedens- und Umweltbewegung in Großbritannien. Eine empirische Untersuchung im Hinblick auf das Konzept der 'Neuen Sozialen Bewegungen'* (Wiesbaden: Deutscher Universitats-Verlag, 1990), pp. 32–36; Detlef Murphy, 'Von Aldermaston nach Greenham Common. Politischer Protest und neue soziale Bewegungen in Großbritannien', in Brand, *Neue soziale*, pp. 140–199, here pp. 163–169.

51 On the British Greens, see Lynn Bennie, 'Greens in the United Kingdom and Ireland. Weak but persistent', in van Haute, *Green Parties in Europe*, pp. 196–216.

nuclear material.[52] Reports from the Office parlementaire d'évaluation des choix scientifiques et technologiques [Parliamentary Office for the Evaluation of Scientific and Technological Choices] for parliament broached these aspects, but without giving them a major sense of urgency.[53] French parliamentary discourse shows a strong consensus in favour of nuclear energy across all political groups.[54] Even the debates sparked by the Chernobyl disaster focused solely on international security systems.[55] Renewable energy was primarily grasped from the perspective of diversifying energy sources, as demonstrated by the use of the term 'new energies'.[56]

References to anti-nuclear arguments were instead found in presidential campaigns and the government's rhetoric.[57] The presidential elections of 1981 were an important moment, as the future socialist candidate François Mitterrand expressed criticism starting in 1979 of the *all-nuclear* path envisioned in France.[58] After his election, he decided to slow the expansion of the nuclear sector, with the creation of the Agence française pour la maitrise de l'énergie [French Agency for the Control of Energy] coming in 1982.[59] Still, French parliamentary debates after the change of government in 1981 – which for the first time revealed certain trends critical of nuclear power – concentrated almost exclusively on the political aspects of the energy

52 The connection between air pollution and energy production was explored in Journal officiel/Assemblée Nationale, 3rd session from 18.12.1979, pp. 12235–12248; 2nd session from 24.6.1980, pp. 2228–2231.

53 See, for instance, Report No. 3192 (Assemblée Nationale, 7th legislature) by Georges Le Baill, appendix to the minutes for the session from 13.12.1985; Report No. 1156 (Assemblée Nationale, 8th legislature) by Jean-Marie Rausch/Richard Pouille, appendix to the minutes for the session from 17.12.1987.

54 See, for example, the very representative commentary on this consensus by the UDF deputy Pascal Clément: Journal officiel/Assemblée Nationale, 1st session from 4.4.1979, p. 2235.

55 On overall French reactions to the Chernobyl disaster, see Karena Kalmbach, *Tschernobyl und Frankreich. Die Debatte um die Auswirkungen des Reaktorunfalls im Kontext der französischen Atompolitik und Elitenkultur* (Frankfurt: Peter Lang, 2011).

56 See, for example, Journal officiel/Assemblée Nationale, 1st session from 4.4.1979, pp. 2220–2235; Minsters' responses to written questions, 14.12.1987, pp. 6782–6784.

57 Between 1988 and 1992, France had a Green (Brice Lalonde) in the ranks of the government, without there being any Green deputies in the Assemblée Nationale.

58 See Mitterrand's signing of the petition 'Pour une autre politique de l'énergie. Pour un débat démocratique sur l'énergie', *La Gazette Nucléaire* **28**: http://gazettenucleaire.org/1979/28.html#-sommmaire (accessed 13 May 2018); Pierre Michel, 'Les socialistes et l'énergie, un témoignage', *L'Économie politique* **56** (2012): 85–95, here 89.

59 See Marc Ambroise-Rendu, *Des cancres à l'Élysée. 5 Présidents de la République face à la crise écologique* (Paris: Jacob Duvernet, 2007), pp. 216–226.

source (centralism, bureaucratism and lack of transparency), without alluding to its environmental implications; there was no question of a possible abandonment of nuclear power.[60]

Beginning in the 1990s, environmental criteria were taken somewhat into account. This can be seen, for instance, in more frequent use of the terms 'green energies' or 'renewables'.[61] When Minister for the Environment, Ségolène Royal, reported in 1992 on the progress of negotiations in advance of the Earth Summit in Rio, she deemed it necessary to first give deputies a fairly basic explanation of the problem of global warming, in order then to claim that it represented a 'genuine civilisational problem'.[62] The Earth Summit and the negotiations that followed regularly sparked debates.[63] This is why the connection between energy and the environment largely came down in French discourse to the problem of CO_2 emissions. In this context, the transpartisan consensus in favour of nuclear energy generally remained intact,[64] as it was seen as beneficial to the environment because it did not contribute to air pollution.[65] The national energy system was considered 'a considerable advantage' for France,[66] and a model for the rest of the world.[67] At the same time, all parties were highly reserved

60 See the Journal officiel/Assemblée Nationale, 2nd session from 6.10.1981, pp. 1504–1520; 3rd session from 6.10.1981, pp. 1522–1547; 1st session from 7.10.1981, pp. 1561–1572; 2nd session from 7.10.1981, pp. 1574–1600.

61 Journal officiel/Assemblée Nationale, 1st session from 13.1.1994, p. 73.

62 Journal officiel/Assemblée Nationale, 2nd session from 3.6.1992, p. 1882. For other examples, see Michel Destot (Socialist), 2nd session from 25.11.1993, p. 6416.

63 See, for example, Journal officiel/Assemblée Nationale, 2nd session from 25.11.1993, p. 6416; 1st session from 15.6.1996, p. 10.

64 This is true despite an entire series of reports by the Office parlementaire d'évaluation des choix scientifiques et technologiques that explored the subject of nuclear energy during the 1990s. See Report no. 1839 (Assemblée Nationale, 9th legislature) by Christian Bataille, appendix to the minutes from the session from 14.12.1990; Report no. 2624 (Assemblée Nationale, 9th legislature) by Jean-Yves Le Déaut, recorded at the Presidency of the Assemblée Nationale on 22.4.1992; Report no. 2689 (Assemblée Nationale, 10th legislature) by Christian Bataille, recorded at the Presidency of the Assemblée Nationale on 27.3.1996; Report nos. 1008, 1496, 1825, 2417, 2651, 2765, 3491 (Assemblée Nationale, 9th – 11th legislatures) by Claude Birraux, recorded at the Presidency of the Assemblée Nationale on 4.12.1991, 10.6.1992, 10.2.1994, 19.12.1994, 19.3.1996, 2.4.1997, 25.3.1999; Report nos. 978 and 1359 (Assemblée Nationale, 11th legislature) by Christian Bataille and Robert Galley, recorded at the Presidency of the Assemblée Nationale on 11.6.1998, 2.2.1999.

65 See, for example, the remarks by the communist deputy Roger Meï, Journal officiel/Assemblée Nationale, single session from 9.4.1998, p. 10.

66 Journal officiel/Assemblée Nationale, 2nd session from 25.11.1993, p. 6399.

67 Ibid., p. 6415; Journal officiel/Assemblée Nationale, single session from 15.6.1996, p. 28.

regarding a possible liberalisation of energy markets.[68]

With much French electricity coming from nuclear power, there was a belief that mobility was the only domain in which action was required. With respect to transportation policy, some socialist deputies posited the need for radical changes during the second half of the 1990s.[69] The first debate in the Assemblée nationale that included an in-depth discussion of the connection between energy and the environment took place in 1996, as part of examining the bill 'on air and the rational use of energy'. This debate largely concentrated on transportation policy.[70]

With seven ecologist deputies joining the Assemblée for the first time in 1997, anti-nuclear voices favourable to deeper changes began to speak out in the Assemblée nationale.[71] This was also the first time that deputies voiced support for opening markets, which was assumed to be favourable to the development of renewable energy.[72]

Although France saw a considerable rise in environmental and anti-nuclear movements during the 1970s,[73] these were unable to exert substantial influence on parliamentary discourse, as the large majority of the French elite continued to support nuclear power, and its hegemonic discourse proved difficult to penetrate. Moreover, in the institutional system of the Fifth Republic, parliament seemingly had no more than a limited capacity to transmit social demands to the decision-making centre of the state. As a result of its relatively weak position, it was seen to a certain extent as being outside the heart of political life. Furthermore, due to the Fifth Republic's election by majority vote, attempts to create a national green party achieved less success

68 See, for example, Journal officiel/Assemblée Nationale, 2nd session from 25.11.1993, pp. 6395–6432; 1st session from 20.6.1994, pp. 3225–3234.

69 Michel Destot, Journal officiel/Assemblée Nationale, 2nd session from 25.11.1993, p. 6416.

70 Journal officiel/Assemblée Nationale, 1st session from 15.6.1996, pp. 10–30; 2nd session from 15.6.1996, pp. 12–53; 2nd session from 19.6.1996, pp. 4–58; 1st session from 20.6.1996, pp. 4–41; 2nd session from 22.11.1996, pp. 41–57, 1st session from 26.11.1996, pp. 3–34; single session from 27.11.1996, pp. 14–46.

71 See, for example, Journal officiel/Assemblée Nationale, single session from 9.4.1998, pp. 7–8.

72 See, for example, Journal officiel/Assemblée Nationale, 3rd session from 17.2.1999, pp. 1596–1597, 1601.

73 On the French anti-nuclear movement, see Tompkins, *Better Active than Radioactive!* and Milder, *Greening Democracy*. On ecological thinking and protest, see Michael Bess, *The Light-Green Society. Ecology and Technological Modernity in France, 1960–2000* (Chicago: University of Chicago Press, 2003); Graeme Hayes, *Environmental Protest and the State in France* (Houndmills: Palgrave Macmillan, 2002).

than in West Germany.[74] Overall, environmental thinking gradually began to spread in French parliamentary discourse due to international policy for climate preservation rather than strong social pressure.

An additional factor that slowed the integration of environmental discourse in the language of political elites was the persistent division between the right and the left in the French political arena – which seems more important than the actual impact of existing social conflicts. By contrast, the economic ideas of liberalisation and privatisation, which were so influential in the UK, did not have a major impact.

A comparison of the three analysed cases first and foremost reveals clear differences between West German, British and French parliamentary discourse:

1. Importance of specific environmental subjects: in the FRG, air pollution was a highly visible and enduring subject in the parliamentary agenda, with very little doubt regarding its importance; it was nuclear energy that nonetheless became the focal point in debates surrounding the connection between energy and the environment, along with encouragement for conceiving radical alternatives for energy policy. In the UK, air pollution was generally recognized as a pressing problem in energy policy, although conflicts surrounding nuclear power were less important than in the FRG, and rejection of this technology did not become the driving force behind a global conception of energy transition. In France, air pollution long played a subordinate role, while the supposed environmental performance of nuclear energy prevented an in-depth formulation of the link between energy and the environment in French parliamentary discourse.
2. Discourse structures corresponding to environmental logic: environmental perspectives pervaded all discourses in West Germany. This led to the formation of two distinct discourses on the problem of integrating environmental considerations in energy policy. In the UK, environmental perspectives were only one factor among others taken into account; the divisions in debates over energy were primarily structured according to economic criteria. In France, discourse was relatively homogenous. The environment was not a structural element before French ecologists joined the Assemblée nationale in 1997, with its impact remaining weak.
3. Capacity attributed to the market: in the FRG, discourse was dominated by the idea that the energy sector should only be partially exposed to free market forces, and

74 On the history of the French Greens, see Pierre Serne, *Des Verts à EELV, 30 ans d'histoire de l'écologie politique* (Paris: Les Petits Matins, 2014); Bruno Villalba, 'From the Greens to Europe ecology – The Greens. Renaissance or more of the same?' in van Haute, *Green Parties*, pp. 92–111.

that a certain degree of state intervention in energy markets was justified; however, supporters of *Energiewende* called for more liberalisation, which was supposed to promote energy conservation and remove market distortions that placed renewable energies at a disadvantage. In the UK, the environmental argument gave conservatives an additional justification for the liberalisation of markets. In France, the self-regulating power of the market was generally seen with scepticism in the field of energy; an alternative perspective emerged only in the late 1990s, when French ecologist deputies joined the parliament.

4. Evolution of discourses over time: in Germany, the 1980s were a particularly virulent phase, whose spirit continued during the 1990s. In the UK, the high-water mark of environmental considerations in matters of energy policy came in the late 1970s, with this dynamic running out of steam in the ensuing decade, followed by a certain resumption during the 1990s. In France, the environment became a part of parliamentary debate on energy only during the 1990s.
5. National and Transnational Impetus: in the FRG, the drive to include the environment in energy policy came in very large part from national forces that perpetuated themselves through their enduring impact on the party system; during the 1990s, international policy also contributed to developments in West Germany. In the UK, intrinsic factors prompted thinking on the environmental effects of energy, but the country did not see the emergence of a Green political force to consolidate this momentum; beginning in the 1990s, it was international policy on climate preservation in particular that provided decisive impulsion. Finally, in France national driving forces were very weak, with UN policy in favour of the climate in the 1990s serving as the primary motor.

The differences are therefore largely predominant, and so extensive as to conclude that a European discursive space regarding the link between energy and environment did not exist on the parliamentary level. The question of a broader European public arena beyond this topic would nonetheless require a study of the relative public debates, notably the social movements of each country. In any event, we can deduce from the differences observed that the permeability of the parliamentary sphere to environmental concerns varied greatly from one country to another.

PART IV.

EUROPEANISING ENVIRONMENTAL POLICIES FROM BELOW?

CHAPTER 13.

RESPONDING TO THE EUROPEAN PUBLIC? PUBLIC DEBATES, SOCIETAL ACTORS AND THE EMERGENCE OF A EUROPEAN ENVIRONMENTAL POLICY

Jan-Henrik Meyer

'The public is the most important ally of the environment!'[1] Officials working in the European Commission's Service for the Environment and Consumer Protection (set up in 1973), and subsequently DG XI (Environment), such as the German official Ludwig Krämer quoted here, were acutely aware of the importance of the public for the new policy area. Indeed, debates in the media and scandals contributed decisively to the rise of environmental policy around the world in the early 1970s. From the late 1960s, environmental issues were increasingly present in the media in many European countries and the United States. Critical experts pointed to problematic developments, such as the deterioration of air and water quality; journalists increasingly specialising in the issue publicised what they described as the scandalous destruction of nature and the pollution of human environments. Sweden and the United States were pioneer countries in this respect.[2] Environmental groups such as the newly founded international non-governmental organisations (NGOs) Friends of the Earth, Greenpeace or the WWF (created by 1961) used the new awareness in the public sphere to campaign for environmental causes using emotionally compelling images of despoliation and animals suffering.[3] At the time, such campaigns often triggered concrete and specific environmental action.

1 Interview with Ludwig Krämer, former head of unit in the European Commission, conducted by Jan-Henrik Meyer, 19 Sept. 2017, Madrid.

2 Adam Rome, *The Genius of Earth Day: How a 1970 Teach-In Unexpectedly Made the First Green Generation* (New York: Hill & Wang, 2013); David Larsson Heidenblad, 'Mapping a new history of the ecological turn: The circulation of environmental knowledge in Sweden 1967', *Environment and History* **24** (2) (2018): 265–284.

3 Frank Zelko, *Make It a Green Peace!: The Rise of Countercultural Environmentalism* (Oxford: Oxford University Press, 2013). On Greenpeace's campaigns in Europe, see Anna-Katharina Wöbse, 'Greenpeace and the Brent Spar campaign. A platform for several truths', in Frank Uekötter (ed.), *Exploring Apocalyptica* (Pittsburgh: University of Pittsburgh Press, 2018), pp. 129–149; Liesbeth van de Grift, Hans Rodenburg and Guus Wieman, 'Milieuactivisten in maatpak: de Europeanisering van Greenpeace International (1987–1993)', *Tijdschrift Voor Geschiedenis* **130** (1) (2017):

The West German 'Waldsterben' debate of the early 1980s is a case in point. Cover stories conjuring up fear-inspiring visions of a treeless Germany put sufficient pressure on the national government to undertake a reform of emissions control legislation for large combustion plants. The public debate helped to overcome opposition from important economic interests and find sufficient political support to implement international commitments, notably the Long-Range Transboundary Air Pollution Treaty of 1979.[4] A technical solution was found for coal-fired power plants: scrubbers were installed to remove the sulphur dioxide from the exhaust. Furthermore, responding to the 'Waldsterben' debate, Christian democratic politicians fearful of the rise of the electorally successful Green Party also committed to introducing stricter limits on car emissions. Catalytic converters, a technological fix that major German car-makers were already using in automobiles produced for export to the United States, would help cleaning the exhaust. In order to avoid a disproportionate burden for domestic industry, the West German government sought to extend the obligation to use this technology to the entire European Communities (EC).[5] This example demonstrates that the impact of the public sphere on environmental policy could sometimes be indirect. National politicians 'uploaded' to the European level issues that were discussed primarily in national public spheres.

Assuming that a European public – directly – contributed to the rise of a European environmental policy seems counterintuitive at first sight. For a long time, one of the key tenets about the European Union (EU) (and its predecessor the EC), has been that it lacked a European Public Sphere. While deemed necessary as a counterpart to the emerging EU political system to enable truly democratic decision making at the European level, it simply did not seem to exist.[6] Even those researchers who have diagnosed a certain progress towards a more integrated, more active and more effective European public sphere have conceded that a European public sphere is indeed fragmented along national lines, 'segmented', uneven and asym-

83–100; Frank Zelko, 'The Umweltmulti arrives: Greenpeace and grass roots environmentalism in West Germany', *Australian Journal of Politics & History* **61** (3) (2015): 397–413.

4 Rachel Rothschild, 'Burning rain: The long-range transboundary air pollution project', in James Rodger Fleming and Ann Johnson (eds), *Toxic Airs: Chemical and Environmental Histories of the Atmosphere* (Pittsburgh: University of Pittsburgh Press, 2014), pp. 181–207.

5 Birgit Metzger, *'Erst stirbt der Wald, dann du!'. Das Waldsterben als westdeutsches Politikum (1978–1986)* (Frankfurt: Campus, 2015).

6 E.g. for the more sceptical view: Dieter Grimm, 'Does Europe need a constitution?' *European Law Journal* **1** (3) (1995): 282–302.

metric in its structures of transnational exchange. Compared to the national model it is much more episodic – i.e. responding to and thus 'existing' only around crucial events.[7]

Thus, the impact of such a weak and haphazard public sphere on a specific policy area, such as European (environmental) policy, would likely be very limited. If the EU indeed continued to be characterised by nationally segmented public spheres, any impact on European policymaking would have to go via national public spheres and member states' 'uploading' to Europe only those issues they cared about.[8]

This chapter seeks to explore the link between the (European) public sphere and European environmental policymaking. It argues that, despite the supposed deficits of a European public sphere, European publics indeed impacted on the rise and the contents of the new policy. They did so in different ways. The national route was only one of the channels available. In fact, if an issue was debated in public spheres in more than one of the member states this increased the likelihood that the issue would make it to the European policy agenda. The nascent European environmental movement was aware of this and cooperated and campaigned transnationally. The chapter also argues that the importance of the public sphere varied across different stages of the policy process – from agenda setting to policy implementation. Empirically, the chapter focuses on the emerging environmental policy of the 1970s. The chapter is organised as follows. First, given the controversy about the European public sphere, I will start with a conceptual clarification. Secondly, I will examine the contribution of the public sphere, its mediation into the institutional system of the EC/EU. For this purpose I will draw on three different cases of the emergent environmental policy of the EC

7 Jan-Henrik Meyer, *The European Public Sphere. Media and Transnational Communication in European Integration 1969–1991* (Stuttgart: Franz Steiner, 2010); Thomas Risse, 'European public spheres, the politicization of EU affairs, and its consequences', in Risse (ed.), *European Public Spheres: Politics Is Back* (Cambridge: Cambridge University Press, 2014), pp. 141–164; Risse, 'No demos? Identities and public spheres in the Euro crisis', *JCMS: Journal of Common Market Studies* **52** (6) (2014): 1207–1215; Ruud Koopmans and Paul Statham (eds), *The Making of a European Public Sphere. Media Discourse and Political Contention* (New York: Cambridge University Press, 2010); Stefanie Sifft et al., 'Segmented Europeanization: Exploring the legitimacy of the European Union from a public discourse perspective', *Journal of Common Market Studies* **45** (1) (2007): 127–155.

8 Conceptually, this would reflect a variant of the simple two-level game model that Andrew Moravcsik claimed to be the essence of EU policy making: national public spheres (instead of the business lobbies Moravcsik held to be all-powerful) impacting on national governments who in turn dominate European policy decisions. Andrew Moravcsik, 'Preferences and power in the European Community: A liberal intergovernmentalist approach', *JCMS: Journal of Common Market Studies* **31** (4) (1993): 473–524.

in the 1970s: first, the introduction of environmental policy between 1969 and 1973; second, the emergence of the EC birds directive; and, thirdly, the conflict at the EC level about the controversial issue of nuclear power. These cases demonstrate how the role of the public sphere varied with a view to its functions, its structure and its impact. Finally, I will assess the impact of the European public sphere in the 1970s on European (environmental) policymaking, as well as the functions it fulfilled.

What is a (European) public sphere?

Historians have frequently used a concept of the European public sphere that encompassed a wide range of cultural as well as political phenomena.[9] At its core, however, the notion of a public sphere is a political concept with strong normative connotations, dating back to the Enlightenment.[10] In the 1960s, the German political philosopher Jürgen Habermas re-inserted the concept into social scientific and political discourse by revisiting its eighteenth-century origins, albeit in a slightly idealised fashion.[11] According to the most widely accepted conceptualisation a public sphere designates a space between state and society, a sphere in which citizens can freely engage with and discuss political issues. From a normative point of view, the existence of such a sphere is indispensable for the functioning of democracy, because it ensures that citizens can exchange their views independently of the state. Only thus can they deliberate, develop and voice their preferences and form opinions. In a well-functioning public sphere, citizens are free to criticise their respective governments and hold them to account.[12]

Researchers have highlighted different functions the public sphere fulfils in democracy. The first and most fundamental function is to ensure transparency, to overcome official secrecy and put information out in the open. Transparency is a crucial precondition for opinion formation, which is the second function. Ideally, in forming opinions, citizens should freely exchange and

9 Robert Frank et al. (eds), *Building a European Public Sphere. From the 1950s to the Present. Un espace public européen en construction. Des années 1950 à nos jours* (Brussels: PIE Peter Lang, 2010).

10 Lucian Hölscher, 'Öffentlichkeit', in Otto Brunner, Werner Conze and Reinhardt Koselleck (eds), *Geschichtliche Grundbegriffe. Historisches Lexikon zur politisch-sozialen Sprache in Deutschland* (Stuttgart: Klett-Cotta, 1978), pp. 413–467.

11 Jürgen Habermas, *The Structural Transformation of the Public Sphere: an Inquiry into a Category of Bourgeois Society* (Cambridge, MA: MIT Press, 1989 [1962]). For a critical reflection see: Craig J. Calhoun (ed.), *Habermas and the Public Sphere* (Cambridge, MA: MIT Press, 1992).

12 Meyer, *The European Public* Sphere, pp. 24-26.

weigh their views in rational discourse. In real-world public spheres, however, rational arguments go hand in hand with emotions and stereotypes. Political actors and the media 'frame' issues to influence the public. Such aspects are crucially important in processes of opinion formation.[13] Thirdly, grievances, claims and demands discussed and assembled in the public sphere are to be inserted back into the political system. This 'input function' is crucial for political 'agenda-setting',[14] the first step in the policy process. At the same time, and this is a fourth function, the public sphere is to hold authorities and governments to account, evaluate and criticise their action and suggest alternative avenues. This function is essential for democracy as a system of checks and balances. A fifth function of the public sphere is its role in forming collective identities. This is also important for democracy, because collective identities underlie any political community. Scholars of identity and nationalism have demonstrated how public spheres shape the formation of imagined communities, instilling a sense of belonging and of solidarity.[15]

In terms of the scope of the public sphere, three levels can be distinguished analytically. First, and most simply, a public sphere of encounters, i.e. face-to-face and often dialogical exchanges between citizens. Such a sphere is very open to participation. Equality between participants is easy to establish, and it is usually very informal. At the same time, encounters rarely have an important impact on politics. Secondly, public spheres of assemblies are more organised and less egalitarian as to who is able to talk, and which issues are selected. Even in protest meetings, there is often a clear distinction between an 'elite' of speakers and audiences. On the other hand, assemblies still allow for some dialogue, covering specialist issues in a more discursive and in-depth manner than mediated debates. Assemblies may have a strong impact on politics. Events are often picked up by media and thus raise further attention in a broader public sphere. The third level, the public sphere of the media, reaches a large number of citizens, via printed or audiovisual media. However, mediated debates are most unequal, limit-

13 Falk Daviter, *Policy Framing in the European Union* (Houndmills: Palgrave, 2011); Juan Díez Medrano, *Framing Europe: Attitudes to European Integration in Germany, Spain, and the United Kingdom* (Princeton, NJ: Princeton UP, 2003).

14 Jan-Henrik Meyer, 'Getting started: Agenda-setting in European environmental policy in the 1970s', in Johnny Laursen (ed.), *The Institutions and Dynamics of the European Community, 1973–83* (Baden-Baden: Nomos, 2014), pp. 221–242; Sebastiaan Princen, 'Agenda-setting strategies in EU policy processes', *Journal of European Public Policy* **18** (7) (2011): 927–943.

15 Meyer, *The European Public Sphere*, pp. 52–57; Thomas Risse, *A Community of the Europeans: Transnational Identities and Public Spheres* (Ithaca: Cornell University Press, 2010).

ing most citizens to a passive recipient role. At least in the days before the internet, the number of speakers (and writers) was severely circumscribed. The selection of issues is left to professional journalists and media corporations (both public and private).[16]

The concept of the public sphere is an ambiguous one. It refers to empirical realities, but it is at the same time an analytical as well as a normative concept. Hence, we can analyse the public sphere on the one hand as a 'structure of communication', by studying media discourses, and on the other hand as a 'sphere of action', as a battleground between different actors struggling for attention and political influence.[17] This essay will consider both aspects. Emphasis will however be placed on the second aspect, namely by enquiring how political actors mobilised and used the (European) public sphere to advance (or limit) European environmental policy.

What makes a public sphere a European one? Three aspects have been highlighted. First, the 'Europeanness' of a public sphere can be defined by its contents, and by the political system it addresses. Any communication directly or indirectly referring to Europe and the EC/EU as a political system can pragmatically be considered part of a European public sphere, because it serves to fulfil political functions for European democracy.[18] Hence, according to this definition, all communication calling for or addressing European level environmental action would count as part of a European public sphere.

Secondly, a European public sphere in the sense of a transnational public sphere must be constituted by cross-border communication. It is a sphere of action in which actors from different European countries interact, discuss, struggle about (European) politics, ideally referring to each other. Thirdly, drawing on Habermas, a European public sphere can be defined as a sphere of synchronous and structurally similar communication across borders. A

16 Jürgen Gerhards and Friedhelm Neidhardt, 'Strukturen und Funktionen moderner Öffentlichkeit. Fragestellungen und Ansätze', in Stefan Müller-Dohm and Klaus Neumann-Braun (eds), Öffentlichkeit, Kultur, Massenkommunikation. Beiträge zur Medienkommunikationssoziologie (Oldenburg: BIS, Bibliotheks- u. Informationssystem der Universität Oldenburg, 1991), pp. 31–89, here 50–56; Meyer, *The European Public Sphere*, pp. 54–55.

17 Bernhard Peters, 'Der Sinn von Öffentlichkeit' *Kölner Zeitschrift für Soziologie und Sozialpsychologie. Öffentlichkeit, öffentliche Meinung, soziale Bewegungen. Sonderheft* **34** (1994): 42–76, here 50–56.

18 This understanding underlies the most recent research projects on the issue, which all draw on key-word searches of large electronic databases: Ariane Brill, *Abgrenzung und Hoffnung: 'Europa' in der deutschen, britischen und amerikanischen Presse 1945–1980* (Göttingen: Wallstein, 2014); Enrico Bergamini et al., Talking about Europe: Le Monde 1944–2018 (2019) Bruegel, http://bruegel.org/2019/03/talking-about-europe-le-monde-1944-2018/# (accessed 25 Mar. 2019).

European public sphere in this sense discusses 'the same issues at the same time using the same criteria of relevance'.[19]

All these criteria seek to conceptualise a European public sphere as a coherent whole. Many researchers have refuted claims that European democracy is impossible without a European public sphere, and that a European public sphere is impossible because of the lack of a common language.[20] Thus they have attempted to demonstrate that indeed there is a European public sphere mirroring the European Union's political system, cutting across linguistic divisions and across nationally organised media systems. Some researchers have argued that one should rather speak of European public spheres in the plural. Varying 'issue publics' are referring to different political topics or problems. They are composed of different actors, and characterised by different media. They may exist – to a varying degree – in different countries.[21] For the purpose of this study, such distinctions are less relevant. Nevertheless, it is worthwhile to use these distinctions to explore the scope and structural features of the European public sphere dealing with environmental issues.

Subsequently, I will explore how different aspects of a European public sphere impacted on the nascent environmental policy of the EC/EU in the 1970s: first, the emergence of the policy in the early 1970s; secondly, the creation of the birds directive as a first element of policymaking in the area of nature conservation, an issue previously considered the prerogative of the Council of Europe; and thirdly, nuclear policy, an issue that was forced into the public sphere by a transnational alliance of anti-nuclear activists, against the backdrop of protest and increasingly critical reporting in the media. These cases not only provide a glimpse into the origins and early development of EC environment policy. They also differ in many ways with a view to the role of a European public sphere.

19 Meyer, *The European Public Sphere*, pp. 26–28, 64–65; quote: Klaus Eder and Cathleen Kantner, 'Transnationale Resonanzstrukturen in Europa. Eine Kritik der Rede vom Öffentlichkeitsdefizit' in Maurizio Bach (ed.), *Die Europäisierung nationaler Gesellschaften* (Opladen: Westdeutscher Verlag, 2000), pp. 306–331, at 315.

20 Grimm, 'Does Europe need a constitution?'

21 W. Lance Bennett, Sabine Lang and Alexandra Segerberg, 'European issue publics online: the cases of climate change and fair trade', in Thomas Risse (ed.), *European Public Spheres: Politics Is Back* (Cambridge: Cambridge University Press, 2014), pp. 108–138.

Transnational public spheres and the creation of a European Environmental Policy

Most textbook introductions to the EC / EU's environmental law and policy attribute the origins of the new policy to what supposedly was an initiative by the heads of state and government meeting in October 1972, when at the Paris summit they gave the go-ahead for the start of an environmental policy.[22] The policy eventually took shape in the first Environmental Action Programme issued in November 1973. Formally, this claim is correct, and reflects the wording of the official documents. Nevertheless, the timing in the aftermath of the Stockholm UN Conference on the Human Environment in 1972 suggests that the heads of state and government responded to broader political debates in a transnational, perhaps global, public sphere.

In fact the preparations for a European environmental policy go almost three years further back, and they started with a transnational scandal. On 19 June 1969, a massive fish kill happened in the Rhine. The river was not only Western Europe's main artery for transport, but also its largest sewer, transporting only partially treated wastewater from Switzerland, West Germany and Eastern France into the Netherlands, where it meets the sea. Being situated downstream, Dutch citizens, agriculture and industry relied on the river for fresh water. The fish kill was caused by the Frankfurt chemical works Hoechst. On a regular basis, and apparently tolerated by the authorities, Hoechst emitted residues of a highly potent insecticide into the river Main, a tributary to the Rhine. In June 1969, low water levels limited the river's capacity to sufficiently dilute this pollution, so that the fish in the river were poisoned and killed. The Dutch authorities found the cause of the pollution relatively quickly, but they had not been warned by their German counterparts upstream.[23]

This contributed to cross-border resentment and public debate on the issue in the countries along the Rhine, notably the Netherlands and West

22 E.g. Stibbe Simont and Monahan Duhot, *Environment and Europe. European Union Environmental Law and Policy and its Impact on Industry* (Deventer: Kluwer, 1994), p. 1; Tom Delreux and Sander Happaerts, *Environmental Policy and Politics in the European Union* (London: Palgrave, 2016), p. 18.

23 Mark Cioc, *The Rhine. An Eco-Biography, 1815–2000* (Seattle: University of Washington Press, 2002), p. 141; idem, 'Europe's river. The Rhine as a prelude to transnational cooperation and the Common Market', in Erika Marie Bsumek, David Kinkela and Mark Atwood Lawrence (eds), *Nation-States and the Global Environment. New Approaches to International Environmental History* (Oxford: Oxford University Press, 2013), pp. 25–42; Christoph Bernhardt, *Im Spiegel des Wassers. Eine transnationale Umweltgeschichte des Oberrheins (1800–2000)* (Cologne: Böhlau, 2016).

Germany.[24] Dutch members of the European Parliament (EP) raised the issue in the Strasbourg assembly. They convinced the EP's Committee on Public Health and Social Affairs to produce an own initiative report, one of the key instruments the unelected assembly had. Even if the EP had no meaningful say in the process of European law-making, it was able to use this instrument to engage in agenda-setting.[25] Against the backdrop of international efforts – notably in the United States, and in various international organisations – to promote environmental policy,[26] and of environmental scandals – such as the *Torrey Canyon* oil spill of 1967[27] – MEPs used this incident to raise a concrete environmental issue. The Rhine was a highly appropriate starting point for such an attempt, because it demonstrated the transnational interdependence and interconnectedness of Europe's economies, ecosystems and livelihoods.[28] In that sense, the Rhine was also the prototype of a European river: except for Italy, all the EC's founding members shared parts of its floodplain. Hence, raising the issue in a European forum seemed more than appropriate. Expanding the issue in the report from the pollution of the Rhine to river pollution more generally was part of the agenda setting that the European Parliament undertook.[29]

The rapporteur, the Dutch Christian Democratic labour unionist Jacob Boersma emphasised the role of the public by inserting a subsection on 'The response of the public to the growing pollution of rivers', which starts out with a description of scandal of 1969, the Thiodan-induced fish kill. The content

24 E.g. NN, 'Flüsse. Rheinvergiftung. Nur ein Sterben', *Der Spiegel* 30 June 1969, 65-66; Sepp Binder, 'Die Ratten verließen den Rhein. Gift in Deutschlands größter Kloake – tote Fische in Richtung Holland', *Die Zeit* 4 July 1969.

25 Meyer, 'Getting started'.

26 Meyer, Appropriating the Environment. How the European Institutions Received the Novel Idea of the Environment and Made it Their Own, *KFG 'The Transformative Power of Europe' Working Paper*: 31 (2011), 1–33, http://userpage.fu-berlin.de/kfgeu/kfgwp/wpseries/WorkingPaperKFG_31.pdf (accessed 25 Mar. 2019); Evanthis Hatzivassiliou, *The NATO Committee on the Challenges of Modern Society, 1969–1975. Transatlantic Relations, the Cold War and the Environment* (London: Palgrave, 2017). See also contributions in Wolfram Kaiser and Jan-Henrik Meyer (eds), *International Organizations and Environmental Protection. Conservation and Globalization in the Twentieth Century* (New York: Berghahn, 2017).

27 Timothy Cooper and Anna Green, 'The Torrey Canyon disaster, everyday life, and the "greening" of Britain', *Environmental History* **22** (1) (2017): 101–126.

28 Cioc, 'Europe's river'.

29 Jacob Boersma, 'Bericht im Namen des Ausschusses für Sozial- und Gesundheitsfragen über die Reinhaltung der Binnengewässer unter besonderer Berücksichtigung der Verunreinigung des Rheins, 11 November 1970', Historical Archives of the European Parliament (HAEP) PEO-AP RP/ASOC.1967 AO-0161/70 (1970).

and wording is remarkably similar to German newspaper reports on the issue from the summer of 1969.[30] Clearly, the rapporteur was aware of the version of the scandal presented in European news media.[31] Indeed, in the subsequent paragraph that stressed the urgency of taking action, Boersma actually invoked the 'European public sphere'. 'The European public sphere was deeply concerned about this catastrophe and voiced its concern about the possibility that accidents of this kind may happen again – at an even larger scale'.[32] At a factual level, such a usage of the term 'European public sphere' describes an actual cross-border sphere of communication, covering those affected by the event. At the same time, it plays on a familiar rhetorical tradition. For the past 200 years journalists and dissident writers have used rhetorical appeals to a 'European public sphere' to scandalise ethically problematic behaviour, as Jörg Requate and Martin Schulze-Wessel have demonstrated.[33]

Furthermore, Boersma's report points to the immediate reaction by the Council of Europe, which drew up a response within a week of the scandal, on 25 June 1969. Such arguments played on the growing competition among international organisations, which at the time tried to stake their claims in the emerging new policy area of the environment.[34] The report also relayed the demands of one of the few European-level societal actors in the European public sphere existing at the time.[35] In the aftermath of the accident, the European consumers' association BEUC (Bureau Européen des Unions de Consommateurs, founded in 1962),[36] had called on the member

30 E.g. NN, 'Flüsse. Rheinvergiftung. Nur ein Sterben'; Binder, 'Die Ratten verließen den Rhein'.

31 Boersma, 'Bericht im Namen des Ausschusses für Sozial- und Gesundheitsfragen über die Reinhaltung der Binnengewässer unter besonderer Berücksichtigung der Verunreinigung des Rheins, 11 November 1970', §28.

32 Ibid., § 29. Translated by the author, here and in the following.

33 On the traditions of the rhetorical invocation of a European Public Sphere as a 'court of appeal', see Jörg Requate and Martin Schulze-Wessel, 'Europäische Öffentlichkeit. Realität und Imagination einer appellativen Instanz', in Requate and Schulze-Wessel (eds), *Europäische Öffentlichkeit. Transnationale Kommunikation seit dem 18. Jahrhundert* (Frankfurt: Campus, 2002), pp. 11–39.

34 Jan-Henrik Meyer, 'Who should pay for pollution? The OECD, the European Communities and the emergence of environmental policy in the early 1970s', *European Review of History: Revue européenne d'histoire* **24** (3) (2017): 377–398.

35 On the concept of societal actors, see Wolfram Kaiser and Jan-Henrik Meyer, 'Beyond governments and supranational institutions. Societal actors in European integration', in Kaiser and Meyer (eds), *Societal Actors in European Integration. Polity-Building and Policy-Making 1958–1992* (Basingstoke: Palgrave, 2013), pp. 1–14.

36 On the consumer associations' activities in the public sphere in the 1970s, see Liesbeth van de Grift, 'Representing European society. The rise of new representative claims in 1970s European politics', *Archiv für Sozialgeschichte* **58** (2018): 263–278.

states to establish a supranational authority to control water pollution and prevent future accidents.[37]

The own initiative report on water pollution and the Rhine of 1970 was followed by another report on air pollution in 1971.[38] Connecting the European institutions to a transnational public sphere of the media on concrete environmental issues, the EP inserted the issue into the EC political process, encouraging the European Commission – the sole proposer of European law – to take action. Given its limited role in law making, the EP itself functioned as a public sphere of assembly, acting as a (self-declared) proxy for the European people. Through its deliberations, it raised attention and pushed the Commission to act, and it used various references to voices and reports in the European public sphere as an argument for European policy action.

Indeed, the Commission subsequently produced a first proposal for environmental policy in 1971. The so-called 'First Communication of the Commission about the Community's Policy on the Environment' picked up the concrete focus on water and air pollution the EP had flagged.[39] At the end of the document's introductory section, which described the issues to be covered by the new policy, the Commission highlighted the urgency of combating the pollution of rivers, notably the Rhine. This demonstrates the relevance of this issue in a European public sphere, and the EP's intermediary role in relaying it to the Commission: 'Lastly, problems like the cleaning-up of rivers and waterways, e.g., the Rhine and its tributaries, passing through a number of member states, ... are immediate and urgent.'[40] Subsequently, the EP commented thoroughly on the Commission's subsequent proposals in 1972[41] and 1973[42] with reports, discussions in relevant committees and in the plenary.

37 Boersma, 'Bericht im Namen des Ausschusses für Sozial- und Gesundheitsfragen über die Reinhaltung der Binnengewässer unter besonderer Berücksichtigung der Verunreinigung des Rheins, 11 November 1970', § 30.

38 Hans Edgard Jahn, 'Bericht im Auftrag des Ausschusses für Sozial- und Gesundheitsfragen über die Notwendigkeit einer Gemeinschaftsaktion zur Reinhaltung der Luft, 15.12.1971', HAEP PE0 AP RP ASOC.1967 0181/71 (1971).

39 European Commission, First Communication of the Commission about the Community's Policy on the Environment. SEC (71) 2616 final, 22 July 1971, Archive of European Integration, http://aei.pitt.edu/3126/1/3126.pdf 12–13 (accessed 26 Mar. 2019).

40 Ibid., 5.

41 European Commission, 'Communication from the Commission to the Council on a European Communities' programme concerning the environment (submitted on 24 March 1972)', *Bulletin of the European Communities. Supplement* **5** (5) (1972): 1–69.

42 Commission of the European Communities, Programme of Environmental Action of the European Communities. Part II: Detailed description of the actions to be undertaken at Community level

'Save the birds' – The European Public Sphere and the birds directive of 1979

The birds directive of 1979 established strict European rules against the hunting of songbirds and introduced the protection of birds' habitats.[43] The European public sphere played an important role in the making of this first concrete project of EC nature protection policy, in two respects: First, references to the 'public' served as an *argument* for supranational institutions, the EP and the European Commission, to argue in favour of establishing bird protection at the European level. Secondly, environmental protest in the European public sphere as a '*sphere of action*'[44] clearly mattered in particular to supranational European policymakers. In the 1970s, the European institutions were intent on bringing Europe closer to citizens, and on demonstrating the 'added value' of European integration.[45] Campaigns in the media and actions of societal actors such as environmental groups proved crucial in mobilising sufficient political support to actually enact relevant environmental legislation.

Calls for EC action on the protection of migrant birds against hunting predated the establishment of the EC environmental policy. In December 1967, Hans Richarts, a German Christian Democrat from Trier and long-serving MEP (1958–1973), raised the issue of nationally different legislation on animal and bird protection and bird hunting. He pointed to 'the interest' this 'aroused' within a 'broader public sphere', and to media reporting about the mass hunting of songbirds for human consumption in some member states. This, he suggested, undermined the efforts at strict protection of these same bird species in other member states.[46] In 1967/68, before the international breakthrough towards the environment as a political issue, the Commission was not convinced that this was a concern for EC political action. Responding to Richarts' question in March 1968, the Commission suggested that it was neither competent to act on the issue nor entitled to

over the next two years. Forwarded by the Commission to the Council, COM (73) 530 final C, 10 Apr. 1973, 1973, http://aei.pitt.edu/5451/01/001084_1.pdf (accessed 11 Mar. 2012).

43 On the origins of the birds directive, see Jan-Henrik Meyer, 'Saving migrants. A transnational network supporting supranational bird protection policy in the 1970s', in Wolfram Kaiser, Brigitte Leucht and Michael Gehler (eds), *Transnational Networks in Regional Integration. Governing Europe 1945–83* (Basingstoke: Palgrave, 2010), pp. 176–198; Andrew L.R. Jackson, *Conserving Europe's Wildlife, Law and Policy of the Natura 2000 Network of Protected Areas* (Abingdon: Routledge, 2018).

44 Peters, 'Der Sinn von Öffentlichkeit', 50–56.

45 Grift, 'Representing European society', 268.

46 Hans Richarts, 'Written Question No. 254/67, 11 December 1967, to the Commission concerning the harmonisation of rules for bird protection', HAEP PE0 AP QP/QE E-0254/67 (1967).

recommend to the member states to harmonise their respective laws.[47]

However, in April 1973, when the Commission presented its draft of an environmental action programme, it justified the inclusion of bird protection by pointing to 'worldwide protests against the countries which allow the trapping of birds', and promised to promote 'joint action by the member states of the Council of Europe and other international organisations' and to study the possibility of harmonising bird protection legislation by the end of 1974.[48] This change of mind may also be attributed to a barrage of Parliamentary questions in 1971–1973 by various MEPs, who urged the EC to take action. In their questions, MEPs explicitly asked for EC action and inclusion of the issue into the Environmental Action Programme. For instance, the German Christian Democrat Hans Edgar Jahn,[49] who had already served as rapporteur on the issue of air pollution and the early Commission proposals, backed up his demand for EC action by reference to 'worldwide protest activities' by 'animal protection groups and action committees' against the killing of '200 million birds in Italy'.[50] At the time, when the environment was still a nascent policy area, the Commission was particularly receptive to ideas for European action.

The presence of the bird hunting issue in transnational public spheres was clearly relevant for the inclusion of bird protection in the Environmental Action Programme. However, it was the activities of societal actors in the European public sphere that convinced the Commission to draft a concrete legal proposal and to eventually getting it into the statute book. Indeed, in the course of the 1970s, transnational networks of bird protection activists cooperating with the European institutions managed to successfully push for a birds directive. Their action involved both cooperating with European institutions and organising protest campaigns in the public sphere.

47 European Commission, 'Answer to Written Question No. 254/67, 1 March 1968, by Mr Richarts, concerning the harmonisation of rules for bird protection', HAEP PE0 AP QP/QE E-0254/67 (1968).

48 Programme of Environmental Action of the European Communities. Part II: Detailed description of the actions to be undertaken at Community level over the next two years. Forwarded by the Commission to the Council, COM (73) 530 final C, 10 Apr.1973, §II.67–68.

49 On Jahn, see Jan-Henrik Meyer, 'A good European. Hans Edgar Jahn – anti-Bolshevist, Cold-Warrior, environmentalist', in Ann-Christina L. Knudsen and Karen Gram-Skjoldager (eds), *Living Political Biography. Narrating 20th Century European Lives* (Aarhus: Aarhus University Press, 2012), pp. 137–159.

50 Hans Edgar Jahn, 'Written Question No. 620/72, 15 February 1973, concerning mass killing of migratory birds in Italy', *Official Journal of the European Communities* **16**: C 39, 7.6.1973 (1972): 12.

This happened in three steps, in three different contexts.[51] First, an incipient network of radical bird protection groups, led by the Dutch Stichting Mondiaal Alternatif, sent a pètition to various international organisations, including the EP. This encouraged Jahn, the deputy chair of the relevant EP committee to produce an own initiative report demanding EC bird protection legislation.

Secondly, informally collaborating with various MEPs, who continued to submit parliamentary questions, bird protection groups engaged in public relations activities, press conferences, media work and organised letter-writing campaigns. When the Commission – after consulting with experts, many of whom were themselves members of traditional bird protection groups such as the British Royal Society for the Protection of Birds (RSPB) – actually submitted a proposal for legislation, the officials justified this step by referring to the about 50,000 letters the Commission had apparently received on the issue.[52]

Thirdly, when the issue lingered in Council, where the proposal required unanimous support of the member state governments, bird protection groups established a European network, the Working Group of Bird Protection Groups (WEBS), in order to better coordinate their work at the national levels. Apart from occasional direct lobbying of governments, notably in the UK where the large and influential RSPB enjoyed access to government, bird protection activists mainly acted via the public sphere. Again, they used letter-writing campaigns – this time to the Council – and cooperated with journalists and media at national levels. For instance, after its most well-known leader, Zoo director and TV show host Bernhard Grzimek had already been involved as an expert in the European bird protection policy project, the first campaign of the newly established West German environmental group BUND (Bund für Umwelt und Naturschutz Deutschland) was called 'Save the Birds'. For this purpose, they collaborated with a widely distributed popular weekly magazine to further press the salience of the issue in the (national) public sphere. These activities took place in a transnational European public sphere as a sphere of action, in which environmental groups tried to promote the same issue at the same time 'using the same criteria of relevance',[53] and clearly contributed to the eventual adoption of the birds directive.[54]

51 For further detail and full references, see Meyer, 'Saving migrants'.

52 Ibid., 185.

53 Eder and Kantner, 'Transnationale Resonanzstrukturen in Europa', 315.

54 Meyer, 'Saving migrants', 188–190.

Nuclear power – Demanding transparency and accountability

The nuclear issue, which was arguably the most contentious environmental issue in the public sphere across Europe in the 1970s,[55] did not feature very prominently in the EC's environmental policy, not least because it fell within the scope of the separate Euratom Treaty. Nevertheless, the issue was present at the European level, right from the start. This reflects the close connection between the European public sphere and European institutions, mediated via various channels.

The nuclear issue already featured in the EP's first report on water pollution and the Rhine, mentioned above. In 1970, when plans for the siting of nuclear installations in many countries became public, the first environmental critique of nuclear plants, however, did not concern radiation, but thermal pollution of rivers. Along the river Rhine, Swiss, French and West German utilities competed for the ample cooling water the large river provided.[56] The 1970 EP report thus flagged the need to address the transnational issue of thermal pollution. The wording of the EP report closely mirrors media reports published in early 1970, which back up this concern with recent research in Germany and experiences from the United States.[57] The issue of thermal pollution continued to feature in discussions by different EC institutions during the 1970s about the need for European rules for the siting of nuclear power plants at intra-Community borders.[58]

In 1975, protests around the nuclear power plant at Wyhl on the Franco-German border, involving participants from the neighbouring Alsace, Switzerland, and West Germany, not only kicked off a series of massive protests

55 Helmuth Trischler and Robert Bud, 'Public technology: Nuclear energy in Europe', *History and Technology* **34** (3–4) (2019): 187–212, here 199–200.

56 Arne Kaijser and Jan-Henrik Meyer, 'Nuclear installations at the border. Transnational connections and international implications. An introduction', *Journal for the History of Environment and Society* **3** (2018): 1–32.

57 Spiegel, 'Tod im Strom. Industrie Kernkraftwerke', *Der Spiegel* 23 Feb. 1970, 46; Theo Löbsack, 'Wenn der Rhein dampft. Zu den geplanten Atommeilern darf nicht geschwiegen werden', *Die Zeit* 24 Apr. 1970, 67; Boersma, 'Bericht im Namen des Ausschusses für Sozial- und Gesundheitsfragen über die Reinhaltung der Binnengewässer unter besonderer Berücksichtigung der Verunreinigung des Rheins, 11 November 1970', 4 §12; 6–7 §11.

58 E.g. Hanna Walz, 'Report drawn up on behalf of the Committee on Energy, Research and Technology of the European Parliament on the Conditions for a Community Policy on the Siting of Nuclear Power Stations taking account of their Acceptability for the Population, doc. 392/75, 26 November 1975', HAEP PE 40.985/fin (1975).

against nuclear power plants in West Germany and France.[59] To a group of young pro-European left-wingers, such transnational protest appeared like the embodiment of the European spirit. Wyhl seemed to epitomise a Europe from below. It looked very different from the Europe of big business the EC seemed to represent and that they resented.[60] Among these young people were members of the Young European Federalists (Junge Europäische Föderalisten, JEF) including the young German Social Democrat Petra Kelly, who worked as an official in the secretariat of the EC's Economic and Social Committee. She was to become one of the founders of the German Green Party and lead candidate in the European elections of 1979.[61] Among them was also Jo Leinen, a young lawyer and alumnus of the College of Europe, who was working for the German Young Socialists at the time. Unlike Kelly, he stayed with the Social Democrats, and served as chairman of the EP's environmental committee until 2014; he continued to be a member until 2019.

From 1974 both of them had collaborated on JEF's magazine *Forum Europa*, which devoted much attention to the new policy issues of the 1970s, such as the environment and nuclear power. JEF organised events to address the lack of a 'democratic European public' and sought to help construct such a space via their own publications and events.[62] In particular, they aimed at encouraging a more open debate on nuclear energy also at the level of the EC, along the lines of public hearings and debates conducted in various European countries at the time.[63] They reasoned that international organisations, and in particular the EC, were important promoters of nuclear energy, through Euratom's research programmes and assistance in funding nuclear projects.

59 Andrew Tompkins, *Better Active than Radioactive! Anti-nuclear Protests in 1970s France and West Germany* (Oxford: Oxford University Press, 2016); Stephen Milder, *Greening Democracy. The Anti-Nuclear Movement and Political Environmentalism in West Germany and Beyond, 1968–1983* (Cambridge: Cambridge University Press, 2017).

60 For a more detailed account, with full references, see Jan-Henrik Meyer, '"Where do we go from Wyhl?" Transnational anti-nuclear protest targeting European and international organisations in the 1970s', *Historical Social Research* **39** (1) (2014): 212–235.

61 Saskia Richter, *Die Aktivistin. Das Leben der Petra Kelly* (Munich: DVA, 2010).

62 'Special Issue: Atomenergie: politischer und sozialer Sprengstoff', *Forum Europa Zeitschrift für transnationale Politik* **6** (3–4) (1976): 2; Josef M. Leinen, 'Protokoll des Forum Europa-Seminares zu dem Thema "Europa ohne demokratische Öffentlichkeit" vom 14./15. Januar in Bonn, Hotel Eden, Teilnehmerliste', *Archiv Grünes Gedächtnis (AGG)* PKA: 2249 (1976): 1–5; 'Umweltschutz Themenheft', *Forum E Bulletin der Jungen Europäischen Föderalisten* **3** (3) (1972).

63 E.g. in Denmark, Jan-Henrik Meyer '"Atomkraft – Nej tak". How Denmark did not introduce commercial nuclear power plants', in Astrid Mignon Kirchhof (ed.) *Pathways Into and Out of Nuclear Power in Western Europe: Austria, Denmark, Federal Republic of Germany, Italy, and Sweden* (Munich: Deutsches Museum 2020), pp. 74–123, here pp. 94–99.

Indeed, in debates on nuclear power in the EP in the mid-1970s, MEPs almost unanimously perceived the expansion of nuclear energy as the only possible response to the oil crisis and ever-growing energy consumption, and routinely disparaged critics of nuclear power as 'motivated by emotional irrationality'.[64]

Collaborating within the European Environmental Bureau (EEB), the federation of environmental groups in Brussels, and with the Brussels-based Euro-federalist group 'agenor', of which they both were members, the JEF first attempted to organise 'hearings' on nuclear energy in Brussels. Their goal was to encourage a more critical public debate about the risks of radioactivity, the impact of nuclear power on the environment, economic, political and ethical issues.[65] However, it proved difficult to convince pro-nuclear politicians and experts to attend an event organised by a civil society group consisting of mainly young people with anti-nuclear credentials. Commissioner for Energy Henri Simonet, who had promised his participation, apparently declined to attend at the very last minute.[66]

The JEF and agenor activists changed strategy. They addressed the new Commissioner for Energy and Research Guido Brunner, a member of the German Free Democrats. At the time, the German liberals presented themselves as the party of the environment. After all, as minister of the interior, party leader Hans-Dietrich Genscher had introduced environmental policy in Germany in the early 1970s, and claimed to have coined the German word term 'Umweltschutz' (i.e. environmental protection).[67] JEF activist John Lambert, who had organised the agenor hearings in Brussels, not only managed to persuade Brunner to hold 'Open Debates on Nuclear Energy' but also instructed Brunner's cabinet on how best to organise such an event and how to achieve maximum credibility and impact, notably by publishing the

64 In the debate on the report by Hanna Walz on the Siting of Nuclear Power Stations (see above), only the Scottish Labour MEP William Winter Hamilton voiced his concern about leaving behind large quantities of nuclear waste to subsequent generations. Hamilton rejected Walz' suggestion that critics of nuclear power were 'motivated by emotional irrationality', believing that they were 'highly qualified academics, highly qualified scientists, highly qualified technologists' (p. 64). Mr Hamilton, 'Speech in European Parliament, 13 January 1976, on Community Policy on the Siting of Nuclear Power Stations', *Official Journal of the European Communities, Annex: Proceedings of the European Parliament* January 1976: 63–64.

65 Agenor, 'Europäische Hearings und Arbeitsgruppen über Atomenergie, Brüssel, 5.-8. November 1975, veranstaltet von Agenor', *Archiv Grünes Gedächtnis (AGG)* Petra Kelly Archiv: 1913 (1975).

66 Agenor, 'Hearings report', *Agenor* **58** (1976): 1–4, at 2.

67 Hans-Dietrich Genscher, *Erinnerungen* (Berlin: Siedler, 1995), pp. 125–138.

results.[68] Brunner's cabinet collected information on national experiences.[69] Brunner managed to convince his fellow Commissioners and the reluctant member states to organise such hearings. They eventually took place in two sessions from 29 November to 1 December 1977 and from 24 to 26 January 1978 at the exhibition centre Heysel in Brussels underneath the iconic Atomium, with Brunner in the chair.

Cooperation between JEF/agenor and Brunner's cabinet on the organisation of the event did not preclude continued controversy on substance. Even if the Commission invited the EEB to suggest suitable counter-experts with anti-nuclear credentials, this did not stop the EEB and its member organisations from publishing critical press releases.[70] In any case, apart from functioning as a European public sphere of an assembly, the event drew substantial attention of an – albeit temporary – European public sphere of the media to European aspects of nuclear energy and the environment. Numerous newspaper articles on the event collected by the European Commission provide evidence of this.[71] Furthermore, the event provided an opportunity for environmental and anti-nuclear organisations from all over Europe, such as the Danish Organisation for nuclear information (OOA)[72] or the German Federation of Citizen Action Groups (BBU), to reach out to this European public sphere with their own messages. They were routinely critical of the event as well as of the Commission's generally pro-nuclear stance. Thus they provided both a structure of communication and a sphere of action at the same time.[73]

68 John Lambert, 'For Guido Brunner: Thoughts on Commission hearings about energy policy, Group Agenor, Brussels, 25 January 1977', *Historical Archives of the European Commission* (HAEC) BAC 144/1987,254 (1977): 84–86. The results were actually published in 1978 as European Commission, *Open Discussions on Nuclear Energy. Held by the European Commission, Brussels, 29/11-1/12/1977 and 24-26/1/1978* (Luxembourg: Office for Official Publications of the European Communities, 1978).

69 E.g. Helmut Hirsch and Helga Nowotny, 'Europe's nuclear debate (1): Austria: a case study. Helmut Hirsch and Helga Nowotny assess Austria's nuclear energy information campaign. Nature Vol. 266, 10 March 1977, 107–108', HAEC BAC 144/1987, 254 (1977): 186–187.

70 EEB, 'Various press releases by the European Environmental Bureau', HAEC BAC 144/1987, 255 (1977-1978): 248-258, 264-265; id., 'Response to the communication on the conclusions drawn by the Commission from the public debates on nuclear energy, Brussels, 14 July 1978', HAEC BAC 144 1985: 250 (1978): 75–77.

71 European Commission, Cabinet Brunner, 'Presse Nuklear-Hearings, 29.11.-1.12.1977, 24.-26.1.1978', HAEC BAC 144/1987, 255 (1977–1978): 16–241.

72 On the OOA, see Meyer '"Atomkraft – Nej tak"', pp. 82–84.

73 BBU, '"Energie-Debatte: Öffentliche Show ohne Konsequenzen oder Beginn einer energiepolitischen Neuorientierung?" Press release by Bundesverband Bürgerinitiativen Umweltschutz (BBU), 1 December 1977', HAEC BAC 144/1987, 255 (1977): 267–268; OOA, 'Press release by the Organisationen til Oplysning om Atomkraft, 1 December 1977', HAEC BAC 144/1987, 266.

This chapter has examined how and to what extent a (European) public sphere was relevant for the emergence and development of a European, that is, EC/EU level, environmental policy. It studied the interaction between debates in the media and the public sphere and the development of European environmental policy and certain environmental measures. Three empirical examples were discussed, which differed with a view to the role of the public sphere.

First, the start of an EC environmental policy can be viewed as a response to a wider transnational public sphere in which environmental issues as issues of public policy increasingly featured at the same time 'using the same criteria of relevance'. The EP proved an important mediator between these transnational and international debates on environmental issues, which often reflected very concrete local impacts, such as the quality of the water of the Rhine in the Netherlands. The EP – itself a European public sphere of an assembly – also fulfilled some of the functions of the public sphere, trying to hold the EC institutions to account.

This role of the EP was very similar in the second case, on bird protection. This issue was discussed in a more fragmented European public sphere: at the same time but not necessary 'using the same criteria of relevance'. In some countries – the Netherlands, West Germany and the UK – bird protection and the fight against bird hunting were much more important. These issues were viewed very differently in Italy or France, where trapping and hunting small birds was not frowned upon, but considered part of local traditions. Italian bird protection activists however, were important players in a transnational public sphere – in the sense of a sphere of action. Together with their partner groups in other EC member states, they played a crucial role in scandalising the issue, using the public sphere to convince national and EC policy makers to enact European legislation. Such pressure – increasingly coordinated across borders – contributed to successfully guiding the issue through the legislative process,which at the time required a substantial threshold, namely the unanimous support of member state governments.

Thirdly, on the environmental impacts of nuclear installations again the European Parliament proved an important mediator of public debates into the Brussels institutions. The EP proved much less receptive to the controversies on nuclear power than to the less divisive bird issue. Most MEPs remained committed to the pro-nuclear elite consensus, even if members worried about the social acceptability of what they considered an indispensable energy source. Nevertheless, through direct lobbying, pro-European anti-nuclear activists convinced the European Commission to hold 'Open Discussions on Nuclear Energy' in 1977/78. Thus they helped create a European public sphere of an assembly which intensely and controversially discussed the issue beyond the usual experts' forums, thus contributing to transparency, opinion-formation and accountability. Even if the wider impact on a broader European public sphere was temporary, this was an important achievement in the face of a clear pro-nuclear consensus within the institutions.

Thus, we can conclude that the European public sphere mattered very concretely with a view to European environmental policymaking – from agenda setting to policy implementation. It even instigated controversial 'open discussions' in a European public sphere on an policy that the EC was committed to by the Euratom Treaty. Thus this concrete study of the role of the European public sphere in policymaking arrives at different results from what media studies found regarding the European public sphere. While media studies have stressed the incipient and fragmented nature, we can conclude that the European public sphere on the environment was able to fulfil some of its most important functions. The European public sphere clearly had an impact on the rise and shaping of European environmental policy. Societal actors, notably environmental groups, used and thereby constructed the European public sphere as a 'sphere of action' and used it in their favour.

The European public sphere on the environment also engaged in European identity formation. Remarkably, most of the voices in the debates in a European public sphere on environmental issues invoked and constructed a European identity. This is also reflected in the various emphatic rhetorical invocations of a European public sphere as a normative benchmark or authority. Three reasons may account for this predominance of European identity constructions: First, the self-selection of those civil society groups, societal and institutional actors who were involved and interested in a European environmental policy; second, for strategic reasons, when demanding European policy action, it made sense to talk the European talk and appeal to familiar federalist convictions shared by those working for the EC institutions. This, thirdly, was clearly more appealing at the time than today. In the 1970s, European policy was considered weak, almost insignificant. European Union was a vision, not a reality. At the time of the supposed 'permissive consensus',[74] demands for more Europe, and emphatic constructions of European identity, were much less controversial than today.[75]

74 Leon N. Lindberg and Stuart A. Scheingold, *Europe's Would-be Polity. Patterns of Change in the European Community* (Englewood Cliffs, NJ: Prentice-Hall, 1970).

75 This finding is well in line with Meyer, *The European Public Sphere*, pp. 291, 297.

CHAPTER 14.

THE MAJOR STAGES IN THE CONSTRUCTION OF EUROPEAN ENVIRONMENTAL LAW

Sophie Baziadoly

The environmental norms of the European Union are among the most substantial and strict in the world. The environment is not a concept defined in the Treaty of Rome, which would have risked fixing a domain that is constantly evolving alongside scientific and technological progress. The first definition was proposed by the European Commission in its first communication regarding the environment: it is 'the combination of elements whose complex interrelationships make up the settings, the surroundings and the conditions of life of the individual and of society'.[1] As this definition is not the only one available, a broader definition of the notion of the environment is needed. One could say that European environmental law corresponds to a series of measures taken to combat pollution and environmental nuisances, in an effort to protect the essential elements of nature (flora, fauna, etc.) and the environment.

European environmental law is strict because it adheres to a high level of environmental protection, and should actually be understood as a law for environmental protection. Procedures ensuring this protection have been available to each individual since the adoption of the Aarhus Convention in 2001 (right to access to environmental information, public participation in the decision-making process and access to justice in environmental matters).

European environmental law was conspicuously absent from the treaties that established the first communities; as a result, the environment was firstly a matter of international law. In fact, despite the Commission's first communications in 1973 on the need to ensure environmental protection, it was in the context of the United Nations Conference on the Human Environment held in Stockholm in June 1972 that 'the starting signal for the environment'[2] was launched, at the initiative of Scandinavian countries, as a

1 First communication of the Commission about the Community's policy on the environment. SEC (71) 2616 final, 22 July 1971.

2 Guy Corcelle, '20 ans après Stockholm, la conférence des Nations-Unies de Rio de Janeiro sur l'environnement et le développement : point de départ ou aboutissement ?' *Revue du marché commun et de l'Union européenne* **365** (1993): 107.

major concern for states and citizens. Against this backdrop, heads of state and of government decided, during the Paris Summit in October 1972, to develop a European environmental law, on account of the advantages offered by Community law, which is a law of integration that notably involves the transfer of competence to institutions independent from the member states.[3]

The silence of the Treaty of Rome that established the EEC meant that the first environmental measures would be taken on the basis of the First Environment Action Programme (EAP), which was launched in 1973 in the form of a declaration. Originally, member states established their own environmental norms; since these internal measures were likely to pose an obstacle to trade in connection with implementation of the Common Market, they were harmonised.

The implementation of the Single European Act (SEA) of 1 July 1987 marked the creation of a new Community environmental law, by introducing within the Treaty of Rome a dense title relating to the environment (current title XX of part 3 of the Treaty on the Functioning of the European Union (TFEU) grouping together articles 191 to 193). This new and specific legal foundation emphasises a series of principles: the prevention principle, which consists of taking measures to avoid the repetition of a risk and/or to limit its consequences; the polluter pays principle, in which damage is paid for by polluters; the principle of subsidiarity, which regulates the division of competence between the Communities and member states; and the integration principle, which enables the environment to become a component of all other policies. This new title also introduced the possibility of adopting environmental measures with a qualified majority, in order to facilitate decision making.

The coming into effect of the Maastricht Treaty on 1 November 1993 was important especially because it inserted environmental policy in article 3 of the treaty, which instituted the European Community. With regard to the environment, it included a corollary principle to that of prevention, namely the precautionary principle, which states that, in the absence of scientific certainty, when an initial scientific evaluation provides sufficient reasons to fear potential harmful consequences for the environment or health, measures must be taken to prevent this risk even if it has not been demonstrated. It also extended the decision-making procedure of a qualified majority to almost all areas of the environment, made the co-decision procedure the

3 See Sophie Baziadoly, *La politique européenne de l'environnement* (Brussels: Bruylant, 2014), pp. 10–16.

standard legal procedure in the domain of the environment, and placed the European Parliament and Council on equal footing.

The implementation of the Treaty of Amsterdam in late 1999 marked a major change in environmental decision-making procedure (expansion of the qualified majority vote, etc.). The environment also had a role in the Treaty of Nice from 2001, as environmental questions were regularly on the European Union's agenda, and played an increasingly important role in the implementation of policy.

The Treaty establishing a Constitution for Europe enshrined the environment as one of the EU's priorities: environmental protection and sustainable development represent the EU's third objective, after peace and free trade. The Charter of Fundamental Rights considers environmental law to be a fundamental right, and the EU's accession to the European Convention for the Protection of Human Rights and Fundamental Freedoms strengthened this protection. The constitutional treaty established an environmental law more in touch with citizens, thanks to institutional advances such as the introduction of a right to petition, which gives one million citizens from a significant number of states the right to ask the Commission to present a bill. The constitutional treaty reinforced the planetary dimension of the issues of European environmental law, especially by emphasising the fight against climate change. As the ratification process for the constitutional treaty did not come to a successful conclusion, the measures relating to environmental protection that it contained were nevertheless implemented through inclusion in a renegotiated treaty, the Treaty of Lisbon.

Hence, it was chiefly the consideration of objectives, principles, and framework conditions that conferred legal competence on the EU to act in all areas of environmental protection. These requirements have led to the adoption of a large body of secondary law texts during the last three decades. Since the 1970s, European environmental law has continued to evolve under the guidance of a key actor of its implementation, the European Commission.

Consideration of objectives, principles and framework conditions

European environmental law was built around objectives to be met. It is based on fundamental principles. It was developed in accordance with framework conditions.

A European Environmental Law built around objectives to be met

There are two kinds of objectives that underpin European environmental legislation: a priority objective in sustainable development, and general objectives.

A priority objective, sustainable development

Sustainable development was defined in 1987 by the United Nations World Commission on Environment and Development as a development 'that meets the needs of the present without compromising the ability of future generations to meet their own needs'.[4] This notion means that, at a strict minimum, the natural systems that sustain life, such as the atmosphere, water, land and living beings, must not be endangered.[5] Also, in order to meet this objective, economic growth and the environment must be balanced. Sustainable development grew out of ecological catastrophes that occurred in the 1980s, such as Bhopal (India) in 1984 and Chernobyl (Ukraine) in 1986.

This objective was established by the Preamble of the Treaty on European Union (paragraph 9), articles 3 and 11 of the TFEU, and article 37 of the Charter of Fundamental Rights.

The concept of sustainable development materialised in different ways. For instance, it served as a basis for the Fifth EAP (1994), which underpins a great deal of legislation. Today it is expressed through the mediation of an effective principle of European environmental law, the principle of integration, which allows the EU to engage in ambitious strategies, such as the one seeking to combat climate change.[6]

The general objectives of European Environmental Law

Article 191, § 1, TFEU provides a list. It is not possible to separate them from the numerous environmental problems that relate to multiple objectives.[7] The establishing of general objectives opens a large field of action.

4 This definition was taken from the Brundtland Report of April 1987.

5 See Philippe Léger (ed.), *Commentaire article par article des traités UE et CE* (Paris-Brussels: Dalloz-Bruylant, 2000), pp. 159–163.

6 See Sophie Baziadoly, *Le droit communautaire de l'environnement depuis l'Acte unique européen jusqu'à la Conférence intergouvernementale* (Brussels: Bruylant-ULB, 1996), pp. 76–77.

7 See Léger, *Commentaire article*, pp. 1328, 1330–1331 and 1334–1335; see also Baziadoly, *La politique européene*, pp. 24–30.

The preservation, protection, and improved quality of the environment represent the primary objectives identified by the EU in taking environmental measures. They allow for approaching all aspects of environmental protection, including the sudden appearance of new and unforeseen problems, or problems detected outside the EU.

Protecting the health of individuals is the second objective, and is connected to the first, as the quality of the environment has important consequences for public health.

The prudent and rational use of natural resources is an objective that flows from the first objective. This third objective targets natural resources in the broader sense, which is to say both those that are renewable (water, etc.), and those that are not (oil, etc.). The concept of prudent and rational use of renewable natural resources entails using renewable resources so that their regular renewal is not compromised; for non-renewable resources, all waste must be avoided.

The fight against climate change is the EU's new priority (Treaty of Lisbon), and has no equivalent. It confers an international dimension to environmental protection, and is in keeping with the broader objective of sustainable development.

A European Environmental Law Based on Fundamental Principles

Article 191, § 2, sentence 2 of the TFEU distinguishes the specific principles that apply only to matters of environmental protection (1), and general principles that involve not just the environment, but all European policies (2).[8]

The specific principles

The precautionary principle is the most recent. It entails adopting appropriate measures of prevention against threats of serious and irreversible harm to the environment and the health of humans, animals or plants, without having to wait for the proof of an immediate danger to be fully reported. It is thus based on uncertain scientific results, unlike the prevention principle, which applies when we know the dangers of a measure. At the outset, prevention and precaution did not exclude one another, for it was because of the shortcomings of the prevention principle that the European Com-

8 See Léger, *Commentaire article,* pp. 1336–1344 and 1350–1355; also Baziadoly, *La politique européene,* pp. 30–36 and 37–41.

mission established the precautionary principle. The precautionary principle was highlighted as part of the fight against climate change, which admitted scientific uncertainty regarding the polluting impact of human activities.

The First EAP in 1973 formulated the prevention, rectification at source of environmental damage and polluter pays principles, which have been constantly reaffirmed since.

The prevention principle, which is a corollary to the precautionary principle, recommends avoiding pollution and environmental nuisances by adopting or adapting measures to eradicate a known risk. The objective of the prevention principle is therefore not to eliminate the pollution and nuisances that occur, but rather to take measures to prevent such events from taking place. It is an essential principle that involves the use of prior environmental impact assessments, which necessitates the study of environmental effects before the implementation of a plan or project, the establishing of systems requiring authorisation for any activity producing a serious effect on the environment, and monitoring and control of authorised emissions.

The principle of rectification at source is a corollary to the prevention principle, in the sense that it involves the rectification of environmental damage at the source, for instance the use of pesticides or fertiliser in agriculture.

The polluter pays principle (PPP), initially developed by the OECD in 1972, is an economic principle that became a legal one with the first treaty amending the provisions of the Treaty of Rome. The instruments for the application of the PPP were identified through a Council Recommendation in 1975 as binding legal norms.[9] These include environmental quality norms that prescribe a level of pollution or environmental nuisance not to be surpassed within a particular environment, in addition to economic and fiscal instruments such as water pollution charges, which offer a way of changing the behaviour of polluters. The PPP signifies that the costs for reducing environmental pollution are to be borne by those who caused the pollution, or who risk causing pollution. It has three facets: it can be applied preventively, which is to say before the damage takes place; it can be applied after environmental damage occurs; and in certain conditions, it authorises controlled pollution, with the polluter paying an ecotax. The PPP can be invoked in connection with the elimination of waste, water pollution, or environmental responsibility.

9 See Council Recommendation 75/436/CEE, from 3 Mar. 1975, regarding cost allocation and action by public authorities on environmental matters, *JOCE* L 194, 25 July 1975.

General principles

The principle of integration is based on the fact that all the EU's policies and actions can have an impact on the environment, especially key sectors of the economy (common agricultural policy, transportation, energy, industry, tourism). It is an innovative principle because it requires legislators to take environmental protection into consideration in all domains, and at all stages, in which they intervene, from design to realisation. This principle has created an integrated policy for energy and the environment, notably in order to fight climate change. The principle of integration prompted a change in working methods within the European administration, with the Commission's Directorates-General (DGs) possessing a department or sector exclusively addressing environmental problems.

The principle of subsidiarity does not apply to the exclusive competence of the EU, which is to say the domains in which member states do not intervene, such as the management of marine natural resources. This principle is applicable to shared competence, that is the domains in which the EU has not yet passed legislation; when it does so, the member states no longer intervene. The objective of the principle of subsidiarity is to determine whether the EU can act within a shared domain of competence, or whether it must allow member states to regulate the matter. Article 5, § 3, TEU states the condition for initiating an action, namely that the member states are unable to sufficiently meet the treaty's objectives, and consequently these objectives can be better achieved on the EU level. The principle of subsidiarity is bolstered by the Treaty of Lisbon, which established measures for control over the subsidiarity carried out by national parliaments. It is, for instance, an effective principle for combatting pollution on the international and European level. It is also a principle whose definition and implementation criteria offer the advantage of less regulation of the environment on the European level.

The principle of international cooperation applies in matters of shared competence, member states and the EU in concluding external agreements relating to environmental protection: this principle enabled, for instance, the conclusion of the Kyoto Protocol in 1990 to combat climate change.

The principle of proportionality is antithetical to the principle of subsidiarity, as it applies only in the context of a normative action conceded to the European Union, and therefore only if subsidiarity does not apply. Provided for by article 5, § 4, TEU, it requires the actions of European institutions not to surpass the limits of what is appropriate and necessary for achieving objectives legitimately pursued by the regulation in question; it is understood that, when there is a choice between a number of appropriate measures, the least restrictive should be used, and that the resulting disadvantage should not be measured based on the established goal.

A European Environmental Law developed in accordance with framework conditions

In developing European environmental law, the EU takes into consideration both framework measures and national mechanisms for dispensation and safeguard.

Framework measures

Framework measures are general conditions that the EU must adhere to in establishing environmental protection measures. Included in the Treaty of Rome by the SEA at the request of multiple member states facing specific national difficulties, they are three in number.[10]

The taking into consideration of available scientific and technical data is due to the United Kingdom, which lamented the lack of a scientific foundation for environmental measures during the mad cow affair; it conveys the idea of a constantly evolving environment that must adapt to technological advances and the evolution of scientific knowledge. In this sense, it combines with the application of the prevention and precautionary principles. For example, this data is systematically taken into consideration in directives granting marketing authorisation for medicine.

The advantages and burdens that can result from action or inaction entail taking environmental protection measures that do not lead to excessive costs, and that subsequently take economic situation into consideration, particularly for companies. It also entails, with regard to member states, not to hinder the need for balanced economic and social development of different regions in the EU. High-performance and non-polluting technologies can prove financially costly for companies, and can lead to major economic and social changes, such as offshoring to third countries that do not have these obligations. With this in mind, the EU engages in a dual balancing, between the positive and negative consequences of its action, and the consequences between its action and inaction.

10 See Léger, *Commentaire article*, pp. 1360–1364; see also Baziadoly, *La politique européenne*, pp. 47–51.

The environmental, economic and social elements of the various regions of the EU (article 191, § 3, TFEU), which less developed countries particularly insisted on (Ireland, Greece, etc.), calls for considering the level of economic and social development of members states in environmental decision-making. As a result, since the Treaty of Amsterdam, this framework condition involves consulting the Committee of the Regions before adopting a measure pertaining to the environment. It notably provides for support mechanisms, so that environmental protection does not represent an obstacle to development.

National mechanisms for dispensation and safeguarding[11]

The minimal protection clause (article 193 TFEU) allows member states such as Germany, which feared having to accept European measures offering little environmental protection, to maintain and even establish stricter national rules for protection in domains in which the EU has not yet acted.

The safeguard clause (article 191, § 2, TFEU) enables member states, as part of the harmonisation of national legislation needed to create the internal market, to apply national provisions when they deem it necessary, and when EU provisions are justified for reasons of the general interest considered to be urgent, such as protection of the environment. These measures must be compatible with the treaties; for example, they must not be disproportionate in relation to the objective being pursued – environmental protection – in view of the obstacles they create. The Commission must be notified of them. Today, given that there are many Europeans texts in environmental protection, member states have little latitude to invoke such dispensations.

A voluminous body of texts of secondary legislation

Normative actions revolving around environmental protection form the primary source of European environmental law for member states, but environmental protection also goes beyond the European norm.

Normative actions, the primary sources of European Environmental Law

Since the 1970s, European environmental law has developed through the creation of environmental action programmes; EAPs are notably supple-

11 See Baziadoly, *La politique européenne*, pp. 51–56.

mented by the working programme that the Commission produces each year, whose content varies according to the presidency of the member states.[12] EAPs are powerful drivers of strong environmental legislation; the adoption of numerous legal acts regulates the various fields.

Environmental action programmes, a powerful driver for strong environmental legislation

The first environmental measures were developed in the form of action programmes. Today there are seven EAPs: the first was adopted by the Council in 1973, and was followed by six others in 1977, 1983, 1987, 1992, 2001 and 2012. The EAPs formulate a philosophy for the protection and improvement of the environment; they also define a calendar of specific actions to undertake during the years covered by these general programmes. With the exception of the first EAP, they are inspired by the results from preceding programmes, their shortcomings and successes and especially the evolution of environmental problems in Europe and the world (climate change, etc.).

The first two EAPs were essentially in the fairly restrictive domain of combatting pollution and environmental nuisances. As the situation had improved thanks to the adoption of legal instruments, the Third EAP changed direction by recommending the development of a global policy strategy for the environment, based notably on the inclusion of environmental concerns within the conception of all EU activity.

The Fourth EAP was innovative in that it accompanied the taking into effect of the SEA, and hence the creation of a new European environmental law. As a result, all while calling for the continuation and acceleration of earlier programmes that had not been executed, it insisted on the need to fix stricter norms, ensure the effective application of directives, and develop more active information and education policy for the environment. The Fifth EAP, entitled 'towards sustainability', gave new momentum to environmental protection by seeking to balance the environment and development.

The Sixth EAP, inspired by the Fifth EAP, covered a period of approximately ten years (2002–2012). It set out from the consideration that our environment would be subject to constant constraints as a result of the global pursuit of growth, prompting the identification of fundamental priorities. In order to have more effective environmental protection, it recommended the use of economic and financial instruments in addition to legislation.

12 See Ludwig Krämer, *EC Environmental Law* (London: Sweet and Maxwell, 2007), pp. 147–148.

The Seventh and final EAP, concluded for the period running from 2012 to 2020, is based on an evaluation of the penultimate programme. Its content seems to be more targeted, as it is in keeping with both the environment and the economy. It highlights the cost, fifty billion euros, of poor application of environmental legislation, and hence the need to improve implementation.

The first five EAPs were not legally binding, as they were adopted in the form of declarations, resolutions and decisions, and did not emanate solely from the Council, but also from member states; they can be seen as policy guiding principles that led to the adoption of many European legal acts.

The TEU established a new legal framework for the adoption of the Sixth EAP and the programmes for the coming years: action programmes henceforth had to be made by co-decision between the Council and European Parliament as part of ordinary legislative procedure (article 294 TFEU); this provision created a legal effect enabling either co-legislator to bring an action for failure to comply against the Commission, if this institution does not develop proposed legislation in the sectors targeted by the EAP.[13]

The adoption of numerous legal acts for regulating various environments

For the most part, European directives enable legislating in various domains. They contain general arrangements that member states can flexibly interpret as they integrate them within their internal legal order. Until the late 1990s, directives took a sectoral approach to environmental protection, and were issued as environmental problems appeared. The development of the integrated approach to the environment prompted the EU to change strategy in order to regulate the different environments.

Water was the first sector in which the EU legislated. The EU has taken an evolving approach to water pollution, as it manages water pollution comprehensively rather than based on one sector. It did so by replacing the directives from the 1970s and 1980s – for instance fixing acceptable levels of pollution for bathing water – by a directive (2000/60/CE) that established a European framework to provide member states with water supplies sufficient in quantity and quality, with a view to the sustainable use of this resource.[14] With this in mind, member states had to establish integrated

13 Ibid., p. 47.

14 See Baziadoly, *La politique européenne*, pp. 97–99; Krämer, *EC Environmental Law*, p. 296; Patrick Thieffry, *Droit de l'environnement de l'Union européenne* (Brussels: Bruylant, 2011), pp. 282–342; Gérard Druesne, *Droit de l'Union européenne et politiques communautaires* (Paris: PUF, 2006), p.

water resource management programmes by creating hydrographical territories, which are areas possessing abundant water (including especially streams, lakes and rivers flowing into the sea); they were also required to establish districts (water agencies of a sort) tasked with managing water resources. Furthermore, the directive reaffirmed the polluter pays principle by including the cost of water pollution in the price of water provided by an economic operator to a consumer.

Air is a sector in which the European Union was late in legislating,[15] with the first directives relating to air pollution dating from the 1980s, as a result of both the 1973 oil crisis and energy constraints weighing on member states. Today there is an integrated approach to air pollution thanks to a 2008 directive (2008/50/CE), which, for instance, requires monitoring pollution from road traffic (PM 2.5) presenting a substantial negative impact on human health (establishing a limit for particles emitted by transportation, harmonisation of air quality monitoring in Europe by establishing days without cars, etc.).

The improvement of air quality is also a global priority in the fight against climate change, as air pollution is responsible for global warming. With this in mind, the EU initiated an ambitious strategy for adapting its economy to this requirement, by planning measures in all sectors that emit polluting gases (carbon dioxide, etc.), such as industry. The European Union's climate strategy has existed since the 1990s, in particular with the Kyoto Protocol and later the Copenhagen Agreement in the 2000s. It led in 2006 to a proposal to develop a new integrated policy, which notably enabled the adoption of the Climate and Energy Package in late 2008, and the issuing of a directive, upgraded a number of times (2009/29/CE), establishing an emissions trading system (authorising companies that have surpassed their pollution quota to purchase the emissions rights of other companies that reduced their greenhouse gas emissions). The definition of climate policy remains a major challenge for the EU. The minimalist Copenhagen Agreement led to new meetings in Paris in 2015. It may appear, for that matter, difficult to reconcile environmental objectives with the preservation of European competitiveness.

537. See also Raphaël Romi, Thomas Dubrueil, Sandrine Rousseaux and Mary Sancy, *Droit international et européen de l'environnement*, 2nd edition (Paris: Montchrestien, 2013), pp. 185–193; Louis Dubouis and Claude Blumann, *Droit matériel de l'Union européenne* (Paris: Montchrestien, 2012), no. 382.

15 Baziadoly, *La politique européenne*, pp. 100–105; Thieffry, *Droit de l'environnement,* pp. 197–279; Druesne, *Droit de l'Union européenne*, pp. 539–542. See also Romi et al., *Droit international*, pp. 201–210; Dubouis and Blumann, *Droit matériel*, no. 380.

Member states produce tons of waste each year, including dangerous waste. The landfilling of this waste is unsatisfactory, especially due to long-term impacts that are poorly understood. The best solution for managing waste emphasises the prevention of waste production, along with reintroducing it in the life cycle of products by recycling the materials of which they are made.[16] EU waste management takes place through the adoption of successively reinforced framework directives.[17] A number of specific measures were taken on the basis of this general framework, such as the implementation of a system for granting ecological labels.

Regarding action for the protection of nature, it is important to begin by citing a binding legal instrument, the Bern Convention on the Conservation of European Wildlife and Natural Habitats, adopted in Switzerland on 19 September 1979, and coming into effect on 6 June 1982. Forty-seven of the signatories, including the European Union, committed to granting special importance to the protection of both natural habitats in danger of disappearing and endangered species (including migratory species). Furthermore, the EU's application of the Bern Convention was primarily carried out by the implementation of Birds and Habitats Directives. With regard to habitats, the requirements of the Bern Convention were satisfied by designating protected areas as part of the Natura 2000 Network.[18]

Manipulation of genetic material, which is increasingly frequent, presents risks for both health and the environment. Since 1990, the European Community and later the European Union adopted directives to supervise the bringing to market of genetically modified organisms; these directives require that trials be conducted in order to proceed with marketing by the member states.

Greater awareness of the harmful effects of certain industrial activities during the 1960s and 1970s – using chemical products that generate major risks – raised constant concerns for environmental protection.[19] For instance, the first directive issued in this domain (67/548/CEE) concerned dangerous

16 See Baziadoly, *La politique européenne*, pp. 106–110; Thieffry, *Droit de l'environnement*, pp. 391–487; Druesne, *Droit de l'Union européenne*, pp. 546–547. See also Romi et al., *Droit international*, pp. 251–257; Dubouis and Blumann, *Droit matériel*, no. 388.

17 See, for example, directive 2008/98/CE of the European Parliament and of the Council from 19 Nov. 2008, on waste and repealing certain directives, *JOUE* L 312, 22 Nov. 2008, pp. 3–30.

18 See Maguelonne Dejeant-Pons, 'Les droits de l'homme à l'environnement dans le cadre du Conseil de l'Europe', *Revue trimestrielle des droits de l'homme* **60** (2004): 861–888.

19 Baziadoly, *La politique européenne*, pp. 114–119; Thieffry, *Droit de l'environnement*, pp. 506–547.

chemical substances. Today the REACH system authorises the bringing to market of existing chemical products.

Serious and environmentally catastrophic industrial accidents, such as those in Seveso (Italy) in 1976 and Toulouse (France) in 2011, prompted the EU to issue directives (Seveso) to regulate the storage of dangerous products. The latest Seveso directive (2012/18/EU) has been in effect since 2015, and provides for informing the public living near industrial installations that produce dangerous chemical products.

The implementation of less normative instruments

The integrated approach to the environment is based on instruments that are not always legislative in nature, and that can change the behaviour of individuals or administrations. These instruments are different from more traditional regulation, as they often have the distinguishing basis of inciting manufacturers and consumers to adopt behaviour more compatible with environmental requirements. These include trans-sectoral directives, as well as economic and financial instruments.

Trans-sectoral directives

Trans-sectoral directives apply to the various sectors in which the EU has legislated. For instance, the impact assessment directive (2001/92/EU) provides for member states to assess, in advance of the European environmental decision, the realisation of public and private projects likely to have a notable impact on the environment. This was the case, for instance, with the construction of freeways or railway lines that could interfere with the protection of habitats for protected birds. It was supplemented by another directive (2001/42/CE) that provided for establishing a system of environmental assessment prior to the planning stage, in other words before decisions are made, insofar as the impact assessments for projects sometimes came after strategic decisions that proved decisive for them.

Impact assessment directives are instruments to help in decision making, although the incomplete nature of impact assessments conducted in advance of environmental decisions makes it difficult to gauge the potential impact of the measures proposed by the Commission.

Due to the impossibility of establishing a general system of environmental responsibility following the failure of the Lugano Convention of 21 June 1993 on civil responsibility for environmental damage caused by dangerous activities, the environmental responsibility directive (2004/35/CE) established minimum common rules for preventing damage to the environment, and ensuring it is repaired by those responsible. The economic operator could thus be held responsible for damage caused to species and habitats protected by Birds and Habitats legislation; once the harm is proven, reparative measures should be taken by the operator (restoration of protected habitats, establishment of an animal or plant species in a space other than the one restored, etc.).

Economic and financial instruments

The Eco-label economic instrument (regulation (CE) no. 66/2010) is granted by a national organisation to products and services that reduce negative impact on the environment (paints and varnishes, etc.). The Eco-label is awarded for five years, and its use is subject to a usage fee. It is recognisable by its logo, a daisy, and is generally awarded to companies that include social and environmental concerns in their activities.

The Eco-label can be supplemented by another economic instrument, EMAS (regulation (CE) no. 1221/2009), a European system of analysis for the environmental practices of companies in all sectors of economic activity (waste management, etc.). The objective is to work towards a more respectful view of the environment. It is expensive to implement, with millions of companies operating in the European Union not using this system.

A number of financial instruments are affected by environmental protection (structural funds, European Bank investment, state aid, etc.), however the Life + programme is the only financial instrument specifically devoted to environmental protection. Funding allocated through co-financing with national budgets has made it possible to finance, for the 2014–2020 period, the implementation of the primary European regulations on the protection of nature (fifty per cent resources from Life); this includes Birds and Habitats directives, and notably creation of the European protected areas network called Natura 2000, which grew out of these two directives and the European Union's action in combatting climate change.

A European environmental law implemented by a central actor, the European Commission

European environmental law is developed and applied by central actors, with the European Commission playing an active role due to its powers and organisation. This European institution also played an essential role in establishing environmental protection measures, and exercised increased control over the application of instruments relating to environmental protection.

An active role in environmental protection thanks to its powers and organisation

The European Commission plays an active role in environmental matters, thanks to its power to propose legislation, as well as its power of control over adherence to European environmental law. For instance, as part of its right of initiative, the Commission makes proposals in which the environment can be combined with other areas of EU action via application of the principle of integration. Integrating the environment can be improved through the Commission's verification of the environmental impact of all proposed legislation. Moreover, as the guardian of treaties, the Commission conducts procedures for infractions and complaints.

The dynamism of the Commission is also based on its composition and functioning. For instance, in order to emphasise that the fight against climate change is an EU priority, the President of the Commission appointed a European Commissioner for Climate Action to go alongside the European Commissioner for the Environment. Furthermore, the Environment DG was created in 1973, at the same time as European environmental policy: it is presided over by a senior European civil servant who provides reports of its environmental action to the European Commissioner for the Environment. It is among the Commission's larger DGs due to the issues it addresses and its human and financial resources.[20]

An essential role in developing environmental protection measures

In broad outlines, within the context of developing environmental protection measures, the Environment DG consults with different actors (representatives of governments, non-governmental organisations such as Greenpeace, interest groups such as industry, technical experts such as engineers, public opinion). The text is then transmitted by adoption to the European College of Commissioners. If the proposed legislation is selected, it is in principle transmitted as part of ordinary legislative procedure to the Council, European Parliament, and the DGs concerned, as the environment is horizontal in nature.

The negotiations and discussions surrounding a proposal of environmen-

20 See Krämer, *EC Environmental Law*, p. 38.

tal legislation can lead to a certain amount of back and forth between the three actors in the institutional triangle, and can notably modify an initial version.[21] Proposals for environmental legislation are regularly added to the Council's agenda in an effort to generate a common policy position on one or more negotiating points. Depending on the results from the working groups, propositions can be submitted one final time to the Council in order to be definitively settled.

Increased control over the application of environmental protection instruments

Concern for better application of European environmental law is present in a number of EAPs, especially the most recent. In order to ensure compliance with instruments relating to environmental protection, the European Commission has measures of control at its disposal in order to take action. Complaints, for instance, represent one of its primary sources of information for the poor application of environmental legislation; they often come from NGOs, the general public, and the European Parliament, a partner of the Commission. To contend with the increasing number of complaints, the Commission has proposed penalising the most serious infractions of environmental protection, for example the dumping of certain substances in water, such as hydrocarbons.[22]

As the increase in environmental legislation was not compensated by an improvement in environmental protection, in 1982 the Commission created, in cooperation with the Dutch presidency of the time, the Impel network, a discussion forum that brings together the environmental organisations of member states a few times a year around problems relating to environmental policy, such as the creation of an environmental inspector corps.

As the guardian of treaties, the Commission can conduct a stern policy to improve the effective application of environmental directives. It initiates numerous failure to comply proceedings before the Court of Justice of the European Union (CJEU), on the basis of articles 258 to 260 of the TFEU, against member states that do not adhere to environmental legislation (for instance member states that do not transpose or are late in transposing a directive into their internal legal order). The CJEU simply records a member

21 See Baziadoly, *La politique européenne*, pp. 154–156.

22 See directive 2008/99/CE of the European Parliament and of the Council from 19 Nov. 2008 on the protection of the environment through criminal law, *JOUE* L 328, 6 Dec. 2008, pp. 28–37.

state's violation of an obligation; it is then up to the member state to acknowledge the consequences and conform to European environmental law. In cases where this is not executed, a new procedure can lead to a second failure to comply decision. Beginning with the first failure to comply decision, the Court can assess dual punishment by imposing the payment of a lump sum and a penalty on member states, following where necessary the amounts established by the Commission. The Commission began actively to pursue recalcitrant member states in the mid-1980s following the Seveso disaster. However, decisions imposing severe financial penalties were issued only in the 2000s. For instance, in 2015 France was required to pay a lump sum of twenty billion euros, and a quarterly penalty of 57.67 million euros, for its insufficient controls of fishing.[23]

Governing bodies were created over the last decades to assist the Commission in accomplishing its work, and to ensure the public is sufficiently informed about the state of the environment. The most important is the European Environmental Agency (EEA), which is based in Copenhagen, in a country that is attentive to environmental protection. The EEA gathers environmental data that is used to implement European legislation; it is also tasked with evaluating progress in the fight against climate change.[24]

The major stages in the construction of European environmental law show that environmental problems were neglected for a long time, as environmental protection was not considered to be an economic factor. It was in 1968 that the decision was made to organise a major international conference on humans and their environment, held in Stockholm in 1972. Today environmental problems occupy an important role in international relations, especially commercial relations. In this respect, the EU has chosen the policy of including sustainable development provisions in the bilateral trade agreements it concludes with its partners, to secure their participation in improving environmental and societal norms.

The simultaneous emergence of global environmental problems such as climate change has led to the negotiation and signing of numerous international agreements, such as the decision to establish binding objectives for

23 See ECJ, 12 July 2005, case C – 304/02, Commission against France, Rec., 2005, p. I – 06263.

24 See Baziadoly, *La politique européenne*, pp. 169–171. See also Krämer, *EC Environmental Law*, p. 45.

the reduction of greenhouse gases, especially as part of the Paris Agreement in 2015. The question of climate change also deserves to be approached from the perspective of security, in the sense that it can threaten growth, prosperity and stability on a global level.

The major stages in the construction of European environmental law demonstrate growing environmental awareness, as well as the development of an environmental movement based on citizens protesting against the destruction of their environment, in parallel to protests held by scientific experts (Stockholm). This environmental awareness has renewed the concept of citizenship, and has also politicised the environmental movement, with the creation of Green and Ecology parties in a number of EU member states, the European Parliament and third countries (the United States, etc.).

CHAPTER 15.

MULTI-LEVEL LEARNING: HOW THE EUROPEAN UNION DRAWS LESSONS FROM WATER MANAGEMENT AT THE RIVER BASIN LEVEL.

Marjolein van Eerd and Duncan Liefferink

Transboundary water management and River Basin Commissions

Rivers function as veins on the world's continents and many human, economic and ecological systems depend on rivers that cross state borders.[1] Initially, water management followed a technical and instrumental approach, often being the exclusive task of technical and state experts. The main presumption was that water resources could be predicted, managed and controlled. Yet, the urgency of cross-border cooperation and changing climatic conditions have triggered a paradigm shift towards more integrated, transboundary water management.

Challenges for water management, such as water quality issues, flood events and the impact of climate change do not stop at man-made, historical, geographical and territorial borders. Governing water resources should therefore be approached from a transboundary perspective. An example of such perspective is the application of a river basin approach, which inherently leads to challenges for existing, conventional and often deeply embedded governance frameworks.[2] Collaboration across borders is complex due to upstream-downstream asymmetries. Upstream and downstream located states are likely to have different interests, discourses, approaches and problems, yet are dependent upon each other for river basin management. In addition, state sovereignty may hinder the development of cross-border cooperation.[3]

1 See M.C.J. van Eerd, , M. Wiering and C. Dieperink, 'Exploring the prospects for cross-border climate adaptation between North Rhine-Westphalia and the Netherlands', *Utrecht Law Review* **10** (2014): 91.

2 See W. Steele, I. Sporne, P. Dale, S. Shearer, L. Singh-Peterson, S. Serrao-Neumann, F. Crick, D. Low Choy and L. Eslami-Andargoli 'Learning from cross-border arrangements to support climate change adaptation in Australia', *Journal of Environmental Planning and Management* **57** (5) (2014): 682–703.

3 See T. Bernauer, 'Explaining success and failure in international river management', *Aquatic Sciences* **64** (2002): 1–19; M.C.J. van Eerd, M. Wiering and C. Dieperink, 'Exploring the prospects for

The current trend of addressing water issues from a river basin management perspective is stimulated by, for example, the European Union (EU).[4] The catchment level plays an important role in today's water management, since it is considered logical to govern water issues along hydrological boundaries. River Basin Commissions (RBCs) can be seen as mediating platforms bridging the gap between various territorially organised organisations concerned with water management at multiple levels.[5] RBCs are unique and functional platforms of collaborative management, involving governmental and non-governmental actors from multiple levels and sectors.[6] In this contribution we assess the International Commission for the Protection of the Rhine River (ICPR) and its role in sparking off learning processes in other institutional contexts, notably the EU. The ICPR has been chosen since collaboration in the Rhine basin has often been referred to as one of the most successful examples of cross-border water management.[7]

Multi-level learning from practical experiences

Actors ranging from the local to the international and supra-national level are concerned with the daily governance of water quality and quantity issues in Europe. Continuous interplay and interaction take place between these levels of governance, for example via the exchange of knowledge, information and expertise. Such exchanges may enable multi-level learning, which occurs when actors adjust their cognitive understanding of, for instance, policies, and modify them in the light of experiences gained elsewhere.[8]

cross-border climate adaptation between North Rhine-Westphalia and the Netherlands', *Utrecht Law Review* **10** (2014): 91.

4 See E. Mostert, 'Conflict and cooperation in international freshwater management: a global review', *International Journal of River Basin Management* **1** (3) (2003): 267–78.

5 See J.R. Warner, 'More stakeholder participation? Multi-stakeholder platforms for integrated catchment management', *International Journal of Water Resources Development* **22** (2006): 15–35.

6 See D. Huitema and S. Meijerink, *The Politics of River Basin Organisations: Coalitions, Institutional Design Choices and Consequences* (Cheltenham: Edward Elgar Publishing, 2014).

7 See T. Bernauer and P. Moser, 'Reducing pollution of the river Rhine: the influence of international cooperation', *Journal of Environment and Development* **5** (1996): 389–415; C. Dieperink, 'From open sewer to salmon run: lessons from the Rhine water quality regime', *Water Policy* **1** (1998): 471–85.

8 See P. Hall, 'Policy paradigms, social learning and the state: the case of economic policymaking in England', *Comparative politics* **25** (3) (1993): 275–96; M. Reed, M.A.C. Evely, G. Cundill,

The practical implementation of policies and their subsequent interaction with implementing agents and target groups generates information about how policies are actually received and work out in practice.[9] This information, which we refer to as *implementation experiences*, is an important resource that can be used to influence the ongoing policy process also in other contexts. It may, for instance, affect processes of setting implementation issues on the agenda, adapting implementation elements or changing policies. Implementation experiences consist of 'all knowledge, expertise and information acquired by actors during or as the result of the practical implementation of … policies'.[10] In the specific context of the EU, the European Commission is concerned with monitoring and enforcing policy implementation, yet EU institutions are not directly involved in executing policy implementation at the domestic level. The European Commission is therefore highly dependent upon the implementation experiences of domestic implementing agents to improve the practicability, workability and legitimacy of EU legislation.[11]

Since RBCs have unique expertise about the management of water resources along hydrological boundaries and concerning the collaboration between multiple water-related actors in an international setting, expertise in RBCs may be assumed to be an important additional knowledge resource for agents concerned with EU water governance. However, little is known about how and under which conditions these RBC experiences feed back from the catchment to the EU level.

Policy feedback and implementation experiences

In this contribution we focus on the process of *policy implementation feedback* from the ICPR to the EU, by which practical implementation experiences gathered at the river basin level are taken up in the EU policy process. Such feedback may lead to a reconsideration of existing policies

I.R.A. Fazey, J. Glass, A. Laing, J. Newig, B. Parrish, C. Prell, C. Raymond and L. Stringer, 'What is social learning?', *Ecology and Society* **15** (4) (2010): 1.

9 See A.R. Zito and A. Schout, 'Learning theory reconsidered: EU integration theories and learning', *Journal of European Public Policy* **16** (2009): 1103–1123.

10 See M.C.J. van Eerd, C. Dieperink and M.A. Wiering, 'Opening the black box of Implementation feedback: an analysis of reloading strategies in EU water governance', *Environmental Policy and Governance* **28** (6) (2017): 426–40.

11 See European Commission, *Better Regulation* (Brussels: European Commission, 2016), http://ec.europa.eu/info/law/law-making-process/better-regulation-why-and-how_en (accessed 16 Dec. 2016).

or their implementation. Considering the literature on policy change, it is unlikely that the feedback of experiences will lead to more radical types of policy change, which are rare anyway. Yet, we assume that implementation experiences are an important factor to consider for understanding incremental policy change.[12]

Both the EU and ICPR are institutional arrangements that interact and exchange knowledge and expertise, including implementation experiences. *Institutional arrangements* are (temporarily stabilised) institutions concerned with the governance of a specific policy field, and can be understood as sets of working rules and procedures determining who is eligible to make decisions in a bounded area, and what actions are allowed or constrained. An arrangement consists of four interdependent dimensions: actors and coalitions, resources and power, rules of the game and discourses.[13]

In this study, the ICPR is studied as the 'sending' institutional arrangement, where actors are acquiring and mobilising implementation knowledge. The EU is seen as the 'receiving' institutional arrangement, which might be affected by these experiences. By reviewing the relevant literature, we found that characteristics of both the sending and the receiving institutional arrangements, the relation between them, and contextual conditions affect the exchange of implementation experiences and policy implementation feedback in general. Examples of relevant characteristics of the sending RBC are: its trustworthiness and perceived success,[14] the role and type of actors involved[15] and their resources, capacities and skills[16] as well as

12 See F.R. Baumgartner and B.D. Jones, *Agendas and Instability in American Politics* (Chicago and London: The University of Chicago Press, 1993); P.A. Sabatier, 'The Advocacy Coalition Framework of policy change and the role of policy-oriented learning therein', *Policy Sciences* **21** (1988):129–68.

13 See B. Arts and P. Leroy (eds), *Institutional Dynamics in Environmental Governance* (Dordrecht: Springer, 2006); in particular D. Liefferink, 'The dynamics of policy arrangements: turning round the tetrahedron', pp. 45–51; and F.G.W. Jaspers, 'Institutional arrangements for integrated river basin management', *Water Policy* **5** (2003): 77–90

14 See C.M. Radaelli, 'Policy transfer in the European Union: institutional isomorphism as a source of legitimacy', *Governance* **13** (1) (2000): 25–43

15 See P. Hall, 'Policy paradigms, social learning and the state: the case of economic policymaking in England', *Comparative Politics* **25** (3) (1993): 275–96.

16 See G. Dudley and J. Richardson, 'Competing advocacy coalitions and the process of "frame reflection": a longitudinal analysis of EU steel policy', *Journal of European Public Policy* **6** (1999): 225–48; D. Pesendorfer, 'EU environmental policy under pressure: chemicals policy change between antagonistic goals', *Environmental Politics* **15** (2006): 95–114.

its information management,[17] organisational structure, problem-solving capacity and bindingness.[18] Explanatory characteristics of the receiving arrangement, here the EU, include its openness and responsiveness to experiences.[19] With regard to the relation between these arrangements, the degree of 'fit' between discourses prevalent in both arrangements[20] and the dependency between them, for instance in terms of rules or resources, are important for policy feedback to occur.[21] In addition, contextual conditions provide opportunities and barriers for policy implementation feedback. For instance, the existing network of actors and venues for exchanging implementation knowledge is important,[22] as well as the timing of the exchange of experiences, since a policy momentum and political or social attention are essential.[23]

Aim and structure of this contribution

The main objective of this contribution is to better understand the interaction and multi-level learning between the river basin and the EU level, by exploring conditions affecting the policy feedback of implementation knowledge between a river basin commission (the ICPR) and the EU. This contribution is based upon earlier studies conducted by the authors, in

17 See P. Huntjens, C. Pahl-Wostl, B. Rihoux, M. Schlüter, Z. Flachner, S. Neto, R. Koskova, C. Dickens and I.N. Kiti, 'Adapative water management and policy learning in a changing climate: a formal comparative analysis of eight water management regimes in Europe, Africa and Asia', *Environmental Policy and Governance* **21** (2011): 145–63.

18 See A. Underdal, 'Conclusions: patterns of regime effectiveness', in L.M. Edward, A. Underdal, S. Andersen, J. Wettestad, J.B. Skjaerseth, E.M. Carlin (eds), *Environmental Regime Effectiveness: Confronting Theory with Evidence* (Cambridge, MA: MIT Press, 2002), pp. 433–65.

19 See M. Howlett, M. Ramesh and A. Perl, *Studying Public Policy: Policy Cycles and Policy Subsystems* (Toronto: Oxford University Press, 2009); B.D. Jones and F.R. Baumgartner, 'From there to here: punctuated equilibrium to the general punctuation thesis to a theory of government information processing', *The Policy Studies Journal* **40** (1) (2012): 1–19.

20 See D. Huitema and S. Meijerink, *The Politics of River Basin Organisations: Coalitions, Institutional Design Choices and Consequences* (Cheltenham: Edward Elgar Publishing, 2014); E. Mostert, C. Pahl-Wostl, Y. Rees, B. Searle, D. Tabara and J. Tippett, 'Social learning in European River-Basin Management: barriers and fostering mechanisms from 10 river basins', *Ecology and Society* **12** (1) (2007): 19.

21 See M.S. Yebra, *Learning, Policymaking and Market* (Cambridge: Cambridge University Press, 2009).

22 See S. Meijerink and D. Huitema, *Water Transitions, Policy Entrepreneurs and Change Strategies: Lessons Learned* (Cheltenham: Edward Elgar Publishing, 2010); S.B. Pralle, 'Venue shopping, political strategy and policy change: the internationalization of Canadian forest advocacy', *Journal of Public Policy* **23** (2003): 233–60.

23 See J. Kingdon, *Agendas, Alternatives and Public Policies* (Harlow: Pearson Education Ltd., 2014).

particular the 'Knowledge for Climate' research project.[24]

We structured this contribution as follows: the next section presents a historical overview of the development and role of the ICPR and describes the key characteristics of this institutional arrangement. After that, the management of water resources at the European level over time is elaborated upon, characteristics of EU water governance are described, as well as the interaction between this policy domain and the ICPR. Learning and interaction between the ICPR and EU are then assessed, based on two cases of policy feedback between the two arrangements. Whereas the first case deals with water pollution issues, i.e. water quality, the second case focuses on flood risk management, i.e. water quantity. The final section consists of our concluding remarks.

The International Commission for the Protection of the Rhine river

A historic overview

The Rhine is one of the largest rivers in Europe, and flows through some of the most populated and industrialised regions of West Europe. This river is used for several functions, such as drinking water supply, navigation and irrigation.

The first international initiatives of collaboration in this basin date back to 1449. Collaboration to deal with water quality issues in 1950 can be seen as the start of the International Commission for the Protection of the Rhine (ICPR). The ICPR got an official mandate by the 1963 Treaty of Bern, which was renewed in 1999. Germany, France, Luxembourg, the Netherlands, Switzerland and the European Community are formal

24 See for example M.C.J. van Eerd, M. Wiering and C. Dieperink, 'Exploring the prospects for cross-border climate adaptation between North Rhine-Westphalia and the Netherlands', *Utrecht Law Review* **10** (2014):. 91; M.C.J. van Eerd, M. Wiering and C. Dieperink, *Possibilities for Transboundary Climate Adaptation Governance: Some Lessons from the Rhine and Danube Commissions* (Nijmegen: Radboud University, Knowledge for Climate research project deliverable 5.2.10, 2014); S. Veenman and D. Liefferink, 'Balanced policy networks: the cases of airport noise', *Journal of Environmental Policy and Planning* **15** (3) (2013): 387–402; S. Veenman and D. Liefferink, 'Transnational communication and domestic environmental policy learning' , *ESSA-CHESS - Journal for Communication Studies* **7** (1) (2014): 147–67; H. Joergens, A. Lenschow and D. Liefferink (eds), *Understanding Environmental Policy Convergence. The Power of Words, Rules and Money* (Cambridge: Cambridge University Press, 2014).

members of the ICPR.[25] Next to delegates of member states, NGOs and other stakeholders participate in this RBC as well. The ministers from the ICPR member states are responsible for water policy, i.e. they determine the RBC's mandate, working programme and political goals. The ICPR is organised in quite a hierarchical structure: it consists of a plenary assembly, a strategy group and several working and expert groups that concern topics like flood and low water, water quality and emissions, and ecology. In addition, a secretariat supports the Commission.

Between 1950 and 1970, the main activities of the ICPR were monitoring of water quality and knowledge gathering and dissemination.[26] During the early years of collaboration, the Rhine Commission's role can be characterised as learning facilitator and connector, as it institutionalised the exchange of information between the Rhine members. Over time, the frequency and intensity of this information exchange increased due to the establishment of a strong network consisting primarily of government representatives and experts. Close interaction between the members of this network led to a considerable degree of socialisation and the development of common views about problems and solutions.[27] The role of the ICPR thus broadened from being solely a facilitator for collaboration to multiple roles: as an expert organisation, educator, mediator and coordinator.[28]

The substantive role of the ICPR has broadened over time as well, which was enabled by article 2 of the Bern Treaty stating that the ICPR is competent for all tasks that Rhine members jointly agree upon.[29] In the early years of collaboration, the ICPR focused on visible pollution problems and tasks were quite narrowly defined. Awareness concerning other water issues increased over time. Combined with the occurrence of a number of shock events – e.g. the Sandoz pollution accident in 1987 and the floods of 1993 and 1995 – this led to expansion of the ICPR's scope to a broader

25 See ICPR, *Organisation ICPR*, https://www.iksr.org/en/international-cooperation/about-us/organisation/index.html (accessed 15 Sept. 2017).

26 See Bernauer and Moser, 'Reducing pollution of the river Rhine'; K. Wieriks and A. Schulte-Wülwer-Leidig, 'Integrated water management for the Rhine river basin, from pollution prevention toe cosystem improvement', *Natural Resources Forum* **21** (2) (1997): 147–156.

27 See Bernauer and Moser, 'Reducing pollution of the river Rhine'.

28 See van Eerd, Wiering and Dieperink, *Possibilities for Transboundary Climate Adaptation Governance.*

29 See M.C.J. van Eerd, C. Dieperink and P. Leroy, 'Building upon implementation experiences? Learning lessons from policy feedback between the Rhine catchment and EU water governance', *Water Resources Management* (under review).

range of issues, such as habitat restoration and water quantity management, aiming at the redevelopment of the Rhine's ecosystem. Examples of important programmes are the Rhine Action Programme against pollution (1987), the Rhine Action Programme on Floods (1998) and the Rhine 2020 programme on the sustainable development of the Rhine (2001).[30] Whereas early programmes had a quite sectoral focus, the latest ones are more integrated, comprehensive plans. Since 2007, topics such as climate change, drought issues and micro pollutants have become important elements of the ICPR agenda as well.[31]

ICPR programmes, however, are not formally binding, as the ICPR does not have sanctioning or legal enforcement powers and decisions are based on consensus between states.[32] Yet, peer and social pressure for compliance with ICPR agreements increased over time. In the early years of collaboration, the process can be characterised as 'gentlemen's consultations' with particularly senior officials participating. During the 1990s, the ICPR became more open and transparent by enabling the participation of international NGOs and business organisations, by organising conferences and workshops and by creating informative brochures and a website.[33]

Over time, the ICPR has been identified as a frontrunner for cross-border water governance and served as a best practice example to inspire the development of new international river basin commissions, such as those for the Oder, Elbe and Danube basins.[34] Currently, as an increasing number of (international) actors are dealing with river basin issues, e.g. the European Union and the Danube Commission, the leading role of the ICPR is becoming less prominent.

ICPR as sending arrangement: Key characteristics

One can conclude that the ICPR has a relatively long tradition of international collaboration on water issues. As a result, a strong network has

30 See Bernauer and Moser, 'Reducing pollution of the river Rhine'; ICPR, *Rhine 2020 – Program on the Sustainable Development of the Rhine*, http://www.iksr.org/en/international-cooperation/rhine-2020/index.html (accessed 27 Mar. 2017).

31 See van Eerd, Wiering and Dieperink, *Possibilities for Transboundary Climate Adaptation Governance*

32 See Bernauer and Moser, 'Reducing pollution of the river Rhine'; van Eerd, Dieperink and Leroy, 'Building upon implementation experiences?'

33 See ibid.; Wieriks and Schulte-Wülwer-Leidig, 'Integrated water management for the Rhine river basin'.

34 See van Eerd, Dieperink and Leroy, 'Building upon implementation experiences?'

been created.[35] This network is relatively open to member state representatives, NGOs, IGOs, experts and societal actors. Due to the strong and inclusive network, participating agents are familiar with each other and readily exchange implementation experiences. The European Commission also directly participates in the ICPR network. Over time, a stable ICPR secretariat has been established, which possesses an important knowledge base concerning river basin management. Other institutional arrangements have great confidence in this expertise.[36]

The ICPR functions as a platform for (sub-)national experts to exchange implementation experiences. The ICPR is technically very advanced and considered as a frontrunner concerning (technical) implementation. In comparison to the EU, the innovative and adaptive capacity of the ICPR is positively affected by the greater distance of its working and expert groups to the political level. Although member state representatives are involved at the ICPR's strategic level, it is fair to say that the political accountability of the ICPR is lower as compared to the EU. Another characteristic affecting the ICPR's freedom to act is its weak bindingness. Although the ICPR is based on the Rhine Treaty (1999) and its policies comprise deadlines and norms, their legal bindingness and enforceability is low. At the end of the day, their impact is based on mutual commitment. According to the actors involved, the lower bindingness in comparison to the EU has enabled the establishment of more ambitious objectives and innovative programmes.[37]

Furthermore, the ICPR's member states have a comparable socio-economic and cultural background and have developed a largely comparable understanding about good water governance. Hence, the ICPR favours consensus relatively easily, which enables policymaking at the ICPR level.[38]

EU Water Management

A historic overview

The broadening scope of the ICPR's focus, i.e. towards high water issues, is not unique and should be placed in a European perspective. Concern-

35 See C. Dieperink, 'Successful international cooperation in the Rhine catchment area'. *Water International* **25** (3) (2000): 347–55.

36 See van Eerd, Dieperink and Leroy, 'Building upon implementation experiences?'

37 See ibid.

38 See ibid.

ing flood risk management, for instance, EU INTERREG projects were conducted parallel to the ICPR's development. These parallel tracks of international river basin management enabled a reciprocal exchange of expertise. For understanding interaction and learning between the EU and ICPR level in the upcoming section, we first summarise key developments in EU water management in this section.

In the EU context, water governance is part of the environmental policy field and thus comes under the responsibility of the Directorate–General for the Environment (DG ENV). Environmental policy has gradually developed in the EU since the early 1970s. It acquired a formal Treaty basis in 1987. Since then, it has evolved into a sophisticated, multi-level governance system, constituting one of the EU's major fields of activity.[39] Water was among the first subsectors of EU environmental policy to be developed. It still is one of its most comprehensive sectors.[40]

Three waves of EU water governance can be identified. The first directives focused predominantly on water quality standards, public health and the protection of surface waters allocated for drinking. The second wave, from 1991, focused not only on setting acceptable water quality standards, but also on controlling emission levels as a means of achieving desired standards. Hence, focus was broadened to pollution control and environmental management.[41] The key policy output in this phase was the Nitrates Directive (Directive 91/676/EEC) which is aimed at protecting ground water and surface water against nitrates from 'diffuse' (i.e. mainly agricultural) sources.

The third wave combines the preceding approaches and seeks to integrate them. It started with the adoption of the Water Framework Directive (WFD) in 2000 (Directive 2000/60/EC). This Directive provides an ambitious and innovative framework for water policy based on a river basin approach. It aims to achieve a good chemical and ecological water status for all water bodies. Hence, a paradigm shift from pollution control to integrated river basin management can be identified.[42] The WFD was

39 See C. Knill and D. Liefferink, *Environmental Politics in the European Union. Policy-making, Implementation and Patterns of Multi-level Governance* (Manchester: Manchester University Press, 2007).

40 See G. Kallis and P. Nijkamp, 'Evolution of EU water policy: a critical assessment and hopeful perspective', *Journal of Environmental Law and Policy* **3** (2000): 301–55; J. Richardson, 'EU water policy: uncertain agendas, shifting networks and complex coalitions', *Environmental Politics* **3** (4) (1994): 139–167.

41 See Kallis and Nijkamp, 'Evolution of EU water policy'.

42 See B. Boeuf and O. Fritsch, 'Studying the implementation of the Water Framework Directive in Europe: a meta-analysis of 89 journal articles', *Ecology and Society* **21** (2) (2016): 19; B. Page

followed in 2007 by the Floods Directive (FD) (Directive 2007/60/EC), which requires Member States to assess risks of flooding and to prepare comprehensive flood risk management plans.[43] Both the WFD and the FD will be introduced and discussed in more detail below.

Hundreds of stakeholders can be identified in the water policy field, ranging from water suppliers and polluting industries to environmental groups and consumer organisations. Therefore, the EU water policy process has been described as 'a rather messy amalgam of interrelationships between non-governmental actors and formal institutions'. A potentially large constituency of European level interest groups is interested in the policy area of water, yet only some, such as EUREAU (European Federation of National Associations of Water Services) and ECPA (European Crop Protection) are continuously participating in the EU policy process.[44] The policy domain of EU water management is often referred to as an open policy system, due to, for instance, the extensive public access to information and the multiple venues and channels in which (new) actors can engage.[45] Another example of open collaboration and exchange of expertise in the field of EU water management is the Common Implementation Strategy (CIS), which is an institution for harmonising and enabling the implementation of EU water legislation. The CIS provides a well-organised network for the exchange of information and expertise.[46]

As issues addressed in water policy are quite technical and complex, and include a high degree of uncertainty and ambiguity, the scientific community has a big influence on the EU water agenda. Experts play a key role in identifying issues and providing technologies and solutions.[47] Hence, the water policy domain is dominated by experts and governmental actors.[48] National ministries are ultimately responsible for the implementation of EU water legislation. Governmental actors at the national level transpose

and M. Kaika, 'The EU water framework directive: Part 2 policy innovation and the shifting choreography of governance', *European Environment* **13** (6) (2003): 328–43.

43 See M.C.J. van Eerd, C. Dieperink and M.A. Wiering, 'A dive into floods: exploring the Dutch Implementation of the Floods Directive', *Water Policy* **17** (2) (2015): 187–207.

44 See Richardson, 'EU water policy'.

45 See T. Moss, 'The governance of land use in river basins: prospects for overcoming problems of institutional interplay with the EU Water Framework Directive', *Land Use Policy* **21** (2004): 85–94; Richardson, 'EU water policy'.

46 Van Eerd, Dieperink and Leroy, 'Building upon implementation experiences?'

47 See Richardson, 'EU water policy'.

48 See Kallis and Nijkamp, 'Evolution of EU water policy'.

these directives to national legislation and set up a framework for practical implementation. In practice, however, regional and local water authorities are concerned with the daily implementation

The EU as receiving arrangement: Key characteristics

Over time, multiple issues have arisen concerning democratic legitimacy in the EU, its so-called implementation deficit and its transparency. Since the early 1990s, the EU has taken steps to increase its transparency and openness.[49] This trend has also affected the policy sector of EU water management, which used to be relatively open anyway (see above). The policy subsystem's increasing openness enables policy implementation feedback. However, it also means that there are many actors competing for influence.[50]

Responsiveness of EU actors to the needs, preferences and experiences of their stakeholders is important for the political legitimacy of EU policies.[51] The relevance of implementation experiences in this regard is increasingly acknowledged by EU institutions. The Commission, for instance, seeks to improve the quality and legitimacy of EU legislation by learning from practical experiences in the ongoing 'Better Regulation' programme. As the EU is not directly involved in the practical implementation of its policies, it is dependent upon other actors to acquire implementation experiences.[52] Hence, EU agents, such as the Commission, are willing to learn from other actors' experiences.[53]

The relationship between the ICPR and the EU

Over the years, a strong relationship was established between the ICPR and EU. Three types of links that favour the exchange of implementation experiences exist between these arrangements.

First, since 1976, the European Community, represented by the EU Commission, has been a formal member of the ICPR. The EU's direct participation in the ICPR put pressure on the establishment of agreements

49 See Knill and Liefferink, *Environmental Politics in the European Union.*

50 See van Eerd, Dieperink and Wiering, 'Opening the black box of implementation feedback'.

51 See M. Bovens, 'New forms of accountability and EU-governance', *Comparative European Politics* **5** (2007): 104–20.

52 See European Commission, *Better Regulation.*

53 See European Commission, Personal communication staff member European Commission DG Environment, Feb. and Sept. 2016; van Eerd, Dieperink and Wiering, 'Opening the black box of implementation feedback'.

in the Rhine RBC, as a supra-national organisation was now looking over the shoulder of the ICPR member states. Furthermore, lessons drawn from the ambitious collaboration agreed upon for the Rhine basin could be used by EU institutions as input for EU legislation, which is – in contrast to ICPR policies – legally enforceable. At certain periods in time and for topics that fit the European interest, the EC has been a very active ICPR member. The EU was particularly pro-active in the period after the Sandoz accident (1986) and preceding the development of the Nitrates Directive (1991).[54]

Second, and inversely, the ICPR also participates in EU working groups and the EU strategic coordination group of the CIS network. Initially, this participation consisted merely of observing EU actions. Yet, over the last decades, the ICPR has actively started to mobilise its expertise at EU workshops, meetings and conferences. Capacity constraints, however, hamper the representation of ICPR staff in all EU CIS meetings.

A third, yet more informal, link for policy implementation feedback are the ICPR and EU's common delegates. In practice, national representatives and experts participating at the EU and ICPR level are often the same people. This overlap and ongoing exchange strengthens the network and enables effective learning.[55]

Understanding multi-level learning: Two cases of policy feedback

In this section, two cases of policy implementation feedback between the catchment and EU level are elaborated upon in order to gain a better understanding of multi-level learning. The first case concerns ecological issues, while the second has to do with flood issues. Selection of these specific cases can be justified since both have been – and still are – key issues on the EU water governance agenda. The WFD and the FD that focus on these issues constitute the cornerstones of EU water management. For both cases, learning and the feedback of implementation experiences from the RBC to the EU level is assessed by applying a process-tracing analysis. Data collection consisted of semi-structured interviews, policy document and scientific literature analysis, and observation research at both the EU and ICPR level.[56]

54 See van Eerd, Dieperink and Leroy, 'Building upon implementation experiences?'

55 See ibid.

56 See ibid.

Case One: Addressing Ecological Challenges

Since its inception, the ICPR has been a frontrunner in water governance. The initial work of the ICPR concentrated on water pollution problems in the Rhine basin.[57] This central concern resulted in the Chemical and Chlorides Convention in 1976. More ambitious goals came in reach after 1 November 1986, when a disaster hit the Rhine as a chemical site at Sandoz, near Basel, caught fire. Chemicals flowing into the Rhine and inadequate handling by the fire brigade caused a toxic wave downstream, killing almost all organisms. This accident triggered a lot of publicity and public concern, which put pressure on the ICPR to formulate transboundary actions to prevent comparable accidents in the future. This resulted in the Rhine Action Programme (RAP) of 1987.[58] The RAP had a broader chemical, biological and ecological scope than earlier ICPR policies.[59] It envisioned redevelopment of the Rhine's ecosystem by stimulating ecology and habitat restoration and the return of previously indigenous species by the year 2000, of which the salmon became a symbol. Concrete pollution reduction goals were included. The riparian states, for instance, agreed on a reduction of at least half of the river's load of heavy metals, organic pollutants and fertilisers. Other actions included the building of fish ladders and improving spawning conditions. Compared to other international agreements at this time, the Rhine policies, and in particular the RAP, were very ambitious, comprehensive and specific.[60] Development of this ambitious programme was enabled by social and political pressure, the ICPR's history of collaboration, the involvement of primarily experts and the ICPR's relatively low bindingness. Although the programme was not legally binding, it yielded an active and significant tackling of pollution issues in the Rhine basin.[61]

After failed attempts to introduce a somewhat more ecological ap-

57 See Dieperink, 'From open sewer to salmon run'.

58 See ibid.; ICPR, *Rhine Action Programme ICPR against Pollution*, Strasbourg, 8th Conference of the Ministers, 1 Oct. 1987, http://www.iksr.org/en/international-cooperation/rhine-2020/index.html (accessed 22 May 2017).

59 See Bernauer and Moser, 'Reducing pollution of the river Rhine'; Dieperink, 'From open sewer to salmon run'.

60 See ibid.

61 See Dieperink, 'Successful international cooperation in the Rhine catchment area'; ICPR, *Rhine Action Programme ICPR against Pollution*; and see ICPR 1998, *Action Plan on Floods*, Rotterdam, 12th Conference of Ministers, 22 Jan. 1998.

proach in EU water policy in the early 1990s, pressure for a fundamental rethink of EU water policies came to a head in 1995, as actors continued their call for deregulation and decentralisation of the complex water policy patchwork. This resulted in the design of an integrated directive.[62] The ICPR's experiences concerning the practical implementation of comprehensive water governance provided significant input during the five-year drafting process of what eventually would become the WFD. More specifically, the ICPR's RAP served as a key model for the WFD, since it was widely acknowledged as an innovative and concrete programme that had proved its success. The European Commission, moreover, was eager to draw upon this programme, as the RAP was, at that time, the only international plan that had been implemented in practice and which covered all aspects of water quality management.[63] Examples of WFD elements based on lessons learned from the ICPR relate to integrated water management, the river basin principle and transboundary cooperation, the involvement of stakeholders and the public, dealing with groundwater issues and using programmes of measures.[64] However, not all RAP elements were taken over in the WFD. For example, the ICPR's work was more advanced with regard to integrated water management since it already combined water quantity and quality measures to create win-win situations, and the WFD focuses less on the return of species.[65]

In this case, the existing links between the EU and ICPR supported the exchange of implementation experiences. Common delegates placed the RAP's measures and actions in the spotlight at EU venues. Representatives of Rhine riparian states, for example, defended the river basin management approach, and this concept eventually became a cornerstone of the WFD. Moreover, the EU representative in the ICPR at this time was a very active participant at the Rhine catchment level while the ICPR's president was a former EU staff member. This established an additional, more direct link enabling the mobilisation of implementation experiences. Finally, ICPR

62 See European Commission, *Introduction to the New Water Framework Directive* (Brussels: European Commission, 2017), http://ec.europa.eu/environment/water/water-framework/info/intro_en.htm (accessed 24 Apr. 2017); Kaika, 'The Water Framework Directive'.

63 European Commission, *Personal communication staff member European Commission DG Environment*, Feb. and Sept. 2016.

64 Ibid.

65 See ICPR, *Rhine Action Programme ICPR against pollution*, Strasbourg, 8th Conference of the Ministers, 1 Oct. 1987, http://www.iksr.org/en/international-cooperation/rhine-2020/index.html (accessed 22 May 2017).

staff members participated in EU workshops throughout the drafting process of the WFD.[66] Important venues for exchanging expertise included EU working and expert groups in the CIS process as well as the ICPR's ecology working group. Learning took place as experiences were repeatedly discussed at these venues. The continuous exchange of implementation experiences in all stages of the policy process enabled decision makers at the EU level to use their knowledge base during the WFD's agenda setting and policy formulation stages. Throughout the process, moreover, the EU Commission remained keen on additional experiences from the ICPR.

Case Two: Addressing Flood Issues

Following the floods of 1993 and 1995 in the Rhine basin, the focus of the ICPR was broadened from water quality to water quantity management. Pressure from the downstream riparians in the Netherlands and Germany triggered the ICPR to also include high water issues in its policies.[67] Policies were relatively easily established due to the ICPR's long history of collaboration, the established trust, network, and the existing ICPR structure. As a result, the Rhine Action Plan on Floods (RAPF) was established in 1998, aiming to improve flood protection by 2020 and to extend and enhance floodplains of the Rhine. Five principles were considered leading: the storage of water, giving space to the river, alignment with other sectors, creating awareness for flood risks, and integrated river basin action in the spirit of solidarity.[68] The plan was a bundle of activities and measures, which acts as a target framework that is continuously adapted. An important outcome was the Rhine Atlas (2001, renewed in 2015), mapping flood risks along the Rhine system, which was and still is important for information management and increased public awareness.[69] The latest evaluation of the RAPF shows that the Rhine riparians successfully implemented integrated flood risk management (IFRM) between 1995 and 2010.[70]

66 European Commission, Personal communication staff member European Commission DG Environment.

67 See van Eerd, Dieperink and Leroy, 'Building upon implementation experiences?'

68 See M. Disse and H. Engel, 'Flood events in the Rhine basin: genesis, influences and mitigation', *Natural Hazards* **23** (2001): 271–90; ICPR, *Action Plan on Floods*. Rotterdam.

69 See ibid.

70 See ICPR, *Evaluation of Measures to Reduce the High Water Levels in the Rhine: Implementation of the Action Plan on Floods 1995–2010 including foresight for 2020 and 2020+*, report 199 (Koblenz:

Parallel to the ICPR's shifting focus towards flood issues, several developments can be identified at the EU level. INTERREG projects focusing on high water management across borders were implemented from the 1990s onwards.[71] In 2004, the urgency of flood issues was formally recognised in an EU Communication, calling member states to express their thoughts about an EU FD. This directive was established after a relatively short drafting period in 2007 (Directive 2007/60/EC).[72] Initiating members were France and the Netherlands. Austria, having the Presidency of the EU Council of Ministers at that time, was also strongly involved in the development of the FD. The expertise of representatives from those countries provided important input and enabled the drafting process, as they were involved in both the EU and ICPR network, had a long tradition with domestic IFRM and had experiences with the RAPF's implementation process in the Rhine basin.[73] Due to capacity constraints and the distance between Brussels and daily water governance, actors involved at the EU level were keen on input from these implementing agents. Because of its direct involvement in the ICPR, the EU was well informed about IFRM measures taken in the Rhine basin. Furthermore, the staff of the ICPR was consulted to present best practices of the RAPF in order to convince reluctant EU member states about the need for a FD. Evaluations of the RAPF's implementation (2000 and 2005), and in particular its clear and ambitious measures and targets (e.g. for flood forecasting and water retention) provided important input for the establishment of the FD. However, the ICPR was not the only source of implementation experiences. Flood risk management practices from the Danube Commission, INTERREG projects and other (domestic) institutional arrangements were also important. Implementation experiences concerning the rigorous and detailed WFD also strongly influenced the drafting process of the FD. These experiences, in combination with the limited EU mandate to

ICPR, 2014); ICPR, *Action Plan on floods 1995–2010: Action Goals, Implementation and Results. Short Term Balance*, report 200 (Koblenz: ICPR, 2014).

71 See M.M. van der Giessen, *Coping with Complexity. Cross-border Cooperation between the Netherlands and Germany*, dissertation (Nijmegen: Radboud University, 2014).

72 See European Commission, *Communication form the Commission to the Council, the European Parliament, the European Economic and Social committee and the Committee of the Regions: flood risk management, flood prevention, protection and mitigation* (Brussels: European Commission, 2004), COM 2004/0472.

73 See van Eerd, Dieperink and Wiering, 'A dive into floods: exploring the Dutch Implementation of the Floods Directive', *Water Policy* **17** (2) (2015): 187–207.

work on spatial governance, resulted in a more flexible and largely procedural directive.[74] Several elements from the RAPF were used in the FD. Both policies are based on the solidarity principle, identify similar phases of IFRM (assessment of risks, mapping, and taking action) and offer a framework setting (ICPR 1998). Yet, overlap is less clear when compared to the ecological case described earlier.

Implementation experiences were exchanged particularly at the working group level, i.e. in the ICPR and EU flood working groups. The EU flood working group falls under the institutional structure of the WFD's implementation guidance institute, the CIS network. As flood issues are strongly expert-based and technical, often the same persons represent the Rhine member states at both the EU and the Rhine catchment level, which enabled learning between these institutions. Both during agenda-setting and policy formulation, as well as during the implementation of the FD, experiences of members played an important role.

The two case studies confirm that implementation experiences of the ICPR affected complex water resource management at the EU level. Key ICPR policies served as models for the development of EU water legislation. This can be explained by the *innovative, ambitious* character of these ICPR policies and their reliance on *clear and concrete* programmes, norms and measures that had *proven to be successful*. The relatively high innovative capacity of ICPR policymaking and implementation can be explained by the RBCs relatively *low political accountability*, the *involvement of primarily expert oriented agencies* and its *lower degree of institutionalisation* and *bindingness* in comparison to the EU setting. Furthermore, the ICPR's *long tradition of collaboration*, its *well-organised and stable secretariat* and *mutually familiar members* facilitated reaching consensus. These institutional characteristics determined the ICPR's freedom to act and explain its front-running position in substantive and regulative precision and coverage of water governance.

In addition, evidence from this study shows that the *institutional context* in which both arrangements are embedded played an important role as well. The *existing network and links* between these arrangements, and in particular

74 See van Eerd, Dieperink and Leroy, 'Building upon implementation experiences?'

the large share of actors who have a *combination of memberships* at both the RBC and EU level, enables the exchange of implementation experiences. *Existing venues*, and in particular the working and expert groups at both the EU and ICPR level, can be seen as important platforms for actor interaction and the exchange, bundling and selection of implementation experiences.

Concerning the condition of timing, we found that policies in both cases were established following *external (shock) events* and related public and political *pressure*, e.g. accidental pollutions or flood events. However, timing appeared less relevant for explaining policy implementation feedback between the studied arrangements, which was observed to entail a more long-term, continuous and reciprocal process of exchange.

Furthermore, the two cases suggest that the chance for feedback of implementation experiences is higher when the receiving arrangement is open and responsive to such expertise. In both cases, the EU appeared as *accessible, open* and *eager to learn* from the expertise of RBCs. However, the *responsiveness* of EU institutions to the ICPR's implementation experiences differs between the cases. With regard to water quality, the ICPR's RAP was the only international and integrated plan that had been implemented in practice at the time. Hence, the European Commission was eager to learn from the unique front-running expertise of the ICPR. With regard to flood risk management and the RAPF, however, more expertise was available, decreasing the responsiveness and lowering the EU's exclusive reliance on the ICPR's knowledge. Furthermore, as the formal authority of the EU and the ICPR differ with regard to flood risk management, and the EU lacks formal competence to work on spatial issues, the ICPR's experiences and policies could not serve as a blueprint so easily (*discursive and institutional misfit*). Hence, variation in the relation between both arrangements, the competition for expertise and the characteristics of the receiving institution explain differences between both cases in the extent to which experiences of the RBC have affected EU water governance.

We can conclude that, in our two cases, learning based on implementation experiences predominantly draws upon informal communication and actor interaction. Mutually familiar members, combined membership, the exchange at working group level and existing networks appear as important explanatory factors. This is in line with findings from other research,[75]

75 E.g. see K. Holzinger, C. Knill and B. Arts, *Environmental Policy Convergence in Europe: the Impact of International Institutions and Trade* (Cambridge: Cambridge University Press, 2008); Joergens, Lenschow and Liefferink (eds), *Understanding Environmental Policy Convergence*; S.A. Veenman,

which also provide evidence of the prevalence of communicative, often informal, channels of exchange over the impact of formal regulation or competitive pressures through the market mechanisms in the transfer of policies – or aspects of policies – from one institutional arrangement to another.

Future EU water governance is not likely to progress without implementation experiences from daily water governance at both the local, regional, national and river basin level. Our analysis has shown that river basin organisations can have a key role in providing these experiences, by acting as best practice examples on river basin management for EU water resources management. Following current developments in the EU water policy domain, we expect that river basin organisations will contribute to new EU policies in the field of climate adaptation, micro pollutants and combating medicines and hormones in the EU's water systems.

Domestic Environmental Policy and Transnational Communication. The Cases of Contaminated Land and Noise around Airports (Delft: Eburon, 2008).

CHAPTER 16.

ENVIRONMENTAL PROTECTION AND THE EVOLUTION OF THE FRENCH AND GERMAN ENERGY SYSTEMS FROM 1973 TO THE 2000s

Christopher Fabre

The French and German energy systems are generally presented in terms of their differences, by focusing to a great extent on the nuclear dimension, and implicitly on a relation to the environment deemed to be distinct. Without challenging these elements of differentiation, this chapter will, on the contrary, determine whether the development of environmental protection contributed to phenomena of convergence.

In order to do so, the present analysis will broaden the approach by examining energy systems as a whole, doing so within a long historical perspective from 1973 to 1990. Energy, in the broad sense of the word, aims to respond to a variety of uses in transportation, industrial processes, heating, lighting, etc. Different types of energy – well beyond the opposition currently emphasised between nuclear and renewable energies – can vary in meeting these needs. Let us recall that in 1973, electricity represented just nine per cent of total consumption in France and 10.9 per cent in Germany, and in 2010 respectively reached just 25 per cent and 22 per cent.[1]

This chapter proposes observing the evolution of energy systems through all the energies of which they consist, in addition to major usage categories and the evolution of consumption. This approach reveals a certain number of economic and technical components, along with those relating to the form of institutional organisation, which formed so many structural constraints that largely determined the evolution of energy systems, and ultimately cast doubt on the degree of latitude available in energy policy. More specifically, this approach identifies the growing role of environmental protection as a structural constraint on the Community and even international level, and subsequently characterises it as a factor of convergence for the energy systems of both countries.

1 International Energy Agency, Energy policies of IEA countries, 2009 review (OECD, 2010); International Energy Agency, Energy policies and programmes of IEA countries, 1988 review (OECD, 1989).

Here the focus will be on the period between 1973 and the 2000s, in other words the period including the two oil crises, during which new energy policies were implemented in an effort to abandon the 'all petroleum' model, followed by a period of low oil prices lasting from the oil counter-shock to the rising prices of the 2000s. The first period was marked by the development of the French and German nuclear programmes, implementation of the first energy-saving measures, and the gradual emergence of environmental protection within energy policy. The second period was marked by the affirmation of environmental protection, and the end of major infrastructure programmes in favour of rationalisation of the energy systems of both countries, within the context of German reunification and the broader liberal resurgence.

This analysis is based on different types of documentary sources. We firstly relied on the reports of major international organisations such as the International Energy Agency (IEA) and the Organisation for Economic Co-operation and Development (OECD). The productions of the French and German Parliaments also provided important material, whether in the form of legislative texts, information reports or parliamentary debates, such as those appearing in the Journal Officiel. The documents of the European Commission were also used, such as general reports and bulletins. This approach was finally supported by articles from the specialised and general press on energy subjects, in addition to analysis by historians, political scientists and economists specialising in energy matters.

1973–1981: The start of a French-German convergence as a result of oil crises and the emergence of the protection of nature

Two differing situations with regard to energy dependence, but measures that were on the whole similar and favourable to the environment

During the *trente glorieuses*, economic growth in France and Germany was initially fuelled by the exploitation of coal resources, and then largely by oil imports. In 1973, oil represented 56.2 per cent of energy supplies in Germany, and 66.5 per cent in France.[2] There was therefore a difference with regard to dependence between France and Germany, one that can largely be

2 International Energy Agency, *Energy Policies and Programmes of IEA countries*, 1988 review (OECD, 1989); International Energy Agency, *Energy policies of IEA countries*, 2007 review (OECD, 2008).

explained by the availability of coal in Germany, which represented nearly 31.6 per cent of supplies, as opposed to just 16.2 per cent in France, and to a lesser degree by natural gas supplies, which represented 9.8 per cent in Germany and 7.5 per cent in France.

It was in this context that the first oil crisis surprised all industrialised states. No slowdown was anticipated in GDP growth, and energy consumption was thought to be linked to economic expansion. Leading to a twofold increase in the price of a barrel, the first oil crisis had a greater impact in France than Germany. France's small coal resources, which moreover were non-competitive, represented a first and essential structural constraint at the time of the oil crisis. This was all the more true given that French dependence went hand in hand with France running behind Germany in terms of electrical equipment for households and industry. Per capita electricity consumption in Germany was almost thirty per cent greater than in France (3,270 kWh per capita compared to 2,400 kWh).[3] In addition, the structure of the German economy, along with its considerable export capacities, allowed it to better absorb the effect of the oil crisis, with Germany maintaining a positive current account (+ 6 Md$), unlike France (- 6,25 Md$).

The response of industrialised countries to the oil crisis was to diversify their energy sources, develop national energies and engage in the first efforts toward energy saving. In this perspective, France did not possess competitive national resources in comparison to Germany, and its range for energy saving was much more restrained. Reports of a 'situation of energy penury'[4] were broadly shared during debates in the Assemblée nationale, at a time when the country was seeking to catch up economically with Germany.

Yet affirmation of the scarcity of energy and the need to save it was in keeping with the theories of environmentalists, especially as they were developed in the Meadows report. Similarly, the desire to develop national energies implicitly led to favouring of energy sources that offered environmental advantages. The oil crisis and the measures taken to contend with it were similar on both sides of the Rhine, and laid the groundwork for a French-German convergence regarding the beginnings of environmental protection.

3 Y. Coudé du Foresto et al., *Rapport d'information (n°8) sur l'ensemble des questions nucléaires* (Sénat, 13 Oct. 1970).

4 J.-F. Pintat, *Rapport d'information (n° 96) à la suite de la mission effectuée à Détroit du 20 septembre au 1er octobre 1974, pour la Xème conférence mondiale de l'*énergie (Sénat, 19 Nov. 1974), p. 19.

From recognition of the protection of nature to its affirmation as a structural constraint

According to Daniel Boullet, environmental protection progressed continually from the 1970s to the 1990s, with two moments of acceleration – 1970 to 1976, and 1983/1984 to 1990 – with the years between them marked by a decrease in this trend due to the crisis. This gradual approach was similar in Germany:

> everything suggests that mobilisation to defend the environment was initiated fairly slowly, albeit surely! People in France were struck by the emergence in Germany of a shift in opinion as clear as the one in the United States, even though German society was of course no stranger to the currents of thought crossing through Europe at the time.[5]

The environmental concerns affecting energy were structured around five key themes: water pollution, air pollution, energy management, the nuclear risk and the development of renewable energies.

Water pollution was an important topic in environmental protection during the 1970s, especially the pollution caused by hydrocarbons following the shipwreck of oil tankers. A number of international agreements were signed after the one in Bonn in 1969, as different maritime catastrophes received a certain amount of media attention (*Torrey Canyon* in 1967, *Olympic Bravery* in 1976, *Boehlen* also in 1976, etc.). This series of conventions and laws enforced in France and Germany led to an inherent increase in the cost of imported oil. What's more, these measures came at a time, before the arrival of the second oil crisis, when the cost of oil was tending to slightly decrease. While states initially baulked at implementing measures to combat black tides, it later became clear that this prevention was not so harmful to imports, and that it even helped maintain a marginal price for oil high enough to ensure the profitability of energy diversification programmes.

With regard to the environmental damage caused by energy use, air pollution most certainly garnered the most attention in studies during the 1970s, with a special focus on sulphur oxide and nitrogen oxide from the burning of coal. The focus was more on understanding the phenomenon (measuring emissions, in addition to health and environmental consequences). The regulatory aspect was not put aside, but remained highly disparate among European states, all while giving rise to growing restrictions that were nev-

5 D. Boullet, *Entreprise et environnement en France de 1960 à 1990 – les chemins d'une prise de conscience* (Geneva: Librairie DROZ, 2006).

ertheless limited to the period. The fight against air pollutants involved a *de facto* marginal increase in the price of fossil energies, both through R&D efforts on the topic and the implementation of technologies and techniques seeking to lower emissions. The initiation of this regulation, which was developed on the Community level, tended to introduce a convergence in terms of constraints.

Energy management inherently contributed to decreased pollution, which is proportional to volume of energy consumed. Yet this environmental role of energy management was not initially assumed in a context of debates surrounding topics of 'degrowth'. This did not, for all that, prevent a certain number of measures from being implemented in 1974 in both France and Germany, notably with regard to the thermal insulation of residences, transportation and industry. The Commission level subsequently took over, with a transition from a 'fight against waste' to a genuine policy of 'energy management'.

The treatment of the three aspects of water pollution, air pollution and energy management was also inscribed, albeit haltingly, within an increasingly multilateral framework beginning in the 1970s, leading to a mechanism of convergence in energy systems impacted by their regulation.

Two initially identical nuclear programmes with two different results

The 1970s are nonetheless frequently presented as the period when the divergence between France and Germany on the nuclear question began. This divergence is often seen as being closely linked to a distinct relation to the environment, which calls for nuancing.

Despite their different energy contexts, in 1973 France and Germany each engaged in nuclear programmes of equivalent scope, of the order of 50 GW by 1985.[6] The scope of these programmes reveals the essential role of price as a factor, which acted as a fundamental structural constraint. Parliamentary reports from the period and the publications of international agencies show a consistency in the economic comparison of different sources of energy. A logic of merit order prevailed in public policy decisions between energies. At the time of the oil crisis, nuclear power affirmed itself as the solution by distinguishing itself in matters of economic competitiveness for the installation of new capacities, as compared to fuel oil and coal.

6 Hearing of M. D'Ornano at the Assemblée nationale, 1976; International Energy Agency, *Energy policies and programmes of IEA countries*, 1977 review (OECD, 1978).

The environmental dimension was nonetheless clearly present, with both the detractors and promoters of this energy, and in a manner that was equally pronounced and varied in both France and Germany. In fact, opposition to nuclear programmes was already diverse by nature, and included scientific, unionist and local criticism, in addition to that of environmental activists. Major protests took place in both countries. Conversely, the environmental advantages of nuclear power – its lack of polluting emissions and the small footprint of power plants in relation to energy produced – were highlighted on both sides of the Rhine. However, decision-makers were suspicious of the association between anti-nuclear arguments and the theories of degrowth referred to by many environmental activists. The divergence of French and German nuclear programmes therefore does not appear, within an initial approach, to reside in each country's different relation to the environment, but rather in the difference of their initial structural constraints.

Beyond the response to the oil crisis, these programmes were more broadly inscribed in scenarios of evolving electricity consumption based on the growth of the *trente glorieuses*. In France, forecasts projected an increase in energy consumption of the order of fifty per cent between 1973 and 1985, and of 200 per cent by 2000. In Germany, capacity needs were evaluated at 160 GW by 1990, compared to 76 GW in 1976.

The role of these scenarios was particularly central to the orientation of energy policies, although the projections notably proved to be overestimated, with GDP growth rates two to three times less than those forecast. As a result, in 1990 Germany finished with 96.5 GW of installed capacities instead of 160 GW; in France, a fifty per cent increase in energy consumption did not occur in 1985, but fifteen years later, in 2000.[7]

In this context, French centralised organisation played an essential role in the implementation of the country's nuclear programme. Conversely, Germany's federal organisation offered numerous communication and legal channels for local opposition, thereby slowing the construction of reactors in Germany. These delays in Germany subsequently helped reveal the overestimation of consumption projections, and the need to revise the nuclear programme downwards in a country already very well equipped with regard to electricity.

On the contrary, in France, many elements structurally promoted the implementation of the nuclear programme. First, voluntarist projections were abandoned quite late, in the mid-1980s. Second, France highlighted

7 International Energy Agency, *Energy policies of IEA countries*, 2008 review (OECD, 2009).

its gap in electricity consumption in comparison to Germany, which was reduced only during the 1990s. Finally, the French programme was based as much on coal, especially imported coal, as it was on nuclear energy, with the latter eventually representing just fifty per cent of electricity production.[8] It was therefore the thermal programme based on coal that served as a genuine downward adjustment.

The essential role of the price factor in reaching environmental objectives

The second oil crisis confirmed the essential role of the price of energy as a key deciding factor in the evolution of the energy system. In fact, it was during the period of the two oil crises that the most important transformation of the energy system took place, both with regard to the evolution of consumption and the reorganisation of production apparatus.

Progress was made toward the objectives identified in the aftermath of the first oil crisis: decreased dependence on oil, increased diversification, a greater share of 'national' resources and relative containment of energy consumption. In other words, there was an improvement in energy efficiency, which was identified as a central pillar of environmental protection, in addition to a reduction in air pollution through decreased use of oil and a shift to less polluting sources (electricity and natural gas). Detailed analysis reveals that these achievements were more the consequence of the price of oil than of the policies put in place, with an essential role played by the second oil crisis in particular.

The evolution of consumption reveals the very high responsiveness of energy consumption and energy intensity to rising prices. These two indicators decreased during the two oil crises, but began to rise at the smallest drop in price during the second half of the 1970s and from 1983 onward, and even more significantly with the oil counter-shock of 1986. The impact of oil crises and the development of national energies nevertheless led to a structural decrease in oil consumption in both countries.

The second crisis also led to a revival of the German and French nuclear programmes. However, the installed nuclear capacity of Germany only served to complement pre-existing coal-based thermal capacities. Finally, natural gas developed in continual and relatively similar fashion in both countries, doubling in volume.

8 Coudé du Foresto et al., *Rapport d'information (n°8).*

In the end, the period extending from 1973 to 1990 revealed highly similar environmental dynamics when observing an energy system as a whole, namely:

- A decrease in air pollution following the decrease in oil and coal consumption, and a shift in uses toward electricity and natural gas, whose consumption increased;
- An improvement in energy efficiency and energy savings following the rise in the price of oil;
- Greater investment in research and development on energies emitting fewer pollutants, notably renewable energies.

It was structural evolution of the energy sector that helped achieve environmental objectives, with the latter in turn justifying the sector's growth.

More generally, it becomes clear that, in 1973, France was running behind Germany with regard both to diversifying its supply and the evolution of its energy consumption. Despite the size of the French nuclear programme in 1990, overall electricity consumption in France remained twenty per cent less than in Germany. Until the 1990s, France pursued a logic of catching up to Germany in the field of energy. It was only in 1989 that the energy independence rate in both countries converged around 45 per cent, while in 1973 it was eighteen per cent in France and forty per cent in Germany.

These gaps in consumption were also present in the transportation sector (49 Mtoe compared to 38.5),[9] industry (46 Mtoe compared to 72) as well as other sectors (58.8 Mtoe compared to 77.7).[10] The evolution followed identical trends for each of these sectors: reduction of consumption by industry in both countries (-11.5 Mtoe in France compared to -12 in Germany), increase in transportation (+15 Mtoe compared to +15.5), and relative stability in other sectors (+2.5 Mtoe compared to – 0.8).[11]

Aside from different starting points in the two countries, all the evolutions reveal numerous parallels on both sides of the Rhine, especially structural constraints that largely surpassed the framework of energy policy. The reduction in energy consumption by industry reflects energy savings resulting entirely from price elasticity and the structural decrease of industry in Europe. Consumption in the transportation sector followed a dynamic

9 Million-ton Equivalent to Petroleum.

10 Residential, service industry, public buildings.

11 International Energy Agency, *Energy Policies and Programmes of IEA Countries*, 1988 review (OECD, 1989); International Energy Agency, *Energy Policies of IEA Countries*, 2007 review (OECD, 2008).

of households equipping themselves with automobiles and the price of oil. Residential consumption responded to the introduction of new uses.

Environmental protection considerations increasingly resonated with these general evolutions, with a subsequent convergence between energy and environmental considerations, until the latter came to represent a powerful lever in the evolution of energy systems.

From 1986 to 2000, the continuation of convergence despite the drop in oil prices

A reworked framework for developing energy policy

While the period from 1973 to 1981 demonstrated the decisive role of the two oil crises in the evolution of French and German energy systems, the ensuing period was marked by very low oil prices following the oil counter-shock of 1986. The context in which the energy policy of the two countries was evaluated was in keeping with a counterincentive framework. Moreover, reunification completely restructured the German energy landscape. Despite these fundamental changes, the dynamics of convergence of large aggregates continued during the period.

Environmental protection was considerably strengthened during the period, thanks to increasing awareness on the part of actors and enhanced understanding of the issues. Environmental measures started to fully become a structural constraint in the determination of energy systems, taking the reins from lower prices. The issue of global warming in particular, which France and Germany quickly came to grips with, emerged during the 1990s as a new framework for coherence for environmental considerations relating to energy. However, this did not prevent the divergences relating to nuclear power from continuing, especially following the Chernobyl disaster, albeit always on the basis of the initial structural differences previously mentioned. Finally, the 1980s and 1990s were characterised by the two new dynamics of liberalisation and decentralisation, which initiated an updated energy framework with the environment as an integral part.

The period between the first and second oil crises revealed the decisive nature of oil prices in the evolution of energy systems, which directly impacted the environmental component of these systems. The situation was inverted with the lasting drop in oil prices following the oil counter-shock, as the price in current dollars remained slightly higher than during the preceding decade, but lower in constant dollars. Furthermore, in correlating price levels to the evolution of purchasing power, the decrease in the cost of oil proved even more pronounced, reaching levels similar to those preceding the first oil crisis.[12] This toppling of the price factor exerted crucial influence on both the evolution of energy systems, and how environmental considerations relating to energy were understood and resolved. In other words, this period confirms, by way of an inverse trend to that of the 1970s, the central nature of energy prices as determinants of energy policy.

Strengthened environmental protection and the crystallising role of the fight against global warming

Environmental protection advanced considerably during this period, albeit in non-linear fashion, and with contrasting approaches depending on the subject. In France and Germany, there was an overall movement of rationalisation and systematisation of environmental policy, which could rely on the improved knowledge of the issue that grew out of research conducted during the 1970s. The measures concerning the primary topics from the preceding decade (air pollution, hydrocarbon pollution, alternative energies) reinforced one another in a highly significant way. Still, efforts to save energy suffered from low prices, with no policy succeeding in taking over in this area.

Despite the reorganisation of the energy mix prompted by the oil counter-shock, low prices of the 1990s, and reunification, the environmental considerations emphasised during the 1970s continued to be handled through a strengthening of measures. While this context yielded contrasting results depending on the subject, there was nevertheless an increasingly integrated and crosscutting approach that promoted a rationalisation of previously sketched-out processes, and increasing awareness among political and private actors as well as the public.

These trends unfolded in similar fashion in France and Germany, reinforcing the previously identified dynamic of convergence, especially with an increased role for European governing bodies. The fight against air pollution contributed to a drop in the share of coal, promotion of nuclear power and expansion of natural gas in both countries. The fight against black tides and water pollution continued, endeavouring to limit the expansion of oil through energy saving measures. However, the two countries struggled to change habits and

12 J.-M. Jancovici, *Comment a évolué le prix du pétrole depuis 1860?*, https://jancovici.com/transition-energetique/petrole/comment-a-evolue-le-prix-du-petrole-depuis-1860/ (accessed 16 May 2020).

reorganise the transportation of merchandise, without succeeding in genuinely reducing consumption, including in the residential and service sectors. Both countries pushed the development of renewable energies, although there were different initial choices of industries, which were largely due to the share of coal in their respective energy mixes. France seems to have put itself in an inferior or following role regarding industrial issues relating to environmental protection, as well as in the development of norms.

Finally, the years that followed the second oil crisis saw the affirmation of global warming, which henceforth tended not to supplant preceding considerations, but to encompass them within a concept offering an overall framework for coherence. Global warming led to a transposition of the various measures already put in place since the first oil crisis to meet objectives in supply security, reducing the energy bill and environmental protection. The fight against climate change distinguished itself as a supplemental stage, a new framework for coherence in the dynamic of convergence between the French and German energy systems. It also consecrated the encompassing and transnational nature of environmental protection, which was established more than ever as a constraint imposed on all, as noted by M. Barnier: 'protecting the environment will never be effective if it is limited to a strictly national framework. Responsibility for the atmosphere or the ocean does not fall to any particular state, but to all of them.'[13] The European Commission nevertheless treated the two topics of energy market reform and environmental protection separately, essentially due to the refusal of states to implement taxation of greenhouse gas emissions (carbon tax). While the measure faltered, notably because of the need to deploy an international framework, it nevertheless emerged that France and Germany agreed on the principle. In fact, the carbon tax amounted, without it being fully expressed at the time, to rationalising the inclusion of the environmental costs of producing goods within the price of these very goods (or internalisation of environmental externalities), which had been done continuously but diffusely since the 1970s.

Germany's engagement in the fight against global warming was deemed 'paradoxical' from the beginning, but could be explained by two primary reasons: 'first, environmentalists wield great electoral weight. Also, the greenhouse effect could make it a little easier to accept the reduction in

13 M. Barnier, *Rapport d'information sur la politique de l'environnement*, no. 1227 (Assemblée nationale, 11 Apr. 1990).

coal production desired by leaders, but extremely painful on the social level'.[14] The rise to power of the SPD-Grünen coalition came shortly after signing of the Kyoto Protocol, in which Germany committed to reducing its CO_2 emissions by eight per cent between 1990 and 2010. Chancellor Schröder initially considered CO_2 emissions and the desire to preserve jobs in the nuclear industry when rejecting calls by the Greens to begin phasing out nuclear power. However, the alternatives of natural gas as a means of transition, and renewables as a long-term solution, were put forward with increasing force in the public debate.[15] The Chernobyl disaster fifteen years earlier had left a lasting impression on German public opinion with regard to nuclear power.

Chernobyl and nuclear differentiation between France and Germany

The Chernobyl disaster occurred at a time when the most recent production plants, whose construction had been decided on during the second oil crisis, came into service and resulted in overcapacity. In other words, the Chernobyl disaster, and the challenges to nuclear power that it sparked, coincided with the realisation that growth in electricity consumption was much lower than estimated. This was a significant change, as the preceding years were largely characterised by the fear of not possessing enough capacity to meet demand. The fact that there was now excess supply reshuffled the deck and promoted challenges to nuclear power.

In both France and Germany, this posed a profound challenge to the industry across fairly similar elements: decrease in R&D budgets and renewed focus on issues of security, strengthened norms, abandonment of breeder reactors, issues connected to long-term storage, etc. However, the disengagement was more pronounced in Germany than in France. Industrial actors acted very early on, as there was no possibility to construct new reactors in the country. It is once again clear that it was the sense of possessing a degree of latitude with national resources, thanks to coal, that made possible the agreement of 14 June 2000 to disengage from nuclear power.[16]

14 B. Derosier, *Rapport d'information sur la politique communautaire en matière d'*énergie, no. 2119 (Assemblée nationale, 13 June 1991).

15 'L'Allemagne ne sait comment réduire sa consommation', *Le Monde*, 21 Jan. 1999.

16 *Energie: la sortie du nucléaire entraîne maints réaménagements*, Allemagne Infos (Centre d'information de l'ambassade d'Allemagne, July 2000).

On the contrary, in France, decision-makers and public opinion did not sense the possibility of nuclear disengagement due to the country's high dependence on this energy, which was itself the result of a lack of available national resources. The perception of nuclear power was nevertheless clearly affected, but this was expressed through a growing desire to stop constructing new nuclear power plants, the reinforcement of security norms, abandonment of breeder reactors and postponing issues relating to storage.

It was hence temporalities in particular that differed between the two countries, on the basis of the previous inertia of the two energy systems. The historically more advanced development of the latter in Germany, by the order of a decade – for instance with regard to the rise in consumption or diversification of the energy mix – is confirmed through the time gap between the two countries' decisions to abandon breeder reactors, and then stabilise the share of nuclear energy.

Liberalisation and decentralisation

While environmental topics were indeed significantly developed on the European scale, the European Commission studied the various issues in isolation from one another. While certain elements of energy policy were implemented on the Community level, they essentially consisted of very general and limited objectives. In fact, the possibility of a more integrated approach was relegated to a later period, for the Commission believed that the alignment of regulatory frameworks through liberalisation and the bringing together of environmental norms were both prerequisites. The European reforms of the energy market implemented in the late 1990s were thus an additional stage in the convergence of French and German energy models, but more as a new and common framework imposed on both countries rather than integrated policy objectives and measures. However, as was the case with the preceding changes to energy systems, Germany was once again a step ahead by very quickly coming to grips with the new rules of the liberalised energy market.[17]

The dynamic of decentralisation, particularly in France, was the final element that promoted convergence of the energy models of the two countries. It was often associated with the development of renewable energies, in addition to a greater awareness of environmental considerations. Decentralised organisation was already a clearly constituent part of German

17 'Deutschland drückt aufs Tempo', *Frankfurter Allgemeine Zeitung*, 14 Mar. 2001.

energy policy,[18] as the federal government had to cooperate with the Länder in order to coordinate it. Energy programmes were largely co-financed by the federal state and the Länder, for instance the programme relating to the thermal efficiency of buildings and urban heating, which received equal funding from both. The trend over the period was the reinforcement of this decentralised aspect of the FRG: the third revision of the German energy programme in 1981, for instance, prompted the federal government to ask local authorities to develop concepts for local energy supplies.[19]

The temptation to develop more decentralised energy systems was not a connection, in that they challenged certain historic monopolies with the liberalisation of markets and the introduction of heightened competition. It was noted during the 1990s that there

> is currently a return to the early decades of electricity use, with decentralised production, and greater interconnection of municipal services such as electricity, gas, and water; in the long-term, 'energy service' rather than 'energy distribution' were the keywords from the viewpoint of fundamental change. These were the results of pressure applied by the European Commission for the opening of markets and competition. A major public debate began, which had to inevitably lead to adjustments and a new conception of energy in harmony with various social and economic needs.[20]

Starting in the 1990s, the Commission implemented energy programmes through the procedures for calls for proposals at the regional level.[21]

In France, the Deferre laws that were passed beginning in 1982 organised an initial decentralisation of French administration, also known as Act I of decentralisation. Decentralisation was fully linked to the rise of environmental topics, as noted by Brice Lalonde in 1990: 'decentralisation, which is sometimes presented as an alternative to the unsuitability of the state, is a method that has already been tested in the field of the environment'.

In 1991 the state implemented regional supervisory bodies for the environment that brought together regional services for water management. At the same time, the government brought together the departmental management authorities for infrastructure and agriculture to develop a single technical department for 'environmental protection'. As a result, the dynamic of decentralisation initiated in France tended, all things considered, to draw its

18 Art. 28 of the Basic Law; 1935 law on energy.

19 International Energy Agency, *Energy Policies and Programmes of IEA Countries*, 1981 review (OECD, 1982).

20 *Bulletin d'histoire de l'*électricité **22** (1993).

21 Commission européenne, *Rapport général sur l'activité des communautés – 1990* **24** (1991).

Jacobin model nearer to that of federal Germany. While Germany did not escape the decentralising trend, it was much smaller in scope than in France due to their distinct initial situations in the matter.

The oil counter-shock of 1986 and low oil prices confirmed, this time in the opposite direction from in the preceding period, the decisive influence of price on the composition of the energy mix. With regard to environmental protection, this period of low prices was characterised by contradictory effects. There was a concurrent falling off of energy saving and an increase in oil consumption, along with a decrease in budgets earmarked for the environment and renewables in the late 1980s, before a subsequent rise. Despite this delay following the oil counter-shock, environmental protection was increasingly seen as being essential, and new significant measures were taken, without their effects being visible due to the declining price of fuel. Environmental protection nevertheless became a key marker of French and German energy policy, and not just a relatively ancillary element. It tended to be affirmed as a fully-fledged element of constraint, and was imposed on both sides of the Rhine on similar terms. In this context of low prices, it especially led to a limitation of – rather than a decrease in – the rise in consumption or the emission of pollutants.

At the end of the process, the share of consumption of different sectors (industry, transportation, residential and commercial) had converged between France and Germany, due in equal parts to reunification and fundamental shifts common to the two countries in the evolution of uses and technological innovation. In like manner, per capita levels of consumption became similar at the time.

In this context, environmental protection was central to regulating the evolution of a system, and less so price constraints as in the preceding period. The environmental approach became rationalised and systematised; the environment was no longer seen as being antagonistic to economic growth, and became even more rooted in mentalities. Efforts from preceding decades to understand pollution bore fruit, allowing for an acceleration and adjustment of measures. And it was broadly within the European framework, particularly following the Single European Act, that these measures were developed, notably through increasingly robust normalisation and regulation.

In the end, the differences between France and Germany stemmed from a historic inertia that resulted especially from the availability of coal resources. European construction, the integration of markets and the development of common norms nevertheless led to the creation of determinants that were less national and more European and global, with environmental considerations henceforth chief among them. This emerging common framework led to a gradual convergence of the French and German energy systems from 1973 to the present.

This dynamic has continued to the present, and was embodied during the 2000s by the development of the first energy-climate package establishing European objectives for 2020 in the reduction of greenhouse gases, increasing share of renewable energies, improvements in energy efficiency and implementation of an emissions allowance trading system. The Clean Energy Package presented in late 2016 by the European Commission was yet another step in this growing integration of environmental and energy considerations on the European level, contributing to the convergence of European, and therefore French and German, energy systems.

CHAPTER 17.

TRAJECTORIES OF EUROPEAN ENVIRONMENTAL GOVERNANCE SINCE THE 1970s

Anthony R. Zito

This chapter studies the trajectory of the state, particularly how states seek to govern within the public sphere. A prevailing proposition in public policy and public administration literature is that the traditional mechanisms used by states to steer society towards particular policy goals have altered, with authority diffusing upwards towards international institutions and networks of global players, downwards to subnational (regional, local, municipal) communities and actors, and horizontally to a wide range of different stakeholders operating at the national level.[1] This chapter offers both macro analysis, focusing on the role of constitutions and political institutions, and micro policy analysis of the tools that carry out state objectives.[2]

I assess the evolution of governance arrangements, in the form of policy instruments, for environmental policy in Germany (one large, federal European Union (EU) Member State) and the Netherlands (one smaller, unitary Member State) from 1970–2017. This comparative analysis over time seeks to explain the dimensions by which governance changes have occurred and why these changes have happened. In asking these questions, I analyse the difference that EU process and outputs have made for EU Member States. The chapter focuses on what changes occur in the types of policy instruments (the tools used to steer public aims) adopted.

The core analytical explanation follows sociological institutionalism, especially the dynamic of process sequencing.[3] It looks to highly critical events where policy actors face recurring choices and have scope to choose substantially differing mechanisms for governing. The framework concen-

1 See, for example, J. Pierre and B.G. Peters, *Governance, Politics and the State* (Basingstoke: Macmillan, 2000).

2 O. Treib, H. Bähr and G. Falkner, 'Modes of governance: Towards a conceptual clarification', *Journal of European Public Policy* **14** (1) (2007): 1–20.

3 J. Haydu, 'Making use of the past: Time periods as cases to compare and as sequences of problem solving', *American Journal of Sociology* **104** (2) (1998): 339–71.

trates its causal analysis on institutional dynamics, the political context and dynamics, ideas and frames for making governing and policy choices. I use a mixture of interviews with members of policy elites including national and subnational civil servants, civil society actors and academic experts (35 interviews and six questionnaire responses for the Netherlands and Germany in 1992, 2000–2003 and 2012) in the case countries, and sixteen interviews with Commission officials in 1992 and 2016–2018. Primary documentation and secondary sources are also utilised to highlight the nature of environmental policymaking across multiple levels of analysis.

I organise the chapter along an approximate chronological order to compare key sequential moments across the countries. The chapter starts the two cases in 1969; this year permits observation of environmental policy at the moment that social movements and decision-makers visualised environmental issues as a distinct policy problem and developed institutional solutions and policy instruments. Critical events are isolated for each environmental policy sector, followed by an assessment of the state of governance arrangements at each interval. Given the huge scope of the environmental sector, particular attention is paid to the subfields of water quality, waste policy and climate change.

The chapter detects three significant shifts and distinct sequences of governance that operate roughly in parallel across the case countries: enshrining of the environmental policy problem in the governance system in the 1970s; incorporation of notions of sustainability and market principles into government priorities and environmental governance; and retrenchment of economic concerns and competitiveness issues in the governance approaches.

Theoretical framework

Dependent variable

As noted above, the assessment of governance involves examining macro arrangements and micro policy instruments. Students of governance tend to isolate particular governance types, specifically hierarchical (associated with government), market and networks.[4] While this typology works as an ideal type, the reality of governance research suggests policy sectors have elements

4 J. Frances et al., 'Introduction', in G. Thompson et al. (eds), *Markets, Hierarchies & Networks: The Coordination of Social Life* (London: Sage Publications, 1991), pp. 1–19.

of all these governance approaches.[5] More differentiation can be found in the study of the actual policy instruments that deliver the policy aims.

Policy instruments are the tools by which governments implement their governance strategies; they are the actual means for governance modes to impact on policymaking.[6] However, we cannot view policy sectors as arenas for single instruments, but rather acknowledge the likely importance of instrument groups and mixes. This paper uses a modified version of Hood's 1983 typology to assess governing resources: information, rules/authority, finance and a residual category.[7] Information focuses on actors collecting data and using communication to steer actors. Rules/authority denotes the legal power and capacity to regulate. Finance encompasses fiscal instruments and other forms of government assets or fungible resources to steer policy actors. The residual category allows for instruments that do not fall neatly in the other categories.

Independent variables and analytical approach

For the independent variables, this chapter offers three dimensions: institutions, politics and ideas. The institutional dimension includes the institutional rules set out in national constitutions and EU Treaties, and the processes and norms of policymaking institutions. The second dimension focuses on politics, examining the role of elections, changes of government, and notable political actors. The third highlights the role of ideas and paradigms.[8] This encapsulates specific policy ideas and ways of framing policy problems, and takes in broader questions of ideologies and societal values that shape political choice and therefore governing choice.

To understand how these dimensions shape governance, I utilise institutionalist theory and sequencing. Institutionalist theory highlights the importance of sequencing, where particular events that occur at the start of a policy trajectory matter more because they help define the likely range of choices and events that will follow over time.[9]

5 See, for example, G. Capano, J. Rayner and A.R. Zito, 'Governance from the bottom up: Complexity and divergence in comparative perspective', *Public Administration* **90** (1) (2012): 56–73.

6 C. Hood, *The Tools of Government* (London: Macmillan, 1983).

7 Ibid.

8 See P. Hall, 'Policy paradigms, social learning and the state', *Comparative Politics* **25** (3) (1993): 275–96.

9 P. Pierson, 'Increasing returns, path dependence, and the study of politics', *American Political Science Review* **94** (2) (2000): 251–67; J. Mahoney, 'Path dependence in historical sociology', *Theory and Society* **29** (4) (2000): 507–48; M. Howlett and J. Rayner, 'Understanding the historical turn in

Process sequencing

Haydu popularised a sequential approach to explaining key causal patterns behind the changes within institutions; Howlett and Rayner term this approach 'process sequencing'.[10] Haydu suggests that one can gain a greater sense of causal connections over time and between events by 'organising events into sequences of problem solving that span different periods'.[11] Haydu argues for the importance of tracking reiterated problem solving, linking facts from different time periods to build 'narratives of historical switch points that are followed by a more or less durable social regime'.[12] Although a solution contained in a given event will set a new historical direction and limit future choices, the process sequencing account differs from path dependent approaches in viewing outcomes at a given switch point as products of the past rather than 'historical accidents'.[13] Solutions at event A may enshrine problems as well as tools and understandings that actors must confront further down the sequence at event C or D. Thus negative/positive feedback about a taken decision will inform the policy debate in a manner that may come to a crisis at the critical juncture. The event itself may occur over a day or much longer if it involves a sustained process over time.

This approach has the advantage of placing actors and agency at the centre of both the explanation and the method since it requires an understanding of an actor's perspective: how 'they define problems, devise solutions and take action'.[14] It becomes important to observe how different actors perceive recurring problems, realising there may be significant clashes in the definition of the problem and the proposed solution(s); with the outcome potentially defining the historical switch point. At these critical junctures, actors may be able to reverse earlier decisions and strike out in an array of potential directions.[15] Each sequence event involves a range of constraints and opportunities which actors can seize upon to block or promote policy change. Daugbjerg contends that reflective policy makers will use feedback

the policy sciences: A critique of stochastic, narrative, path dependency and process-sequencing models of policy-making over time', *Policy Sciences* **39** (1) (2006): 1–18, here 8.

10 Haydu, 'Making use of the past'; Howlett and Rayner, 'Understanding the historical turn'.

11 Haydu, 'Making use of the past', 341.

12 Ibid., 349.

13 Ibid., 354.

14 Ibid., 355.

15 C. Daugbjerg, 'Process sequencing', in E. Araral et al. (eds), *Routledge Handbook of Public Policy* (London: Routledge, 2012), pp. 473–83.

to learn.[16] In most cases actors choose those precedents and steps that are a logical progression as shaped by the previous event in the policy sequence. Actors tend towards a gradual approach even when there is a significant gap between the current design of policy and the perceived ambition. However, if the context and perceptions of disjuncture lead to a policy crisis, this may generate a more radical and abrupt policy change.

With this approach, the chapter makes a qualitative assessment as to whether the three dimensions reflect a substantive and substantial change in comparison to the previous point in time. I examine the degree of turnover in the actors wielding governing power, the relationships between actors, and paradigm shifts in values and ideologies.

Framing the problem: 1969–1972

A combination of critical pollution events and a growing environmental movement in Organisation for Economic Co-operation and Development (OECD) states triggered the creation, in the 1970s, of environmental institutions and public policy across Northern European countries and the European Communities (EC).[17] Seizing on growing political awareness, national governments in Germany and the Netherlands put forth environmental platforms to distinguish themselves from their domestic political opponents.

The EU level

In the early 1970s, the EC was responding to the pressure of environmental movements in member countries such as the Netherlands and Germany. The Commission harnessed this concern in a 1971 memorandum about this agenda, proposing both a strategy and legislation.[18] In 1972, the original six Member States and three prospective enlargement members started to create national environmental policies in the aftermath of the Stockholm Conference.[19] In the face of environmental concerns and views that unilateral

16 Ibid.

17 The name of the EU before the Maastricht Treaty came into effect in 1993.

18 J. McCormick, *Environmental Policy in the European Union* (Basingstoke: Palgrave, 2001); E. Rehbinder and R. Stewart, *Environmental Protection Policy*, Vol. 2 *Integration through Law: Europe and the American Federal Experience* (Berlin: De Gruyter, 1985).

19 M.S. Andersen and D. Liefferink, 'Introduction: the impact of the pioneers on EU environmental policy', in M. Andersen and D. Liefferink (eds), *European Environmental Policy: the Pioneers* (Manchester: Manchester University Press, 1997), pp. 1–39.

environmental protection at member state level could act as trade barriers, the EC process ratified various measures. The increasing activity of the most energetic European states helped create a self-reinforcing logic, ratcheting up levels of environmental protection. As public and political concern grew, the institutional dynamics of the EC process shaped the policy outcome creating a strategy and regulations to harmonise member state efforts and protect the common market.

The European Court of Justice upheld the validity of these measures as being an implied power (and later an essential objective).[20] This constitutional evolution led environmental policy to fall within the Common Market's scope, and therefore the EC supranational institutions, with states retaining control of implementation.

Germany

The 1969 election of a new German federal government, formed by a coalition of the Social Democratic Party (SPD) and the Free Democratic Party (FDP), triggered the move towards a federal environmental policy. An important distinction, compared to other OECD states such as the United States, is that this political movement happened without a particularly strong recognition by the German population of the environment as a political problem.[21] The coalition won the national elections on the basis of a reform agenda. The FDP leadership, including the FDP Minister of Interior Genscher, saw an opportunity of carving out its own broader reform agenda and change legacy, matching the SPD and Chancellor Brandt's Ostpolitik.[22] Policy activities in the United States and United Nations also had a stimulating role in pushing German federal government action.[23]

The SPD-FDP coalition installed a number of institutional and policy changes through 1969–1974. Most importantly, in 1972 it amended the Basic Law, the German Federal Republic (BRD) constitution, to include

20 I. Koppen, 'The role of the European Court of Justice', in D. Liefferink, P. Lowe and A. Mol (eds), *European Integration and Environmental Policy* (London: Belhaven Press, 1993), pp. 126–49, here p. 133.

21 H. Pehle, 'Germany: Domestic obstacles to an international forerunner', in Andersen and Liefferink (eds), *European Environmental Policy*, pp. 161–209.

22 H. Weidner, '25 years of modern environmental policy in Germany. Treading a well-worn path to the top of the international field', *WZB Discussion Paper* FS II 95–301 (1995).

23 H. Pehle and A. Jansen, 'Germany: The engine in European environmental policy?', in K. Hanf and A. Jansen (eds), *Governance and Environment in Western Europe: Politics, Policy and Administration* (Harrow: Longman, 1998), pp. 82–109.

waste and air pollution. In terms of policy output, 1970 saw the government create its first action programme, followed up by the first comprehensive federal programme in 1971. The Federal government produced two key pieces of legislation: the Waste Disposal Act 1972 and the Federal Emission Control Act 1974. Importantly, these laws focused on guiding industry in accordance to previous industrial regulations, rather than promoting particularly ambitious norms.[24] Thus, political factors predominated in the push to frame an environmental policy, but the BRD's institutional legacy structured the nature of how it was framed.

The Netherlands

During the period 1968–1972, a number of environmental incidents occurred, generating strong environmental interest by the Dutch public and leading to the creation of critical environmental movement bodies, such as Stichting Natuur en Milieu (1972). A shift in this period occurs from perceiving many of the same issues as public health and energy policies to conceptualising a distinctive policy sector with a separate political agenda. In this same period, the Christian Democratic (CDA) Party built a more progressive alliance with some of the smaller parties, such as Democrats '66.[25] The government's general policy attitude retained an emphasis on building consensus with a variety of actors, through consultation and cooperation, including extensive consultation with civil society groups about environmental legislation.[26] Public opinion and external events combined with an evolution in the political dynamics, encouraging the Dutch political class to focus upon this new policy area. The formal and informal structures of how the Dutch created political and policy consensus constrained the nature of how this was done.

The Dutch selection of policy instruments reflected concerns about the implications of certain environmental problems: there was a focus on targeting specific polluter activities in particular societal sectors and environment media with an idea of prohibiting specific pollution and taking remedial steps. Given that Dutch and European environmental policy was

24 J. Hucke, 'Environmental policy: The development of a new policy area', in K. von Beyme and M. Schmidt (eds), *Policy and Politics in the Federal Republic of Germany* (New York: St. Martin's Press, 1985), pp. 156–75.

25 K. Gladdish, *Governing from the Center: Politics and Policy-making in the Netherlands* (Dekalb: Northern Illinois University Press, 1991), pp. 59–61.

26 C. Middendorp, *Ideology in Dutch Politics: The Democratic System Reconsidered, 1970–1985* (Assen/Maastricht: Van Gorcum, 1991).

often located in a public health setting, much of the initial environmental approach reflected the notion of physical regulation of individual sectors. This led to a 'first generation' of policy instruments focused on regulation.[27]

Critical junctures: 1972–2017

Having laid out the starting point for environmental policy, I now highlight the subsequent core critical junctures. Two strong discursive notes were struck across the two political systems. In the 1972–2017 period, there is a governance focus on competitiveness and market solutions and a growing concern about conditionality: i.e. the institution of environmental protection had to be linked to other countries (whether at a regional or global level or both), establishing sufficient levels of protection to balance economic costs and environmental responsibilities.

The EU

Throughout the 1970s–1990s, the EU arena was an important source for environmental governance and instruments for EU Member States. The nature and ambition of this governance push changed in the 1990s, culminating in 1999 with a critical juncture. Although the creation of the Euro had been planned long before, it was only with the currency's 1991 adoption that the political and administrative requirements of this new project became clearer. Although EU environmental goals continued to be touted loudly and sometimes very prominently in such negotiations as Kyoto, after 1999, the priority of safeguarding national economies and their competitiveness started displacing the environment lower down the political and policy agenda of the EU and its Member States. The Lisbon Process enshrined sustainability in its goals in 2000, but it is noteworthy how Lisbon's evolution and ten years of development led to an increasing focus on certain economic achievements, with the environment gradually featuring less prominently.[28] 1999 also was the year that the EU Commission President Santer and his fellow Commissioners resigned. These resignations did not directly reflect upon EU environmental policy, but did dent the body's prestige and created a leadership vacuum for moving policy forward.[29]

27 J. van Tatenhove, *Beleidsvoeringsprocessen in het Nederlandse Milieubeleid in de Periode 1970–1990* (Wageningen: Agricultural University, 1993), p. 118.

28 Interview, Commission official, 10 Jan. 2017.

29 M. Cini, 'Political leadership in the European Commission: The Santer and Prodi Commissions, 1995–2005', in J. Hayward (ed.), *Leaderless Europe* (Oxford: Oxford University Press, 2008), pp. 113–30.

More importantly, the resignation was a culmination of questioning about EU integration ambitions by Member State governments and national populations that has subsequently reshaped EU environmental governance and broader EU integration. The EU had significantly expanded environmental policy in the 1970s–1980s, but in doing so raised issues of the effectiveness of Member State and EU environmental policy implementation.[30]

This question of effectiveness and merit was compounded by questions about the legitimacy of European integration and at what level the power to govern should appropriately lie (the *subsidiarity* debate). A very public signal of this was the 1992 Danish referendum result concerning the Maastricht Treaty, forcing the Treaty's revision. The 1990s also witnessed the British and French attempt to repatriate certain EU legislation back to Member States. Of the 24 EU legislative items on this regulatory target list, seven were environmental measures.[31] Eventually, the EU kept all this legislation but the repatriation bid contributed to an attitude change towards environmental integration, emphasising consolidating public policy rather than an expansion or roll back.[32] The EU Commission had a strong inducement to propose less intrusive legislation such as framework directives, giving greater scope for Member State implementation and non-legislative instruments based on the principle of shared responsibility.[33]

Germany

Germany operated through the 1980s as frontrunner in pushing environmental policy forward. The 1990 German Re-unification changed this outlook. Unification challenges were swiftly followed by the most serious post-war German economic downturn. The level of environmental degradation in the former German Democratic Republic was combined with a difficult economic situation, particularly in the Eastern *Länder*. These realities reshaped German ambition about how to fund and support its governance of the unified territories

30 A. Jordan, 'The implementation of EU environmental policy: A policy problem without a political solution?', *Environment and Planning C: Government and Policy* **17** (1) (1999): 69–90.

31 R. Wurzel, *Environmental Policy-making in Britain, Germany and the European Union* (Manchester: Manchester University Press, 2002).

32 L. Krämer, 'Recent developments in EC environmental law', in J. Holder (ed.), *The Impact of EC Environmental Law in the UK* (Chichester: John Wiley & Sons, 1997), pp. 9–26.

33 A. Jordan, 'Editorial introduction: The construction of a multilevel environmental governance system', *Environment and Planning C: Government and Policy* **17** (1) (1999): 1–17.

and build expertise.[34] The solution to this enormous challenge was to keep the BRD administrative and legal structures while amending regulation to speed the process and restrict debate.[35] Lees also notes reunification's longer term political impact of increasing the number of *Länder* concerned about their economic wellbeing and making the *Länder* governments less inclined towards progressive environmental solutions that they feared they could not meet.[36]

In this context, the German government's environmental policy focus evolved, giving higher priority to concerns about how environmental regulations would create costs affecting the German economy's economic competitiveness. This pushed back against the 1980s ideological argument that enhanced environmental protection was compatible with growth.[37] The Christian Democratic Union/Christian Social Union (CDU/CSU)-FDP coalition government of 1982–98 adopted a 1991 coalition agreement, formally giving preference to voluntary agreements (VAs) over environmental regulations.[38] It is also notable that less than half of this coalition agreement's environmental measures were implemented.[39] Reunification changed the way that Germany was governed as a whole, reflecting new economic and political realities.

Nevertheless, the nature of this direction for German environmental policy was later defined by the second critical juncture: the SPD-Green/Alliance 90 coalition election in 1998. The new coalition agreement involved both more ambitious environmental policy targeting and experimentation with different policy instruments. It explicitly laid out an ecological modernisation objective (i.e. that an ecological standards focus can promote economic growth and productivity, and environmental protection).[40] This idea was

34 D. Graham, 'Study shows high cost of German reunification: report', *Reuters* 7 Nov. 2009, http://www.reuters.com/article/us-germany-wall-idUSTRE5A613B20091107 (accessed 7 Mar. 2016).

35 Weidner, '25 years of modern environmental policy in Germany'.

36 C. Lees, 'Environmental policy: The law of diminishing returns?', in S. Green and W. Patterson (eds), *Governance in Contemporary Germany: The Semisovereign State Revisited* (Cambridge: Cambridge University Press, 2005), pp. 212–39.

37 A. Weale et al., *Environmental Governance in Europe* (Oxford: Oxford University Press, 2000).

38 R. Wurzel et al., 'From high regulatory state to social and ecological market economy? "New" environmental policy instruments in Germany', *Environmental Politics* **12** (1) (2003): 115–136.

39 Pehle, 'Germany'.

40 SPD (Sozialdemokratische Partei Deutschlands) et al., *Aufbruch und Erneuerung – Deutschlands Weg ins 21. Jahrhundert. Koalitionsvereinbarung zwischen der Sozialdemokratischen Partei Deutschlands und BÜNDNIS 90/DIE GRÜNEN*, https://www.gruene.de/fileadmin/user_upload/Bilder/Redaktion/30_Jahre_-_Serie/Teil_21_Joschka_Fischer/Rot-Gruener_Koalitionsvertrag1998.pdf (accessed 12 Mar. 2016).

not new; a (CDU) environment minister in the previous coalition, Klaus Töpfer, had publicly backed the idea.[41]

The change was not an ideological shift as much as a determination of a heavily Green Party infused government to promote substantial changes in environmental practices. Due to space constraints I do not cover perhaps this government's most significant act: the move away from nuclear power. More significant for this chapter are its environmental initiatives. Ecological tax reforms were fully embraced: the idea that ecological taxation and incentives could be used to generate revenue to help the economic sector with welfare and other costs.[42] There were also initiatives to protect natural sites and efforts to improve environmental standards in the production process in a way that also reduced waste.

The Netherlands

The two critical junctures identified in the Dutch case focus on changes of government and the ideological consequences to environmental policy. The Dutch political right has had a decisive shaping role.

The first juncture occurred in 1982 with the assumption of power of the Christian Democratic Appeal Party-Liberal (VVD) coalition. This Lubbers government coalition made fundamental governance changes to the environmental policy sector between 1982–1986, in the context of reassessing the fundamental role of government.[43] This political dynamic found fertile ground in the growing political dissatisfaction with, and increased ecological understanding of, many aspects of Dutch environmental policy and other direct consequences of the 1970s environmental decisions and governance.[44] The government platform included the ideational aim of reducing the scope of government responsibility and regulation while increasing societal/economic actors' responsibility. Simplifying or reducing regulations in all areas would reduce an overloaded government's burden and increase efficiency.

41 K. Töpfer, 'Ecological modernisation of the industrialised state: A federal perspective', in T. Ellwein et al. (eds), *Yearbook on Government and Public Administration* (Baden-Baden: Nomos, 1989), pp. 89–520.

42 SPD, *Aufbruch und Erneuerung*.

43 V. Lauber, 'The political and institutional setting', in A. Mol, V. Lauber and D. Liefferink (eds), *The Voluntary Approach to Environmental Policy* (Oxford: Oxford University Press, 2000), pp. 32–61.

44 H. Bressers, 'Environmental policy instruments in Dutch practice', in European Parliament DG for Research (ed.), *Economic and Fiscal Incentives as a Means of Achieving Environmental Policy Objectives* **16** (1990): 79–96.

This was a core VVD tenet, in keeping with the Reagan/Thatcher period of rightist, neo-liberal governments seeking to redefine the state/society relational boundaries.[45]

VVD politician Pieter Winsemius became environment minister in 1982, bringing entrepreneurial energy and the ability to articulate a coherent ideational vision incorporating VVD concerns and the need for environmental protection. VVD and the business sector believed that contemporary environmental regulations could burden industry, so the government created an interministerial working group to propose regulatory streamlining.[46] Winsemius and the government shifted civil society's role to more closely follow the philosophy of self-responsibility.[47]

By 1986 this had been articulated into two core approaches operating under the rubric of a socialisation strategy: *internalisation* and *target group strategy*. Internalisation expressed the importance of closing the gap between government and societal actors by getting groups to accept their social responsibility while acknowledging that the government had to pay attention to the concerns of the regulated.[48] Target group policy emphasised designing a policy focusing on a relatively homogenous group of actors engaged in activities affecting the environment and modifying this group's behaviour.[49] In line with the government's self-responsibility approach, these target groups acted as joint partners (with government) shaping policy design and implementation. The sustainable development focus in the target group policy suggested a more fundamental effort to deal with problems at the source of production. Taking a preventative, source-orientated strategy gave strong impetus to securing the active involvement and commitment of target groups. Internalisation involved creating environmental policy instruments that would induce societal actors to take responsibility and internalise government policy objectives into the groups' value systems and economic processes. While the focus of the govern-

45 M. van Vliet, 'Environmental regulation of business: Options and constraints for communicative governance', in J. Kooiman (ed.), *Modern Governance* (London: Sage, 1993), pp. 105–18.

46 K. Hanf, 'Deregulation as regulatory reform: The case of environmental policy in the Netherlands', *European Journal of Political Research* **17** (2) (1989): 193–207.

47 K. Hanf and E. van de Gronden, 'The Netherlands: Joint regulation and sustainable development', in K. Hanf and A. Jansen (eds), *Governance and Environment in Western Europe: Politics, Policy and Administration* (Harlow: Longman, 1998), pp. 152–80, here p. 165.

48 D. Liefferink, 'The Netherlands: A net exporter of environmental policy concepts', in Andersen and Liefferink (eds), *European Environmental Policy*, pp. 210–50.

49 K. Hanf and I. Koppen, *Alternative Decision-making Techniques for Conflict Resolution: Environmental Mediation in The Netherlands* (Berlin: WZB, 1994), p. 10.

ance approach changed, these approaches remained in line with Dutch policy practices of the 1970s and before, in terms of neo-corporatism and the close relationship between the state and core societal groups.[50]

October 2010 saw the second critical juncture, when VVD and CDA formed a short-lived minority government, supported by the anti-immigrant Freedom Party. Before unravelling in April 2012, this government shifted away from highlighting environmental and policy issues and focused instead on reducing budgets and boosting growth. This approach was not new: the Balkenende centre-right government in 2002 proposed a budget which substantially shifted away from environmental priorities, cutting a range of environmental taxations and subsidies (but from which the successor Christian Democratic/Social Democratic coalition, 2007–2010, later rowed back).[51]

Although the 2010 coalition found it politically difficult to produce decisive changes in direction, the environment's lower priority was decisive in itself for the Netherlands' current environmental governance positioning.[52] The coalition agreement offered repeated references to a 'level playing field in Europe' on the subject of pollution emissions.[53] This political orientation has shaped the Dutch governments' approach to policy innovation and ambition, linking the possibility of additional domestic environmental targets to other countries' targets. This framing of environmental questions did not include new ideas. Rather, its focus was reducing regulatory burdens and emphasising conditionality of Dutch efforts, with greater responsibility placed on international and EU arenas to take initiatives making significant reductions. The 2018 Climate Change Agreement reflects this focus on international co-operation while also pursuing various agreements with core Dutch sectors

50 D. Liefferink and M. Wiering, 'The Netherlands: An integrated participatory approach to environmental policymaking', in A. Breton et al., *Environmental Governance and Decentralisation* (Cambridge: Cambridge University Press, 2007), pp. 558–83.

51 ENDS Europe, *Dutch 2003 Budget Shifts Focus from Environment*, 18 Sept. 2002, www.endseurope.com/article/7249/dutch-2003-budget-shifts-focus-from-environment (accessed 7 Mar. 2016); D. Liefferink and K. Birkel, 'The Netherlands: A case of "cost-free leadership"', in R.K.W. Wurzel and J. Connelly (eds), *The European Union as a Leader in International Climate Change Politics* (London: Routledge, 2010), pp. 147–62.

52 D. Liefferink and M. Wiering, 'Environmental pioneers in retreat? The cases of the Netherlands and Denmark', paper presented at the 6th ECPR General Conference, University of Iceland, Reykjavik, 25–27 Aug. 2011.

53 Government of the Netherlands, *Freedom and Responsibility: Coalition Agreement*, VVD-CDA, 30 Sept. 2010, article 7, http://mae.ro/sites/default/files/file/2013/pdf/coalition-agreement_olanda_2010.pdf (accessed 7 Mar. 2016).

to voluntarily agree emissions targets and achievement strategies.[54]

Evolution of the governance approaches

Having outlined the critical junctures that shaped environmental governance in the case countries, I now evaluate how environmental governance modes changed over time. We see a governance approach and specific sets of instruments that have been rendered more complex but not overturned in the case countries.

EU Policy developments

Throughout the 1970s, the EC's process created a number of directives (in areas such as waste and water). These gave Member States some scope to tailor how they achieved the Directive objectives to their specific domestic legal and policy circumstances. As with Member States, the EC had a similar focus on specific media regulations, as seen in waste policy; here the rise in waste policies in Germany and elsewhere led the Commission and Member States to agree on the need for a common response, and forced states such as the Netherlands to implement the directives.[55]

From its initial starting point, the EU continued to adopt environmental legislation at a gradually increasing rate towards 1991. It suffered a drop, before rising significantly, then starting a long decrease in 2002.[56] From 1972–2002, the dominant instruments of EU governing continued to be various forms of regulation and hierarchical governance. In the 1970s–1980s, this focused on producing standards to limit emissions. From the late 1980s procedural laws (such as the Directive on Environmental Impact Assessment (85/337)) supplemented the environmental standards stipulated in earlier EU legislation.[57] The limited EU repertoire at this time was notable. Information campaigns were negligible, with rare exceptions such as the ecolabel Blue Flag. In terms of funding dedicated to protecting the environment, EU

54 Government of the Netherlands, *Climate Agreement*, 2019: https://www.government.nl/ministries/ministry-of-economic-affairs-and-climate-policy/documents/reports/2019/06/28/climate-agreement (accessed 19 Jan. 2020).

55 A. Zito, *Creating Environmental Policy in the European Union* (Basingstoke: Palgrave, 2000).

56 N. Haigh (ed.), *Manual of Environmental Policy: The EC and Britain* (Harlow: Cartermill Publishing, 2011).

57 R. Wurzel, A. Zito and A. Jordan, *Environmental Governance in Europe: A Comparative Analysis of the Use of New European Environmental Policy Instruments* (Cheltenham: Edward Elgar, 2013).

funding efforts (e.g. the LIFE programme) have been microscopic compared to the rest of the EU budget.[58]

Although the EU has continued to generate considerable legislative output, greater emphasis has been put on more flexible tools in the environmental arena since 2000. Broad framework laws are more typical: specifying only the most crucial environmental objectives, such as the EU Water Framework Directive (2000/60). Since the mid-1990s, all Commission environmental proposals must incorporate a cost-effectiveness statement.[59] The 2002 Commission Communication renewed the Commission approach to 'avoid making its legislative proposals unwieldy, in accordance with the Protocol on the application of the principles of subsidiarity and proportionality'.[60] The Commission created large consultation meetings involving numerous stakeholders; it also published Green and/or White Papers before proposing legislation, including the Emissions Trading Scheme (ETS Directive). The ETS is the central plank of the EU approach to climate change and involved a major instrumental addition within the administrative context of both Germany and the Netherlands.[61] Some Member States, such as Germany which preferred detailed environmental laws using the best available technology principle for domestic environmental policy, experienced considerable adaptation pressures.[62] Nevertheless, for the Commission and the EU a great deal of consideration has been given to the question of cost-effectiveness and transparency considerations, and the Juncker Commission strongly reflected a more limited approach.[63]

Policy developments in Germany

After the initial establishment of the environmental sector, the 1970s witnessed an extension of the regulatory portfolio (e.g. amending the Federal

58 McCormick, *Environmental Policy*.

59 Wurzel et al., *Environmental Governance in Europe*.

60 CEC, *Communication from the Commission. Action Plan 'Simplifying and Improving the Regulatory Environment*, COM(2002)278 final of 05.06.2002 (Brussels: Commission of the European Communities, 2002), p. 12.

61 Wurzel et al., *Environmental Governance in Europe*.

62 R. Wurzel, 'Germany: from environmental leadership to partial mismatch', in A. Jordan and D. Liefferink (eds), *Environmental Policy in Europe. The Europeanization of National Environmental Policy* (London: Routledge, 2004), pp. 99–117.

63 A. Bürgin, 'The impact of Juncker's reorganization of the European Commission on the internal policy-making process: Evidence from the Energy Union project', *Public Administration* (2018), https://doi.org/10.1111/padm.12388 (accessed 19 Jan. 2020).

Water Act and creating the Federal Nature Conservation Act and Washing Agents Act), but with a lessened intensity.[64] The Waste Water Charges Act was also passed, creating a financial tool.[65] The rest of the decade was spent in an environmental policy rear-guard action against the political pressures, but this action was ultimately successful. The slowdown did not reverse the environmental policy trajectory; it added some vital dimensions to the German approach for the future. The 1974 Emissions Control Act articulates a key German policy idea, the principle of precaution (*Vorsorgeprinzip*: the idea of preventing pollution from occurring, necessitating an active state role) as the basis for a robust environmental policy.[66]

The 1970s–1980s governance mode was hierarchical, with legislation creating general goals and principles; these statutes were implemented through highly detailed regulations and administrative directives (including technical guidelines) that were binding on the state (*Länder*) authorities to implement.[67] These hierarchical tools do not focus on procedural issues but on specifying technical standards and legislative substance. The focus was on enhancing administrative control and using the precautionary principle to pursue more stringent standards than would otherwise be necessary.[68]

The Federal Republic gradually started an increase in new environmental regulations by the 1980s. The quintessential regulation in this era was the 1983 Large Combustion Plant Ordinance, a regulation coming out of the 1974 Emissions statute.[69] At the heart of this increase was the rise of the Green environmental movement and the Green Party, and their electoral threat to traditional German mainstream parties; the 1986 Chernobyl disaster emphasised this move. However, such regulations as the 1983 Ordinance triggered opposition from German industry and its representatives. In particular, these actors argued that German environmental policies and

64 H. Weidner, 'Reagieren statt Agieren: Entwicklungslinien staatlicher Umweltpolitik in der Bundesrepublik Deutschland', *Politische Ökologie* **9** (23) (1991): 14–22.

65 BRD, *Gesetz über Abgaben für das Einleiten von Abwasser in Gewässer (Abwasserabgabengesetz -AbwAG)* (1976), http://www.ecolex.org/ecolex/ledge/view/RecordDetails;DIDPFDSIjsessionid=3685FBC99527096C834331778AE5FD21?id=LEX-FAOC035872&index=documents (accessed 7 Mar. 2016).

66 Weale et al., *Environmental Governance in Europe.*

67 D. Kelemen, *The Rules of Federalism: Institutions and Regulatory Politics in the EU and Beyond* (Cambridge, Mass.: Harvard University Press, 2004)

68 Weale et al., *Environmental Governance in Europe.*

69 J. Newig, 'Symbolic environmental legislation and societal self-deception', *Environmental Politics* **16** (2) (2007): 276–96.

regulatory stringency, compared to those in other countries, damaged economic competitiveness.[70]

The post-reunification Kohl government grappled with the consequences of reunification and economic recession. The reconstruction of the East German *Länder* led the federal level governance focus to favour standard economic growth policies and various planning/infrastructure policies such as road building.[71] In the wake of reunification and by 1993, industrialists argued that environmental and other German regulations were stifling competitiveness. Although German environmental policymakers refuted this stance and the ecological argument remained (as did the production of high-end ecological goods), a number of measures were scrapped.[72]

Despite these concerns, regulations have remained the key governance instrument, reinforced by EU legislative requirements. One significant instrument development was the increase in voluntary agreements in the early 1990s. The Kohl government implemented a coalition agreement giving preference to VAs over traditional regulation;[73] this federal move also reflected the substantial use of VAs by the SPD-Green coalitions in various *Land* governments of the 1980s and 1990s.[74] German VAs are not legally binding, unlike the Dutch covenants, but many of them were adopted under the 'shadow of the law', i.e. recognition that regulation could be the next alternative step.[75] When the Green Party came to power in the 1998 elections, one of the critical coalition demands was for ecological tax reform (i.e. the shift of the national taxation burden towards incentivising better environmental performances). Successor governments have continued this agenda, with substantial budgetary adjustment to cut environmentally damaging subsidies and enhance extant eco-taxes.[76] Thus hierarchical governance remains, but with greater emphasis towards information and market-based incentives in support.

70 Weale et al., *Environmental Governance in Europe.*

71 J. Anderson, *German Unification and the Union of Europe: The Domestic Politics of Integration Policy* (Cambridge: Cambridge University Press, 1999).

72 Pehle, 'Germany'.

73 J. Knebel, L. Wicke and G. Michael (eds), *Selbstverpflichtungen und normersetzende Umweltverträge als Instrumente des Umweltschutzes, Berichte des Umweltbundesamtes* 5 (Berlin: Umweltbundesamt, 1999), p. 30.

74 Lees, 'Environmental policy'.

75 Wurzel et al., *Environmental Governance in Europe.*

76 Umweltbundesamt (ed.), *Quantifizierung der Effekte der Ökologischen Steuerreform auf Umwelt, Beschäftigung und Innovation*, Hintergrundpapier (Berlin: Umweltbundesamt, 2004).

Policy developments in the Netherlands

The 1970s Dutch governance strategy was to use framework legislation to define the broad lines of responsibility in both the policy response and the range of potential instruments.[77] Specific regulations were issued via executive decrees with powers delegated to various government authorities. This was command-and-control legislation (e.g. the Chemical Wastes Act 1976 and the Noise Abatement Act 1979) focused on prohibiting activities threatening the environment. The 1970s legislation gave provinces significant licensing powers, targeting those environmental cases involving technically complex pollution processes.[78] Given the technical demands required in implementation, the Dutch policy operated a consensus-orientated system where multiple levels and groupings of public/private actors are the norm.

In light of the various criticisms raised about the Dutch regulatory approach and its implementation and coordination problems,[79] Dutch policy actors began to articulate a new approach in the 1970s.[80] This learning about environmental governance moved the Dutch thinking towards integrating environmental responses across sectors. Another strand of learning occurred with respect to increasing the role for other policy instruments. The Dutch had considerable tax instrument experience, notably the levy contained in the 1969 Surface Water Pollution Act.[81]

With the critical juncture in 1982, the Dutch government implemented several long-term governance changes, on integrated thematic approaches and integrated multi-year strategic plans.[82] The policies focused on specific pollution sources and on societal actor groups (including consumers, industry, farmers), to target particular policy packages. As regards instruments, the Dutch government's strategy shifted to streamlining regulation and rethinking the nature of policy instruments while increasing business and industry's responsibility in a self-governing process.[83] The governance tool

77 Hanf et al., 'The Netherlands'.

78 Ibid.; Liefferink, 'The Netherlands', pp. 219–20.

79 Bressers, 'Environmental policy instruments in Dutch practice'; Liefferink, 'The Netherlands'; M.S. Andersen, *Governance by Green Taxes: Making Pollution Prevention Pay* (Manchester: Manchester University Press, 1994).

80 G. Bennett, 'Policy planning in the Netherlands', in N. Haigh and F. Irwin (eds), *Integrated Pollution Control in Europe and North America* (London: Conservation Foundation, 1990), pp. 209–39.

81 Andersen, *Governance by Green Taxes*, pp. 148–49.

82 Hanf et al., 'The Netherlands'.

83 van Tatenhove, *Beleidsvoeringsprocessen*.

box was extended to information, education and *covenants* as well as eco-taxation, in order to promote the concepts of shared and self-responsibility on the part of target groups and consumers.[84]

Covenants are a notable Dutch governance innovation; they are negotiated agreements where the parties set environmental targets in writing.[85] Agreements normally include the government (but not always the national government) and industrial groups and organisations but also potentially other actors, such as provincial governments.[86] Both officials and the targeted groups recognise that negotiated settlements are an attractive alternative to ensure higher environmental standards than stricter command-and-control regulation.[87] Covenants remain linked to and are effectuated by the licensing system; covenants serve as guidance for licensing if the groups do not achieve their goals.[88]

At the point of the 2010 critical juncture, the Netherlands was ranked relatively high in certain categories, e.g. fourth in terms of transportation taxes and second in pollution/resource taxes.[89] The Netherlands also had one of the highest proportions of revenue derived from eco-taxes. These realities led the Secretary of State to argue that further Dutch eco-tax efforts must be made dependent on other European countries increasing their environmental tax efforts and revenue.[90] The 2010–2012 rightist coalition went further, actively dismantling part of the environmental taxation structure, such as the waste, groundwater and packaging taxes; aiming instead for 'solid and simple taxes'.[91] The move towards less ambitious

84 H. Bressers and T. De Bruijn, 'Environmental Voluntary Agreements in the Dutch context', in E. Croci (ed.), *The Handbook of Environmental Voluntary Agreements: Design, Implementation and Evaluation Issues* (Dordrecht: Springer, 2005), pp. 261–81; van Tatenhove, *Beleidsvoeringsprocessen*.

85 H. Bressers et al., 'Negotiation-based policy instruments and performance: Dutch covenants and environmental policy outcomes', *Journal of Public Policy* **31** (2) (2011): 187–208, here 189.

86 P. Glasbergen, 'Partnership as a learning process. Environmental covenants in the Netherlands', in P. Glasbergen (ed.), *Co-operative Environmental Governance* (Dordrecht: Kluwer Academic, 1998), pp. 133–56, here p. 133.

87 Bressers et al., 'Negotiation-based policy instruments and performance'.

88 P. Börkey and F. Lévèque, *Voluntary Approaches for Environmental Protection in the EU*, ENV/EPOC/GEEI (98) 29/final (Paris: OECD, 1998): p. 13; Bressers et al., 'Negotiation-based policy instruments', p. 190.

89 CEU, *Taxation Trends in the European Union: Data for the EU Member States, Iceland and Norway* (Luxembourg: Commission of the European Union, Office of Official Publications, 2010), pp. 232–24.

90 Interview, Ministry of Finance official, 2011.

91 Ibid.

environmental governance is made manifest in the design and calibration of the policy instruments.

Current environmental governance approaches

Table 1 summarises the governance approaches in the case countries and the EU. What stands out clearly is the dominance of the hierarchical mode in the environmental policy sector. The 1970s starting point was a focus on hierarchical governance that remains the legacy today. There are important nuances, however. The focus on building a consensual relationship with business is more suggestive of a network mode with a relatively exclusive set of players involved in delivering the policies.

From an early point in the history of environmental governance, the EU operated surprisingly detailed and systematic hierarchical legislation that was agreed and implemented by Member States. Although experimentation has happened with the market mode, essentially the EU has operated its governance as a regulatory state.[92] This has reinforced the hierarchical approach in our two EU Member State cases. The rules of the Single Market place restrictions on certain subsidies and other instruments that Member States can utilise. Nevertheless, the EU's lower ambition in the environmental arena is the main change, with a greater focus on better, more efficient implementation of current legislation.[93] The Dutch and German central governments have had greater scope to create environmental legislation and various forms of instruments, and have arguably done so on a more systematic basis. Nevertheless, both states have sought to experiment with their governance, and both have an underlying principle of building consensus with lower levels of government (where the implementation occurs) and the different core representations of civil society.

The two case countries and the EU have seen some experimentation but, bar the Dutch negotiated agreements and taxation, and German eco-labelling, this has tended to take a secondary role focused within particular sectors (see Table 2). In terms of efforts to change the nature of the governance mode, the EU has facilitated very limited experimentation despite the huge size and diversity of its membership. If one excludes the emissions trading system, the degree of EU instrumental innovation is relatively small: a limited

92 G. Majone, 'The Rise of the Regulatory State', *West European Politics* **17** (3) (1994): 77–101.

93 C. Burns, P. Eckersley and P. Tobin, 'EU Environmental policy in times of crisis', *Journal of European Public Policy* **27** (1) (2010): 1–19.

number of VAs with mixed success, a relatively low-profile eco-label and so forth. The German success with its eco-label and negotiated agreements and the Dutch experimentation with taxation and covenants reflect internal dynamics rather than EU ones.

Table 1.
Governance trajectories for the case countries and the EU.

Country/System	Governance approach before critical junctures	Governance shift after critical junctures
EU	Hierarchical with Member States implementing	Hierarchical but has become more flexible about regulation and attempted to use market and informational instruments
Germany	Hierarchical and more specified but operating on a consensual basis	Remains hierarchical and consensual but with elements of the other modes added
The Netherlands	Hierarchical and more specified but operating on a consensual basis	Remains hierarchical and consensual but with other governance elements added

Table 2 demonstrates the reality that diverse policy instruments are available. The basket of instruments and the available tools has increased over time for each of the case countries and the EU. The EU and Member States have adopted certain instruments belonging to the three analytical types examined here (regulatory, economic and information), but the reality is that regulation remains the main instrument choice.

Table 2.
The policy tools.

Country/System	Environmental Instrument
EU	1. Regulations over a diverse range of emissions and for environmental protection 2. Emissions Trading Scheme 3. Certain VAs with particular industries 4. Informational instruments including an eco-label and an environmental management scheme 5. Limited environmental-focused funding
Germany	1. Regulations over a diverse range of emissions and for environmental protections 2. Eco-taxes and charges 3. EU Emission Trading Scheme 4. Energy subsidies 5. Negotiated agreements under the shadow of the law 6. Eco-labelling 7. EU environmental management system (EMAS) plus other forms of information provision and reporting 8. Research funding
The Netherlands	1. Regulations over a diverse range of emissions and for environmental protections 2. Covenants 3. Energy and environmental taxation and charges 4. Emissions trading schemes (Dutch scheme for nitrous oxygen, EU scheme for climate change emissions) 5. Grants and subsidies for energy saving, soft loans 6. Use of international environmental management systems (EMAS/ISO) 7. Eco-labelling and other forms of information provision

Reliance on regulation, limited experimentation with new instruments and framing of environmental policy in the context of conditionality all hint at similar trajectories. A finer grained analysis however suggests some important nuances.

The EU, with its extremely limited budget and its protection of the single market, has focused on being a regulatory state.[94] This focus, along with the monitoring of state aid and other aspects of competition law in the EU context, has given a strong regulatory content to the EU law that Germany and the Netherlands implement. Dutch and German institutional norms of collaboration and consensus building across the vertical levels of government, as well as horizontally between the state and society, also remain strong determinants of environmental policy without the intervention of the EU process.

Moving away from institutional to ideational analysis, we see that various versions of the neo-liberal approach to governance and the move away from a hierarchical state governance focus have occurred in the three political systems, although it is critical to emphasise in all cases that regulation, albeit calibrated in different ways, remains the predominant governance tool. Although certain critics would depict the EU as an ongoing neo-liberal project, the reality is that the EU's focus on environmental protection and regulation maintained momentum until essentially after 2000. Conditionality dynamics also occurred in the EU context, with Germany, the Netherlands and others interested in their fellow Member States taking equal environmental burdens.

Turning more explicitly to the dependent variable, I have already noted that hierarchy and regulation remain the prominent approaches for environmental governance. The consensus-orientated politics of Germany and the Netherlands, and the presence of EU legislation, have led to environmental protection that is more specific in detailing key targets and means of implementation. One of the remarkable episodes of EU integration was the detailed and prescriptive regulation in the areas of air, water, waste and wildlife in the 1970s–1980s (without, for most of the period, an explicit treaty mention of the environment). This shows considerable entrepreneurship not easily

94 Majone, 'The rise of the regulatory State'.

explained by focusing on single market dynamics.[95] However, the nature of EU regulation has notably changed to give more flexibility in the means.

Equally, as more intractable problems such as climate change come onto the scene with diffuse sources of pollution, the EU and its Member States have all resorted to more governance innovation, in the form of financial/market instruments and more instrument mix. Nevertheless, EU governance, barring climate change, has essentially used regulatory sticks as the key instrument. This is less the case with Germany and the Netherlands with their negotiated instruments, eco-labels and taxation instruments, but their overriding steering mix is backed by regulation and hierarchy.

Noteworthy also is the critical role of the election of new governments with particular ideological and ideational views towards governance and the role of the state in environmental politics. Ideational change and external events have played a substantial role in shaping environmental governance in the case countries, but the election of new governments often provides the dominant switch point. For both critical junctures in the Netherlands, a shift in the politics of government made the critical difference. The 1982 and 2010 coalition governments pointed Dutch environmental protection down particular paths. Recent Dutch governments have generally moved away from ambition in environmental governance, citing issues of competitiveness and the global economic situation, but the way these linkages have been framed reflects the outlooks of those who have gained power. In Germany the 1969 and 2008 coalition governments gave particular impetus to environmental protection although there is no escaping the impact German Reunification had on environmental governance. The two case countries reinforce the importance of understanding the electoral connection and regime change in bringing changes in ideological perspective and perception. The EU has built on and reinforced the tendency towards a regulatory approach, but equally the influence of neo-liberal thinking and approaches, and the prioritisation of economic and competitiveness values, have crucially shifted the EU role.

95 A. Weale, 'European environmental policy by stealth: The dysfunctionality of functionalism?', *Environment and Planning C: Government and Policy* **17** (1) (1999): 37–51.

INDEX

Index

Index

Index

CPSIA information can be obtained
at www.ICGtesting.com
Printed in the USA
BVHW040957210323
PP14759700001B/1

9 781912 186143